MAGNETOFLUID DYNAMICS FOR ENGINEERS AND APPLIED PHYSICISTS

MAGNETOFLUID DYNAMICS FOR ENGINEERS AND APPLIED PHYSICISTS

Kenneth R. Cramer

Adjunct Professor of Mechanical Engineering,
Air Force Institute of Technology
Wright-Patterson Air Force Base
Dayton, Ohio

Shih-I Pai

Research Professor
Institute for Fluid Dynamics
and Applied Mathematics
University of Maryland

SCRIPTA PUBLISHING COMPANY
Washington, D.C.

McGRAW-HILL BOOK COMPANY
New York St. Louis San Francisco Düsseldorf Johannesburg
Kuala Lumpur London Mexico Montreal New Delhi
Panama Rio de Janeiro Singapore Sydney Toronto

Library of Congress Cataloging in Publication Data

Cramer, Kenneth R.
Magnetofluid dynamics for engineers and applied physicists.

1. Magnetohydrodynamics. I. Pai, Shih-I, joint author. II. Title.
QC718.5.M36C7 538'.6 73-9936
ISBN 0-07-013425-1

1234567890 KPKP 79876543

The supervising editor for this book was Glenda Hightower, the designer was Victor Enfield, and its production was supervised by Joseph Egerton. It was set in Century by Scripta Technica, Inc.
It was printed and bound by The Kingsport Press.

CONTENTS

PREFACE

The purpose of this book is to prepare the advanced undergraduate and graduate student in engineering or physics for professional activity in the field of magnetofluid dynamics. Sufficient material is presented for a two-semester course but, by proper selection of topics, the essential features of magnetofluid dynamics can be presented in one semester.

The subject is introduced with simple descriptions of the primary magnetofluid-dynamic phenomena before the detailed presentation of the fundamental equations in the second chapter. While the development of the equations is extensive, it is required to provide the student with a firm basis for understanding the later chapters. Experience has demonstrated that students may be adequately prepared in electromagnetics, but that they are weak in fluid dynamics. The reverse condition occurs with equal frequency. The specific purpose of the other chapters is described at the end of the first chapter. However, the fourth chapter contains a variety of applications to demonstrate the versatility of magnetofluid-dynamic theory. The discussion of an individual topic is restricted to an elementary analysis to provide training in the application of the theory and to stimulate the student's imagination for new applications. The last chapter, not required in an introductory course, was added to present a basis for understanding the limitations of the theory and advanced applications of the magnetofluid-dynamic principles. Additional references, indeed the major sources, are listed in the appendices which discuss the subject in greater depth and present additional applications.

This book evolved from class notes presented to graduate students in the Mechanical Engineering and Physics Departments of the Air Force Institute of Technology at Wright-Patterson Air Force Base, Ohio. (Kenneth Cramer's primary affiliation is at the Thermomechanics Branch, Flight Mechanics Division of the Flight Dynamics Laboratory, Air Force Systems Command, Wright-Patterson Air Force Base.) Their pointed questions and wide-ranging interest in applying the theory contributed significantly to this project. The contributions of the many authors of the cited references are also acknowledged.

Kenneth R. Cramer

Shih-I Pai

MAGNETOFLUID DYNAMICS FOR ENGINEERS AND APPLIED PHYSICISTS

Chapter 1

INTRODUCTION

1.1 INTRODUCTION

Magnetofluid dynamics is the study of flow of electrically conducting fluids in electric and magnetic fields. It unifies in a common framework the electromagnetic and fluid-dynamic theories to yield a description of the concurrent effects of the magnetic field on the flow and the flow on the magnetic field. Magnetofluid dynamics (MFD) deals with an electrically conducting fluid, whereas its subtopics, magnetohydrodynamics (MHD) and magnetogas dynamics (MGD), are specifically concerned with electrically conducting liquids and ionized compressible gases.

There are many natural phenomena and engineering problems susceptible to magnetofluid dynamic analysis. It is useful in astrophysics because much of the universe is filled with widely spaced, charged particles and permeated by magnetic fields, and so the continuum assumption becomes applicable. Again, geophysicists encounter MFD phenomena in the interactions of conducting fluids and magnetic fields that are present in and around heavenly bodies. Engineers employ MFD principles in the design of heat exchangers, pumps, and flow meters; in solving space vehicle propulsion, control, and reentry problems; in designing communications and radar systems; in creating novel power generating systems; and in developing confinement schemes for controlled fusion.

Nothing illustrates the current importance of MFD better than the controlled fusion problem. Here, successful confinement of the plasma, partially an MFD problem, will free mankind of the specter of energy shortage due to the predicted depletion of the world's fuel sources and the rapidly increasing importance of environmental pollution by smokestack effluents and waste energy discharged into streams and lakes. Very large MFD-augmented power stations, for example, 2,000 to 5,000 megawatts, would operate at 15 to 20 percent higher efficiency than steam turbine or nuclear plants, significantly reducing thermal pollution. MFD plants would produce large quantities of commercially valuable oxides of nitrogen and the process of extracting these byproducts would produce very clean smokestack gases. Also, such plants could be fueled by char, a byproduct of the gasification of coal, which will become a major source of gas for heating houses as the demand outstrips our national resources in the late 1970s.

The principal MFD effects were first demonstrated in experiments of Faraday and Ritchie. In 1832, Faraday[1] experimented with flow of mercury in glass tubes placed between poles of a magnet, and discovered that a voltage was induced across the tube by the motion of the mercury across the magnetic field, perpendicular to the direction of flow and to the magnetic field. Faraday observed that the current generated by this induced voltage interacted with the magnetic field to slow down the motion of the fluid, and he was aware of the fact that the current produced its own magnetic field that obeyed Ampère's right-hand rule[2] and thus, in turn, distorted the field of the magnet. He conjectured[1] that similar phenomena could occur in nature, and in 1832 attempted to measure the voltage induced by the flow of the Thames through the earth's magnetic field, but was unsuccessful because the riverbed short-circuited the resultant voltage. However, his theory was confirmed in 1851 by Wollaston[3] who did measure an induced voltage in the much deeper English Channel.

Ritchie,[4] a contemporary of Faraday, discovered in 1832 that when an electric field was applied to a conducting fluid perpendicularly to a magnetic field, it pumped the fluid in a direction perpendicular to both fields.

Faraday[5] also suggested that electrical power could be generated in a load circuit by the interaction of a flowing conducting fluid and a magnetic field.

The practical applications of Faraday's ideas came with Smith and Slepian's invention of an instrument for measurement of ship's speed[6] and with Williams' MFD flow meter,[7] which was based on the principle that the induced voltage is proportional to the flow rate. Young, Gerrard, and Jevons[8] were the first (in 1920) to study tidal motions with an induced-voltage device, a technique since then widely used in oceanography. Further fundamental work in this area was done by Hartmann[9] and Hartmann and Lazarus,[10] who, like Faraday, studied channel flow of mercury.

The first astronomical application of the MFD theory occurred in 1899, when Bigalow[11] suggested that the sun was a gigantic magnetic system. It remained, however, for Alfvén[12] to make a most significant contribution by discovering MFD waves in the sun. These waves are produced by disturbances which propagate simultaneously in the conducting fluid and the magnetic field. The analogy that explains the generation of an Alfvén wave is that of a harp string plucked while submerged in a fluid. The string provides the elastic force and the fluid the inertia force, and they combine to propagate a perturbing wave through the fluid and the string.

More recently, Ahlstrom's experiments[13] demonstrated that disturbances generated by an aerodynamic body could propagate in the magnetic field and the fluid to produce both upstream and downstream wakes.

Even though the important MFD effects were demonstrated long ago, most of the practical work in this field has been done since 1950 because of the era's interest in high-temperature gases, nuclear engineering, and space technology. Among the conspicuous applications of MFD theory are meters for liquid-metal heat-transfer systems employed in nuclear reactors. Since 1960, MFD power generation has been the object of a worldwide research-and-development effort.

In summary, MFD phenomena result from the mutual effect of a magnetic field and a conducting fluid flowing across it. Thus, an electromagnetic force is produced in a fluid flowing across a transverse magnetic field, and the resulting current and magnetic field combine to produce a force that resists the fluid's motion. The current also generates its own magnetic field which distorts the original magnetic field. An opposing, or pumping force on the fluid can be produced by applying an electric field perpendicularly to the magnetic field. Disturbance in either the magnetic field or the fluid can propagate in both to produce MFD waves, as well as upstream- and downstream-wake phenomena. The science of magnetofluid dynamics is the detailed study of these phenomena, which occur in nature and are produced in engineering devices.

1.2 MAGNETOFLUID-DYNAMIC PARAMETERS

To the nondimensional parameters of ordinary fluid dynamics—the Mach, Reynolds, and Prandtl numbers—must be added two new parameters to characterize the two primary MFD effects. The first parameter is the magnetic pressure number $R_H = \mu_e H_0^2/\rho U^2$ which is the ratio of the magnetic pressure to the dynamic pressure, and is a measure of the effect of the magnetic field on the fluid. Only when R_H is the order of unity will the flow be influenced noticeably by the magnetic field and if it is very small, all the magnetic effects can be disregarded. The second parameter is the magnetic Reynolds number $R_\sigma = UL/(1/\mu_e\sigma)$ which is the ratio of the fluid flux to the magnetic diffusivity and is one of the more important parameters in MFD. The magnetic Reynolds number determines the diffusion of the magnetic field along the streamlines similar to the ordinary Reynolds number $R_e = UL/\nu$ which determines the diffusion of vorticity along the streamlines. The magnetic Reynolds number is, then, a measure of the effect of the flow on the magnetic field. If it is very small compared to unity, the magnetic field is not distorted by the flow. When it is very large, the magnetic field moves with the flow and is called *frozen-in*.

In engineering problems, it is difficult to obtain a R_σ greater than unity because of the low electrical conductivity of the useful fluids, whereas R_H can be easily varied from unity to ten. In astrophysical problems, R_σ is very large due to characteristically great lengths, and the electrical current is controlled by the magnetic field rather than by the electrical conductivity when the R_σ is very small.

Other parameters, which may be more appropriate for special problems, can be obtained from combination of the two new MFD numbers and the ordinary fluid-dynamic numbers. Three have been extensively used. The Hartmann number

$$R_h = \mu_e H_0 L\left(\frac{\sigma}{\mu}\right)^{1/2} = (R_e R_H R_\sigma)^{1/2}$$

is the ratio of the magnetic force to the viscous force and was introduced by Hartmann to describe his experiments with viscous MHD channel flow. The magnetic number

$$R_m = \mu_e H_0 \left(\frac{\sigma L}{\rho U}\right)^{1/2} = (R_H R_\sigma)^{1/2}$$

is the ratio of the magnetic force to the inertial force, and when R_σ is very small R_m is also used to measure the electromagnetic effects on the flow. The magnetic Prandtl number

$$P_m = \frac{R_\sigma}{R_e} = \frac{\nu}{\nu_H} = \frac{\mu\sigma\mu_e}{\rho}$$

is the ratio of the viscous diffusivity to the magnetic diffusivity and is similar to the ordinary Prandtl number which compares the viscous and thermal diffusivities. The P_m, which is generally small, is a measure of the relative magnitude of the fluid boundary layer thickness to the magnetic boundary layer thickness.

In ordinary inviscid gas dynamics, there is only one mode of wave propagation, a sound wave, which depends only on the local temperature of the fluid, while in magnetogas dynamics there exist three modes. These new waves are the Alfvén wave, which disturbs the magnetic field and fluid in the transverse direction, and the fast and slow magnetoacoustic waves whose propagation speeds depend on the temperature of the medium as well as on the magnitude and direction of the magnetic field. Consequently, the magnetic Mach number $M_m = v/V_H = R_H^{-1/2}$ is formed with the fluid and Alfvén velocity and is not an additional parameter, since it is defined with the magnetic pressure number R_H. MGD flow patterns are different when M_m is greater or less than unity and are called *super-Alfvén* or *sub-Alfvén flows*.

Before describing some of the other newly recognized MFD effects, a few sample values of R_σ, R_H, and V_H are calculated to provide a feeling for their size in problems of interest. If we take a high value of 100 mho/m for the electrical conductivity of an ionized gas, then $R_\sigma = 4\pi \times 10^{-5} UL$. For a space flight problem, we take a length of 1 m and a velocity of 10^4 m/sec and then $R_\sigma = 1.26$; in astrophysical problems, an appropriate length would be 10^6 m, and $R_\sigma = 1.26 \times 10^6$. If the space flight occurs at a 100-km altitude ($\rho = 2 \times 10^{-6}$ kg/m^3), and we use a magnetic field of 1/10 weber/m^2, which is the strength of a good toy magnet, then $R_H = 40$. Comparing this with a laboratory experiment with the flow of mercury ($\rho = 13.6 \times 10^6$ kg/m^3) at 1 m/sec, $R_H = 0.59$. The Alfvén-wave velocities for these examples would be 6.35×10^4 m/sec for the space flight and 0.77 m/sec for the laboratory experiment. In both cases, the magnetic Prandtl number would be small since for mercury P_m is 0.15×10^{-6} and for a 10,000°K plasma P_m is 2.8×10^{-5}. Apparently, in MFD engineering problems, the magnetic Reynolds and Prandtl numbers will be small, while the magnetic pressure number and Alfvén velocity can vary over a considerable range.

1.3 MAGNETOFLUID-DYNAMIC EFFECTS

The diffusion phenomena in MFD are characterized by the magnetic diffusivity ν_H along with the ordinary gas-dynamic diffusion coefficients μD and k. If the ordinary and magnetic Reynolds numbers are large, along with the Peclet number $P_e = \Pr R_e$, the corresponding diffusion phenomena are limited to thin layers near boundaries and are called *boundary layers*. If they are not large, then the diffusion phenomena spread over the entire flow field. In many problems, R_e and P_e are large and R_σ is small and the fluid is considered inviscid, non-heat conducting, with a small or finite electrical conductivity. If R_σ is also very large due to an infinite electrical conductivity, then the fluid is called an *ideal plasma*. When the magnetic and flow boundary layer thicknesses are approximately the same size, R_σ and M_m have a significant influence on the boundary-layer flow and new phenomena occur. Consider the two-dimensional flow of a uniform stream over a body with a magnetic field in the flow direction. If the magnetic pressure number is greater than unity, there will be an upstream as well as a downstream wake due to the combined magnetic- and fluid-dynamic disturbances.

Studies of MFD boundary-layer flow over flat plates and in the stagnation region of bodies with transverse magnetic fields demonstrate that the magnetic field reduces the skin friction and heat transfer and increases the shock detachment distance. However, in practical cases, the effects are small because of the low electrical conductivity of the fluid. Usable magnetic field strengths for flight applications are also limited due to the large weights of magnetics.

One of the most interesting MGD problems is the use of the electromagnetic force for the control of the flight of an aircraft or a space vehicle. A thin body in a uniform stream produces gas-dynamic forces that are a few percent of the dynamic pressure of the free stream. Now $R_m{}^2 = (R_\sigma R_H)$ should be approximately 0.1 for the magnetic force to be the same order of magnitude as the gas-dynamic force. If we take the length and velocity of our previous space flight example, but with an electrical conductivity of 10 mho/m, then $R_\sigma = 0.1$ and for $R_H = 1$, to produce the desired magnetic number, the magnetic field strength should be 0.015 weber/m^2. Earth and interplanetary magnetic fields are much smaller, but the magnetic field could be provided by a magnet in the vehicle. Since electric and magnetic fields can be controlled, flight in a conducting gas can be controlled by changing the fields inside the vehicle.

The most important application of MFD is the generation of electrical power with the flow of an electrically conducting fluid through a transverse magnetic field. Recently, experiments with ionized gases have been performed with the hopes of producing power on a large scale in stationary plants with large magnetic fields (about 5 weber/m^2). Cryogenic and superconducting magnets are required to produce these very large magnetic fields. Generation of MFD power on a smaller scale is of interest for space applications.

1.4 PERISCOPE

Mainly, this book considers MFD theory as it relates to special problems important to aerospace engineers and applied physicists. In general, laminar MFD theory has been well developed; unfortunately, there is little experimental verification of these phenomena. True understanding of MFD flow problems is limited and combined theoretical and experimental studies of simple flows, given all possible boundary conditions, are required. In ordinary fluid dynamics, the knowledge of turbulent flow is poor and thus, little progress is possible in MFD turbulent flow theory. Experiments will provide the main understanding through semiempirical correlations.

In Chap. II, the MFD equations are derived from the physical laws to provide an understanding of the equations and pertinent terms. Chapter III considers one-dimensional MFD flow in detail because of its importance in applications; the calculation of the associated heat and pressure losses is the main subject of Chap. IV. Chapter V describes the new wave phenomena, important in communication and astrophysical problems. The application of MFD theory to engineering problems such as power generation, pumps, and accelerators for space propulsion is contained in Chap. VI. The extensions and improvements of MFD theory are described in the last chapter.

Appendix E lists references concerning important aspects of MFD, and Appendix C lists property data for selected fluids.

REFERENCES

1. M. Faraday, Experimental Researches in Electricity, *Phil. Trans.*, vol. 15, p. 175, 1832.
2. L. B. Loeb, "Fundamentals of Electricity and Magnetism," John Wiley & Sons, Inc., New York, 1947.
3. C. Wollaston, Tidally Induced E.M.F.'s in Cables, *J. Soc. Tel. Engrs.*, vol. 10, p. 50, 1881.
4. W. Ritchie, Experimental Researches in Voltaic Electricity and Electro-Magnetism, *Phil. Trans. Royal Soc.* (*London*), vol. 122, pp. 279–298, 1832.
5. S. I. Pai, "Magnetogasdynamics and Plasma Dynamics," Springer-Verlag, New York, 1962.
6. C. G. Smith and J. Slepian, Electromagnetic Ships Log, United States Patent 1, 249, 530, December, 1917.
7. E. J. Williams, The Induction of E.M.F.'s in a Moving Fluid by a Magnetic Field and Its Application to an Investigation of the Flow of Liquids, *Proc. Phys. Soc.* (*London*), vol. 42, p. 466, 1930.

8. F. B. Young, H. Gerrard, and W. Jevons, On Electrical Disturbances Due to Tides and Waves, 6th series, *Phil. Mag.*, vol. 40, p. 149, 1920.
9. J. Hartmann, Hg-Dynamics, part I, *Kgl. Danske Videnskab. Selskab, Mat.-Fys. Medd.*, vol. 15, no. 6, 1937.
10. J. Hartmann and F. Lazarus, Hg-Dynamics, part II, *Kgl. Danske Videnskab. Selskab, Mat.-Fys. Medd.*, vol. 15, no. 7, 1937.
11. V. C. A. Ferraro and C. Plumpton, "Magneto-Fluid Mechanics," Oxford University Press, London, 1961.
12. H. Alfvén, On the Existence of Electromagnetic-Hydrodynamic Waves, *Arkiv F. Mat. Astro. O. Fysik. Bd.*, vol. 29B, no. 2, 1942.
13. H. G. Ahlstrom, Experiments on the Upstream Wake in Magnetofluid Dynamics, *J. Fluid Mech.*, vol. 15 (part II), pp. 205–222, 1963.

Chapter 2

THE MAGNETOFLUID-DYNAMIC EQUATIONS

2.1 INTRODUCTION

The MFD equations are presented here as the basis for describing the various phenomena contained in succeeding chapters. The viewpoint adopted for the major portion of this book is to provide continuum or macroscopic equations; microscopic or particle considerations are offered in the last chapter. The magnetoelectro-dynamic equations (Maxwell equations) are obtained as physical laws describing experiments that were conducted in the late 1700s and the early 1800s, a time of fantastic discovery in the physical and mathematical sciences. The equation for the propagation of electromagnetic disturbances is then derived from the Maxwell equations and is included for later comparison with MGD wave phenomena and for its historical importance as a root for the theory of relativity. The equation for the diffusion of a magnetic field through a permeable medium is derived to demonstrate the consequences of large and small magnetic Reynolds numbers. The equations of linear and angular momentum for an electrically conducting fluid are derived from the requirement of local rotational equilibrium and D'Alembert's principle. These equations are used with a generalized Hooke's law for small deformations to obtain the final form of the Navier-Stokes equations. The energy equation is derived from an energy balance for a fluid element and the First Law of Thermodynamics.

The assignment of numerical values to quantities requires the specification of reference or unit quantities. Many systems have been proposed for use in electromagnetics and the details of their development may be found in Refs. 1 and 2. The universally recognized system used in this book was introduced by Giorgi in 1901 and is called the *MKS system*. In this system, the length is measured in *m*eters, mass in *k*ilograms and the time in *s*econds. Another basic quantity must be defined for electric and magnetic quantities. The choice is the coulomb for the unit of electrical charge or ampere for the unit of current which is the flow of charges. Any other electromagnetic or fluid flow quantity can be expressed in terms of these units from their defining relations. The potential is measured in volts, resistance in ohms, and the power in watts. The units for the various quantities and their dimensions are contained in Appendix A. One choice remains between the "rational" and "irrational" system of units. Our choice is the rational system which defines the proportionality constant k, introduced in Eq. (2.1) in the next section, as $1/4\pi$.

2.2 ELECTROSTATICS[3,4,5]

In 1785, Coulomb performed experiments to determine the behavior of the force between electrically charged bodies. His conclusions for the force between two charged objects that are small compared to the distance separating them are:

1. The magnitude of the force is proportional to the product of the magnitudes of the charges.
2. The magnitude of the force is inversely proportional to the square of the distance separating the charges.
3. The direction of the force is along the line connecting the charges.
4. The force is attractive if the charges are not alike, and repulsive if they are alike.
5. The force depends on the electrical permittivity ϵ of the medium in which the charges are placed.

These observations when expressed in mathematical form, which contains a proportionality constant k, are called Coulomb's law.

$$\mathrm{F}_{12} = \left(\frac{kq_1q_2}{\epsilon R^2}\right)\mathbf{a} = \left(\frac{q_1q_2}{4\pi\epsilon R^2}\right)\mathbf{a} \tag{2.1}$$

where F_{12} is the force exerted by charge q_2 on charge q_1. This law can be extended to describe the force on a charge q_0 due to a number of other charges q_1, q_2, q_3, and so forth, by superimposing the individual force vectors to obtain the total vector sum.

$$\mathrm{F}_{0i} = \frac{q_0}{4\pi\epsilon}\sum_{i=1}^{n}\frac{q_i\,\mathbf{a}_i}{R_i^2} \tag{2.2}$$

Coulomb's law is then an "action-at-a-distance" law. The same result could be obtained by superimposing the fields of each charge. The force of an electrostatic field depends only on its strength and not on its origin, and exists even though a charge q_0 is not present to detect it. An electrostatic field E is defined by the force that is exerted on a unit charge in the field and is a vector in the same direction as the force.

$$\mathrm{E} = \lim_{\Delta q \to 0}\frac{\mathrm{F}}{\Delta q} \tag{2.3}$$

The limit is required so that the charge Δq doesn't disturb the field and the smallest charge is an electron or positron with the strength 1.60×10^{-19} coul. In terms of both forms of Coulomb's law, the electrostatic field definition becomes

$$\mathrm{E} = \frac{\mathrm{F}}{\Delta q} = \frac{q\mathbf{a}_R}{4\pi\epsilon R^2} \tag{2.4}$$

and

$$\mathrm{E} = \frac{1}{4\pi\epsilon}\sum_{i=1}^{n}\frac{q_i\,\mathbf{a}_{Ri}}{R_i^2} \tag{2.5}$$

If there are a large number of charges in a finite volume element, then a charge density can be defined as

$$\rho_e = \lim_{\Delta V \to 0} \frac{\Delta q}{\Delta V} \tag{2.6}$$

The limit is approached before the charge Δq approaches an electron and by this limiting process, the charge density can be represented by a continuous function of position. The total charge in a given volume is then

$$Q = \int_v \rho_e \, dV \tag{2.7}$$

The electrostatic field definition can now be restated as

$$\mathbf{E}(x, y, z) = \frac{1}{4\pi\epsilon} \int_V \frac{\rho_e(x', y', z')}{R^2} \mathbf{a}_R \, dV' \tag{2.8}$$

The point (x, y, z) is any point in the electrostatic field due to the charge density in the volume that is a function of the points (x', y', z'). The electrostatic field can also be represented by a flux emanating from a point source. The total flux of $\mathbf{E}$ due to a charge q is determined by integrating the normal component of the field vector $\mathbf{E} \cdot d\mathbf{s}$ over a surface enclosing the charge q. Now

$$\mathbf{E} \cdot d\mathbf{s} = \frac{1}{4\pi\epsilon} \frac{q}{R^2} \mathbf{a}_R \cdot d\mathbf{s} \tag{2.9}$$

and the solid angle $d\Omega$ subtended by the surface element $d\mathbf{s}$ at q is

$$d\Omega = \frac{\mathbf{a}_R \cdot d\mathbf{s}}{R^2} \tag{2.10}$$

then

$$\mathbf{E} \cdot d\mathbf{s} = \frac{q \, d\Omega}{4\pi\epsilon} \tag{2.11}$$

The required integral can now be evaluated as

$$\int_S \mathbf{E}\cdot d\mathbf{s} = \frac{q}{\epsilon} \tag{2.12}$$

since $\int_S d\Omega = 4\pi$. If the surface encloses a number of charges then

$$\int_S \mathbf{E}\cdot d\mathbf{s} = \sum_{i=1}^{n} \frac{q_i}{\epsilon} = \frac{Q}{\epsilon} \tag{2.13}$$

and is called Gauss' flux theorem. This theorem states that the net outflow of the flux of E through a closed surface equals the total charge contained by the surface reduced by the permittivity of the medium. Flux lines originate on positive charges and terminate on negative charges and the number of lines depends on the source strength. Gauss' flux law for a volume of distributed charges is

$$\int_S \mathbf{E}\cdot d\mathbf{s} = \frac{1}{\epsilon}\int_V \rho_e \, dV \tag{2.14}$$

or by using Gauss' divergence theorem

$$\int_S \mathbf{E}\cdot d\mathbf{s} = \int_V \nabla\cdot\mathbf{E}\, dV = \frac{1}{\epsilon}\int_V \rho_e \, dV \tag{2.15}$$

which can be combined to give

$$\int_V \left(\nabla\cdot\mathbf{E} - \frac{\rho_e}{\epsilon}\right) dV = 0 \tag{2.16}$$

If the integrand is continuous in the volume and on its boundary, then

$$\nabla\cdot\mathbf{E} = \frac{\rho_e}{\epsilon} \tag{2.17}$$

which states that the flux lines leaving a volume element is due only to charges. In the absence of electric polarization of the medium, the electric displacement is

$$\mathbf{D} = \epsilon \mathbf{E} \tag{2.18}$$

and then

$$\nabla \cdot \mathbf{D} = \rho_e \tag{2.19}$$

which is the usual form of the charge continuity equation. This equation can be written in another familiar form by defining a scalar potential ϕ as

$$\phi = \frac{q}{4\pi\epsilon R} \tag{2.20}$$

then

$$\nabla \phi = \frac{q}{4\pi\epsilon} \mathbf{a} \frac{\partial (1/R)}{\partial R} = -\frac{q\mathbf{a}_R}{4\pi R^2 \epsilon} = -\mathbf{E} \tag{2.21}$$

or

$$\mathbf{E} = -\nabla\phi \tag{2.22}$$

The charge continuity equation becomes

$$\nabla \cdot \mathbf{D} = -\epsilon \nabla^2 \phi = \rho_e \tag{2.23}$$

or

$$\nabla^2 \phi = -\frac{\rho_e}{\epsilon} \tag{2.24}$$

which is called Poisson's equation. An interesting property of the electrostatic field can be determined by taking the curl of Eq. (2.22).

$$\nabla \times \mathbf{E} = \nabla \times (\nabla\phi) = 0 \tag{2.25}$$

This relation demonstrates that an electrostatic field is irrotational and conservative since the curl of the vector field E is zero.

2.3 ELECTROSTATIC ENERGY

The absolute potential due to a single charge was defined by Eq. (2.20) with R as the distance between a field point and the charge q. The energy required to bring a charge q' from infinity to a distance R from the charge q is then

$$W_e = \frac{qq'}{4\pi\epsilon R} \tag{2.26}$$

and is the potential energy of the charges, which is a function of the final position. The energy in a field of N charges can be calculated by superimposing the individual energies. After the kth charge has been moved into the presence of the previous charges then

$$\left(W_e\right)_k = \frac{1}{4\pi\epsilon} \sum_{i=1}^{N}{}' \frac{q_k q_i}{R_{ik}} \tag{2.27}$$

where the primed summation means that $i = k$ is not included. This equation is only valid for the last charge moved into the region. Thus the energy to assemble N charges is

$$W_e = \frac{1}{2} \sum_{k=1}^{N} \sum_{i=1}^{N}{}' \frac{q_k q_i}{4\pi\epsilon R_{ik}} \tag{2.28}$$

The factor $\frac{1}{2}$ is required since in summing over k each charge energy will be counted twice. Defining the scalar potential due to N charges as

$$\phi_k = \frac{1}{4\pi\epsilon} \sum_{i=1}^{N}{}' \frac{q_i}{R_{ik}} \tag{2.29}$$

then the potential energy in the medium is

$$W_e = \frac{1}{2} \sum_{k=1}^{N} q_k \phi_k \tag{2.30}$$

For a continuous charge distribution, it becomes

$$W_e = \frac{1}{2}\int_V \rho_e \phi \, dV \tag{2.31}$$

and ϕ is the potential distribution due to all the charges and is a continuous function of position. This equation can also be expressed in terms of the electrostatic field with Eq. (2.17),

$$W_e = \frac{\epsilon}{2}\int_V (\nabla \cdot \mathrm{E})\phi \, dV \tag{2.32}$$

Now since

$$\nabla \cdot (\phi \mathrm{E}) = \phi \nabla \cdot \mathrm{E} + \mathrm{E} \cdot \nabla \phi \tag{2.33}$$

then

$$W_e = \frac{\epsilon}{2}\int_V \nabla \cdot (\phi \mathrm{E}) \, dV - \frac{\epsilon}{2}\int_V \mathrm{E} \cdot \nabla \phi \, dV \tag{2.34}$$

or

$$W_e = \frac{\epsilon}{2}\int_S \phi \mathrm{E} \cdot ds + \frac{\epsilon}{2}\int_V \mathrm{E}^2 dV \tag{2.35}$$

and the first integral, when integrated by parts, goes to zero, i.e.,

$$\lim_{R \to \infty} \int \phi \mathrm{E} \cdot ds = \left[\phi \int \mathrm{E} \cdot ds\right]_{R \to \infty} - \left[\int \left(\int \mathrm{E} \cdot ds\right) d\phi\right]_{R \to \infty} = 0$$

since as

$$R \to \infty, \quad \int \mathrm{E} \cdot ds \to 0 \quad \text{and} \quad \phi \sim \frac{1}{R} \to 0$$

If the original volume extends to infinity, then s extends to infinity, i.e., as $R \to \infty$ then $\phi \sim 1/R, E \sim 1/R^2, s \sim R^2$, and the first integral is proportional to $1/R$ which goes to zero. The energy in an electrostatic field in all space is then

$$W_e = \frac{\epsilon}{2}\int_V E^2 dV \tag{2.36}$$

A similar expression can be derived for charges distributed continuously in a dielectric medium, i.e.,

$$W_e = \frac{1}{2}\int_V \mathbf{E}\cdot\mathbf{D}\, dV \tag{2.37}$$

The integrand $\mathbf{E}\cdot\mathbf{D}$ is termed the *energy density* in the sense that its volume integral produces the total potential energy in the field since it is impossible to localize the total energy of the field.

2.4 ELECTRODYNAMICS

If we define a current density $\mathbf{J}$ as the flow of charges across a unit cross-sectional area per second, then the total current flowing across a surface is

$$I = \int_S \mathbf{J}\cdot d\mathbf{s} \tag{2.38}$$

The net flow of current into a volume element must produce an increase in the charge in the volume. Thus,

$$-\int_S \mathbf{J}\cdot d\mathbf{s} = \frac{\partial}{\partial t}\int_V \rho_e\, dV = \int_V \frac{\partial \rho_e}{\partial t}\, dV \tag{2.39}$$

The interchange of the limiting process is permissible since ρ_e is a continuous function of position including the limit that defines ρ_e.

Now using Gauss' divergence theorem and combining

$$\int_V \left(\nabla \cdot \mathbf{J} + \frac{\partial \rho_e}{\partial t} \right) dV = 0 \tag{2.40}$$

and if the integrand is continuous over the volume and on its boundaries, then

$$\nabla \cdot \mathbf{J} + \frac{\partial \rho_e}{\partial t} = 0 \tag{2.41}$$

and is called the *current continuity equation*. In stationary flow $\partial \rho_e / \partial t = 0$ then

$$\nabla \cdot \mathbf{J} = 0 \tag{2.42}$$

Thus the current density is solenoidal and must form current loops since the divergence of the vector field $\mathbf{J}$ is zero.

In 1819, Oersted discovered that when an electrical circuit was closed, a nearby magnetic needle was deflected and his investigations showed that a magnetic field is produced about a conductor carrying a current. The nature of the field was discovered by Ampère, who stated that the direction of the magnetic field is the same as the fingers of the closed right hand when the thumb points in the direction of the current. In 1820, Ampère observed that the force on a magnetic pole m in the neighborhood of a current-carrying wire (Fig. 2.1) is proportional to the length of wire taken perpendicular to the line joining the element and the point being considered, proportional to the current, and inversely proportional to the square of the distance from the current, i.e.,

$$f = \frac{midl}{R^2} \tag{2.43}$$

This "action-at-a-distance" law gives a magnetic definition to the current which is expressed as

$$dH = \frac{idl}{R^2} \tag{2.44}$$

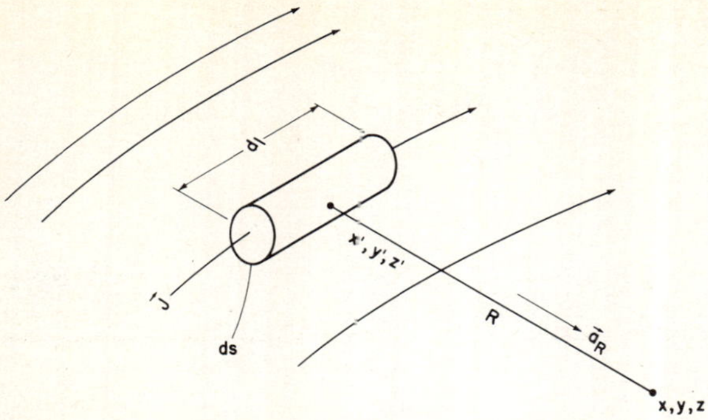

Fig. 2.1 Current produced magnetic field.

where the magnetic field strength H is defined to be the force exerted on a unit magnetic pole.

A field equation by Biot and Savart was formulated to replace Ampère's relation. In the absence of magnetic polarization of the medium, the magnetic induction is

$$\mathbf{B} = \mu_0 \mathbf{H} \tag{2.45}$$

The partial field $d\mathbf{B}$ at (x, y, z) due to the current flow-tube element of length dl at (x', y', z') is

$$d\mathbf{B} = \frac{\mu_0}{4\pi R^2} \mathbf{J}(x', y', z') \times \mathbf{a}_R \, dl \, ds \tag{2.46}$$

and is called the *Biot-Savart law*. In Eq. (2.46)

$$R = [(x - x')^2 + (y - y')^2 + (z - z')^2]^{1/2} \tag{2.47}$$

and the total current is $\mathbf{J}\,ds$. The magnetic field due to the current in a volume v is then

$$\mathbf{B}(x, y, z) = \frac{\mu_0}{4\pi} \int_v \frac{\mathbf{J}(x', y', z') \times \mathbf{a}_R}{R^2} dv' \tag{2.48}$$

Defining a vector potential $\mathbf{A}(x, y, z)$ as

$$\mathbf{A}(x, y, z) = \frac{\mu_0}{4\pi} \int_v \frac{\mathbf{J}(x', y', z')}{R} \, dv' \tag{2.49}$$

and since

$$\frac{\mathbf{a}_R}{R^2} = -\nabla\left(\frac{1}{R}\right) \tag{2.50}$$

then

$$\mathbf{B}(x, y, z) = -\frac{\mu_0}{4\pi} \int_v \mathbf{J}(x', y', z') \times \nabla\left(\frac{1}{R}\right) dv' \tag{2.51}$$

but

$$\nabla \times \left(\frac{\mathbf{J}}{R}\right) = \frac{1}{R} \nabla \times \mathbf{J} - \mathbf{J} \times \nabla\left(\frac{1}{R}\right) = -\mathbf{J} \times \nabla\left(\frac{1}{R}\right) \tag{2.52}$$

since $\nabla \times \mathbf{J}$ is zero for stationary flow. Therefore,

$$\mathbf{B}(x, y, z) = \frac{\mu_0}{4\pi} \int_v \nabla \times \left[\frac{\mathbf{J}}{R}(x', y', z')\right] dv' = \nabla \times \frac{\mu_0}{4\pi} \int_v \frac{\mathbf{J}(x', y', z')}{R} \, dv' \tag{2.53}$$

or

$$\mathbf{B}(x, y, z) = \nabla \times \mathbf{A} \tag{2.54}$$

The interchange of ∇ and the integration in Eq. (2.53) is permissible since ∇ is on (x, y, z) and $\mathbf{J}$ and dv' are defined for (x', y', z'). An essential property of the magnetic field is determined by taking the divergence of the magnetic field vector.

$$\nabla \cdot \mathbf{B} = \nabla \cdot (\nabla \times \mathbf{A}) = 0 \tag{2.55}$$

or

$$\nabla \cdot \mathbf{B} = 0 \tag{2.56}$$

This is the continuity equation for the magnetic field which is solenoidal, continuous, and forms closed loops. The equation is in agreement with the fact that single magnetic poles have never been observed. All magnetics have both north and south poles.

2.5 AMPÈRE'S LAW

We have determined the magnetic field strength at a given point due to the flow of charges at some point or region and we wish to determine the magnetic field strength due to a current at the same point. Remembering that the magnetic field has a zero divergence, the current producing the field must then be related, somehow, to the curl of the magnetic field. This relation can be determined as follows. Taking the curl of Eq. (2.53)

$$\nabla \times \mathbf{B}(x, y, z) = \nabla \times \nabla \times \frac{\mu_0}{4\pi} \int_v \frac{\mathbf{J}(x', y', z')}{R} \, dv' \tag{2.57}$$

Using the vector relation

$$\nabla \times \nabla \times (\;) = \nabla\nabla \cdot (\;) - \nabla^2(\;) \tag{2.58}$$

and since ∇ is defined on (x, y, z) and the integral is defined on (x', y', z') then,

$$\nabla \times \mathbf{B}(x, y, z) = \frac{\mu_0}{4\pi} \int_v \left[\nabla\nabla \cdot \left(\frac{\mathbf{J}(x', y', z')}{R}\right) - \mathbf{J}(x', y', z')\nabla^2\left(\frac{1}{R}\right)\right] dv' \tag{2.59}$$

Now

$$\int_v \nabla\nabla \cdot \frac{\mathbf{J}(x', y', z')}{R} \, dv' = \nabla \int_v \nabla \cdot \left(\frac{\mathbf{J}(x', y', z')}{R}\right) dv' \tag{2.60}$$

$$= \nabla \int_v \left[\mathbf{J} \cdot \nabla\left(\frac{1}{R}\right)\right] dv' \tag{2.61}$$

$$= \nabla \int_v \left[-\mathbf{J} \cdot \nabla' \left(\frac{1}{R} \right) \right] dv' \tag{2.62}$$

since

$$\nabla \left(\frac{1}{R} \right) = -\nabla' \left(\frac{1}{R} \right) \tag{2.63}$$

Thus,

$$\int_v \nabla \nabla \cdot \left(\frac{\mathbf{J}(x', y', z')}{R} \right) dv' = -\nabla \int_v \left[-\nabla' \cdot \left(\frac{\mathbf{J}(x', y', z')}{R} \right) \right] dv' \tag{2.64}$$

since

$$\nabla' \cdot \left(\frac{\mathbf{J}(x', y', z')}{R} \right) = \frac{1}{R} \nabla' \cdot \mathbf{J}(x', y', z') + \mathbf{J}(x', y', z') \cdot \nabla \left(\frac{1}{R} \right) \tag{2.65}$$

and

$$\nabla' \cdot \mathbf{J}(x', y', z') = 0 \tag{2.66}$$

for stationary currents. Using Gauss' divergence theorem

$$\int_v \nabla \nabla \cdot \left(\frac{\mathbf{J}(x', y', z')}{R} \right) dv' = -\nabla \int_s \left(\frac{\mathbf{J}(x', y', z')}{R} \right) \cdot d\mathbf{s}' \tag{2.67}$$

where s' is any surface surrounding v'. Now the stationary current is in a finite region and we can choose s' so large that all the current is inside since it must form loops. Thus

$$\int_s (\mathbf{J}(x', y', z')) \cdot d\mathbf{s}' = 0 \tag{2.68}$$

on s' and

$$\int_v \nabla \nabla \cdot \frac{\mathbf{J}(x', y', z')}{R} dv' = 0 \tag{2.69}$$

Equation (2.59) becomes

$$\nabla \times \mathbf{B}(x, y, z) = -\frac{\mu_0}{4\pi} \int_v \mathbf{J}(x', y', z') \nabla^2 \left(\frac{1}{R}\right) dv' \tag{2.70}$$

Now $\nabla^2(1/R)$ is zero for all finite R except that it has a singularity at $R = 0$ which can be integrated to give a finite nonzero value. Near $R = 0$ the current density doesn't deviate much from its value at $R = 0$, i.e., $x' = x$, $y' = y$ and $z' = z$. Since the integrand is zero everywhere except at $R = 0$, we can take the volume integral over a small sphere and can let the current density be a constant throughout the sphere. Since

$$\nabla^2 \left(\frac{1}{R}\right) = \nabla'^2 \left(\frac{1}{R}\right) = \nabla' \cdot \nabla' \left(\frac{1}{R}\right) \tag{2.71}$$

then

$$\nabla \times \mathbf{B}(x, y, z) = -\frac{\mu_0}{4\pi} \mathbf{J}(x, y, z) \int_v \nabla' \cdot \nabla' \left(\frac{1}{R}\right) dv' \tag{2.72}$$

or

$$\nabla \times \mathbf{B}(x, y, z) = -\frac{\mu_0}{4\pi} \mathbf{J}(x, y, z) \int_s \nabla' \left(\frac{1}{R}\right) \cdot d\mathbf{s}' \tag{2.73}$$

The element area in spherical coordinates is $\mathbf{n} R^2 d\Omega$, where $d\Omega = \sin\theta d\phi d\theta$ and

$$\nabla' \left(\frac{1}{R}\right) = \frac{\mathbf{a}_R}{R^2} \tag{2.74}$$

Therefore,

$$\nabla' \left(\frac{1}{R}\right) \cdot d\mathbf{s}' = \mathbf{n} \cdot \mathbf{a}_R \, d\Omega = -d\Omega \tag{2.75}$$

since $\mathbf{a}_R$ is radially inward. Now

$$\nabla \times \mathbf{B}(x, y, z) = \frac{\mu_0}{4\pi} \mathbf{J}(x, y, z) \int_s d\Omega = \mu_0 \mathbf{J}(x, y, z) \tag{2.76}$$

or

$$\nabla \times \mathbf{B} = \mu_0 \mathbf{J} \tag{2.77}$$

which is the original form of Ampère's circuital law and was derived for steady current flow. Maxwell discovered the following inconsistency between Eq. (2.77) and the current continuity equation, Eq. (2.41). Since the divergence of the curl is zero for any vector then

$$\nabla \cdot (\nabla \times \mathbf{B}) = \nabla \cdot (\mu_0 \mathbf{J}) = \mu_0 \nabla \cdot \mathbf{J} = 0 \tag{2.78}$$

But the current continuity equation is

$$\nabla \cdot \mathbf{J} = -\frac{\partial \rho_e}{\partial t} \neq 0 \tag{2.79}$$

which contradicts the requirement of Eq. (2.78). Maxwell concluded that Ampère's law was incomplete and required another current to satisfy the vector requirement. This additional current can be determined as follows. Let

$$\frac{1}{\mu_0} (\nabla \times \mathbf{B}) = \mathbf{J} + \mathbf{J}_d \tag{2.80}$$

then

$$\frac{1}{\mu_0} \nabla \cdot (\nabla \times \mathbf{B}) = \nabla \cdot \mathbf{J} + \nabla \cdot \mathbf{J}_d = 0 \tag{2.81}$$

and

$$\nabla \cdot \mathbf{J}_d = -\nabla \cdot \mathbf{J} = \frac{\partial \rho_e}{\partial t} \tag{2.82}$$

Now the electric displacement field equation, Eq. (2.19), was derived for stationary fields but should also apply for time-dependent fields. Thus,

$$\frac{\partial(\nabla \cdot \mathbf{D})}{\partial t} = \nabla \cdot \left(\frac{\partial \mathbf{D}}{\partial t}\right) = \frac{\partial \rho_e}{\partial t} \tag{2.83}$$

and then

$$\nabla \cdot \mathbf{J}_d = \nabla \cdot \left(\frac{\partial \mathbf{D}}{\partial t}\right) \tag{2.84}$$

or

$$\mathbf{J}_d = \frac{\partial \mathbf{D}}{\partial t} \tag{2.85}$$

The general form of Ampère's law is then

$$\frac{1}{\mu_0}(\nabla \times \mathbf{B}) = \mathbf{J} + \frac{\partial \mathbf{D}}{\partial t} \tag{2.86}$$

and Maxwell called the additional current a *displacement current density*. At the time he derived this result, there was no direct experimental evidence of such a current. The main supporting evidence is the agreement of its consequences with experiment. The existence of the displacement current is now well founded and can be interpreted as follows. If the charge density in the current tube being considered is permitted to change with time, then this produces a time change in the electric displacement field in the medium. This is equivalent to a current and must be considered when computing the induced magnetic field with Ampère's law. In stationary flows $\partial \mathbf{D}/\partial t$ and $\partial \rho_e/\partial t$ are zero and then the vector requirement is satisfied. Thus, in order to neglect $\partial \mathbf{D}/\partial t$ the value of $\partial \rho_e/\partial t$ must be zero and conversely if $\partial \rho_e/\partial t$ is zero, then $\partial \mathbf{D}/\partial t$ must be zero.

2.6 AMPÈRE'S FORCE LAW

As a part of his experiments, Ampère determined the force exerted by one current loop on another current loop. He also replaced the one circuit by a magnet and demonstrated that the force was magnetic. Ampère deduced from the experiments that the force F_{21} (Fig. 2.2) produced by C_1 on C_2 is

$$\mathbf{F}_{21} = \frac{\mu_0}{4\pi} \oint_{C_2} \oint_{C_1} \frac{I_2 d\mathbf{l}_2 \times (I_1 d l_1 \times \mathbf{a}_R)}{R^2} \tag{2.87}$$

and is an action-at-a-distance law. This law can be restated as a field law similar to the field concepts in electrostatics. Thus, let

$$\mathbf{B}_{21} = \frac{\mu_0}{4\pi} \oint_{C_1} \frac{I_1 d\mathbf{l}_1 \times \mathbf{a}_R}{R^2} \tag{2.88}$$

from the Biot-Savart law, represent the magnetic field. The force on the circuit C_2 due to the magnetic field $\mathbf{B}_{21}$ is then

$$\mathbf{F}_{21} = \int_{C_2} I_2 d\mathbf{l}_2 \times \mathbf{B}_{21} \tag{2.89}$$

and the force on an element of C_2 is

$$\mathbf{F} = I d\mathbf{l} \times \mathbf{B} \tag{2.90}$$

Now a moving charge q is equivalent to a current element and the

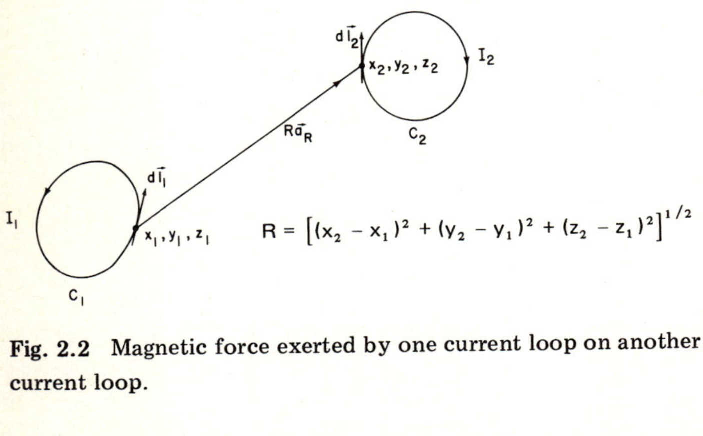

Fig. 2.2 Magnetic force exerted by one current loop on another current loop.

force on a charge moving in a magnetic field with a velocity $\mathbf{v}_p$ is

$$\mathbf{F}_p = q\mathbf{v}_p \times \mathbf{B} \tag{2.91}$$

which is called the *Lorentz force*. The combined force due to the Lorentz force and an electric field moving with the velocity $\mathbf{v}_p$ is then

$$\mathbf{F}_p = q(\mathbf{E} + \mathbf{v}_p \times \mathbf{B}) \tag{2.92}$$

This equation can be interpreted as the force on a charge due to an electric field which is moving with the charge velocity $\mathbf{v}_p$ through a stationary magnetic field. An observer moving with the charge would interpret this force as an electrostatic field. Thus,

$$\mathbf{E}' = \frac{\mathbf{F}_p}{q} = \mathbf{E} + \mathbf{v}_p \times \mathbf{B} \tag{2.93}$$

2.7 MAGNETOSTATIC ENERGY

Since it requires a force to move a charge across a magnetic field, then work is performed by a current moving in the circuit C_1 through the magnetic field produced by the circuit C_2. The energy required is

$$dw_m = I_1 \mathbf{B}_2 \cdot d\mathbf{s}_1 \tag{2.94}$$

where $d\mathbf{s}_1$ is normal to the surface s_1 enclosed by l_1. The total energy is then

$$w_m = I_1 \int_{s_1} \mathbf{B} \cdot d\mathbf{s}_1 = I_1 \int_{s_1} (\nabla \times A_2) \cdot d\mathbf{s}_1 \tag{2.95}$$

or

$$w_m = I_1 \oint_{l_1} \mathbf{A}_2 \cdot d\mathbf{l}_1 \tag{2.96}$$

Now

$$\mathbf{A}_2 = \frac{\mu_0 I_2}{4\pi} \oint_{l_2} \frac{d\mathbf{l}_2}{R} \tag{2.97}$$

and then

$$w_m = \frac{\mu_0 I_1 I_2}{4\pi} \oint_{l_2} \oint_{l_1} \frac{d\mathbf{l}_1 \cdot d\mathbf{l}_2}{R} \tag{2.98}$$

In a single circuit

$$w_m = \frac{\mu_0}{8\pi} \oint_{l} \oint_{l} \frac{I d\mathbf{l} \cdot I d\mathbf{l}'}{R} \tag{2.99}$$

when $d\mathbf{l}$ and $d\mathbf{l}'$ are different elements of the circuit and R is their separation distance. The current densities are

$$\mathbf{J} dv = I d\mathbf{l} \tag{2.100}$$

and

$$\mathbf{J}' dv' = I d\mathbf{l}' \tag{2.101}$$

Thus, the energy is

$$w_m = \frac{\mu_0}{8\pi} \int_v \int_v (\mathbf{J} \cdot \mathbf{J}') \frac{dv dv'}{R} \tag{2.102}$$

But

$$\mathbf{A} = \frac{\mu_0}{4\pi} \int_v \frac{\mathbf{J}' dv'}{R} \tag{2.103}$$

and

$$\nabla \times \mathbf{H} = \mathbf{J} \tag{2.104}$$

for steady current flow and then

$$w_m = \frac{1}{2}\int_v (\nabla \times \mathbf{H}) \cdot \mathbf{A}\, dv \tag{2.105}$$

or

$$w_m = \frac{1}{2}\int_v \mathbf{H} \cdot (\nabla \times \mathbf{A})\, dv - \frac{1}{2}\int_v \nabla \cdot (\mathbf{A} \times \mathbf{H})\, dv \tag{2.106}$$

Applying Gauss' divergence theorem to the second integral

$$\int_v \nabla \cdot (\mathbf{A} \times \mathbf{H})\, dv = \int_s (\mathbf{A} \times \mathbf{H}) \cdot d\mathbf{s} \tag{2.107}$$

Since $\mathbf{A} \sim 1/R$, $\mathbf{H} \sim 1/R^2$ and $d\mathbf{s} \sim R^2$ then as $R \to \infty$ the surface integral in Eq. (2.107) goes to zero. Equation (2.106) when taken over all space is then

$$w_m = \frac{1}{2}\int_v \mathbf{H} \cdot (\nabla \times \mathbf{A})\, dv = \frac{1}{2}\int_v (\mathbf{H} \cdot \mathbf{B})\, dv \tag{2.108}$$

The total energy in an electromagnetic field can now be expressed as

$$w = w_e + w_m = \frac{1}{2}\int_v (\mathbf{E} \cdot \mathbf{D} + \mathbf{H} \cdot \mathbf{B})\, dv \tag{2.109}$$

2.8 FARADAY'S LAW OF ELECTROMAGNETIC INDUCTION

Ampère's law describes the magnetic field produced by a current and it was suspected that a magnetic field could induce a current. Investigators searched for an experimental demonstration and it was found by Faraday in England and J. Henry in America in 1831. Faraday wound two separate coils on an iron ring and discovered that when the magnitude of the current changed in one coil, an induced current would flow in the other circuit. He found that a current could also be produced by moving a magnet in the vicinity of the coil. The experiments were conducted with circuits and wires but the results were assumed to hold in space even when no wires were present to detect the effect. Faraday's experiments took 10 days over a 3-month period and the present knowledge of this phenomenon is little more than his 10 days of discovery.

The direction of the current flow was deduced from the conservation of energy and is called *Lenz's law*. It states that "the current, induced by a magnetic field change in the region of a conductor, flows in the direction that will produce a magnetic field which opposes the change, or results in motion that opposes the change." The current must flow in that direction since a flow in the opposite direction would produce a magnetic field that would increase the changing magnetic field and thus generate a larger current at the expense of no additional energy. Faraday's law can be stated as "for any closed conducting path fixed in space (i.e., stationary), that is subjected to a changing magnetic field, the induced voltage around the path is equal to the negative time rate of change of the total magnetic flux through the closed path." This is an experimental law. The induced voltage around a path c is

$$V_{\text{ind}} = \oint_c \mathbf{E} \cdot d\mathbf{l} \tag{2.110}$$

The magnetic flux through c is

$$\psi = \int_s \mathbf{B} \cdot d\mathbf{s} \tag{2.111}$$

where s is any surface bounded by c. Therefore, Faraday's law is

$$\oint_c \mathbf{E} \cdot d\mathbf{l} = -\frac{\partial}{\partial t} \int_s \mathbf{B} \cdot d\mathbf{s} \tag{2.112}$$

Applying Stokes' law to convert the line integral into a surface integral, then

$$\oint_c \mathbf{E} \cdot d\mathbf{l} = \int_s (\nabla \times \mathbf{E}) \cdot d\mathbf{s} = -\frac{\partial}{\partial t} \int_s \mathbf{B} \cdot d\mathbf{s} \tag{2.113}$$

Since the surface is stationary,

$$\int_s (\nabla \times \mathbf{E}) \cdot d\mathbf{s} = -\int_s \left(\frac{\partial \mathbf{B}}{\partial t}\right) \cdot d\mathbf{s} \tag{2.114}$$

and

$$\int_s \left(\nabla \times \mathbf{E} + \frac{\partial \mathbf{B}}{\partial t}\right) \cdot d\mathbf{s} = 0 \tag{2.115}$$

If the integrand is continuous over the surface and on its boundary, then

$$\nabla \times \mathbf{E} = -\frac{\partial \mathbf{B}}{\partial t} \tag{2.116}$$

is the usual form of Faraday's law of electromagnetic induction. The electric vector field induced by a change of a magnetic field is not the same type vector field as an electrostatic field which has a zero curl. The induced electric field is rotational or solenoidal and is not conservative.

2.9 POYNTING VECTOR

We have just computed the total energy stored in a static electromagnetic field as potential energy. But we require the energy balance when the fields change with time. With Eqs. (2.45), (2.86), and (2.116), we compute

$$\mathbf{H} \cdot (\nabla \times \mathbf{E}) - \mathbf{E} \cdot (\nabla \times \mathbf{H}) = -\mathbf{H} \cdot \frac{\partial \mathbf{B}}{\partial t} - \mathbf{E} \cdot \mathbf{J} - \mathbf{E} \cdot \frac{\partial \mathbf{D}}{\partial t} \tag{2.117}$$

or

$$\nabla \cdot (\mathbf{E} \times \mathbf{H}) = -\mathbf{H} \cdot \frac{\partial \mathbf{B}}{\partial t} - \mathbf{E} \cdot \mathbf{J} - \mathbf{E} \cdot \frac{\partial \mathbf{D}}{\partial t} \tag{2.118}$$

then

$$\int_v \nabla \cdot (\mathbf{E} \times \mathbf{H})\, dv = -\int_v \mathbf{E} \cdot \mathbf{J}\, dv - \frac{1}{2} \int_v \frac{\partial}{\partial t} (\mathbf{E} \cdot \mathbf{D} + \mathbf{H} \cdot \mathbf{B})\, dv \tag{2.119}$$

or

$$-\frac{\partial w}{\partial t} = \int_v \mathbf{J} \cdot \mathbf{E}\, dv + \int_s (\mathbf{E} \times \mathbf{H}) \cdot d\mathbf{s} \tag{2.120}$$

by using Gauss' divergence theorem. This is the required energy balance which states that the time rate of decrease of the stored electromagnetic energy is equal to that dissipated by heating of the medium, called *Joule heating*, and the instantaneous flow of electromagnetic energy across the confining surface. The vector $\mathbf{E} \times \mathbf{H}$ is called the *Poynting vector* and represents the rate of flow of electromagnetic energy per unit area through the surface. The direction of flow of this energy density is in the direction of $\mathbf{E} \times \mathbf{H}$ although it is not possible to localize the energy. The energy balance is made only for the volume and its surface.

2.10 OHM'S LAW

In 1827, Ohm deduced from experiments that the resistance in a part of a circuit is equal to the ratio of the potential drop across that part of the circuit and the current produced in that part of the circuit. His law can be generalized for elements of a conducting medium and is

$$R = \frac{\mathbf{E}}{\mathbf{I}} = \frac{1}{\sigma} \tag{2.121}$$

In Sec. 2.6, we computed the Lorentz force on a single charge moving in a magnetic field and stated that the combined force, due to this motion and a moving electric field, was equivalent to a moving electrostatic field. Similarly, a moving electrostatic field is produced in a conducting medium moving with the velocity, v, in a magnetic field which produces a current called the *conduction current*. The total current in the medium is the vector sum of this conduction current and the convection current due to charges moving with the fluid. Applying Ohm's law for the conduction current,

$$\mathbf{J} = \mathbf{J}_v + \mathbf{J}_c = \rho_e \mathbf{v} + \sigma(\mathbf{E} + \mathbf{v} \times \mathbf{B}) \tag{2.122}$$

is the equation for the current density. This is the generalized form of Ohm's law used in most of this text, and it was derived for a conducting medium whose current is controlled by the resistance of the medium. This restriction applies when the magnetic field diffuses easily through the medium and when the frequency of collision of the charged particles is large compared to their frequency of rotation about the magnetic field lines. If these conditions are not fulfilled, then additional terms will appear in the effective voltage in Eq. (2.122). The Hall current is produced by such a voltage term which has the direction of $-\mathbf{J} \times \mathbf{B}$ and will be introduced in Chap. 7. The current density computed with Eq. (2.122) is the conduction current.

2.11 MAXWELL STRESSES

Faraday originated the idea that an electric field could be interpreted as a stress pattern in the medium. The lines of force crowding together at some points and spreading out at others represent a medium under stress and the forces felt by the medium will be generated by a stress tensor which can be generalized to include magnetic fields. This tensor is the electromagnetic or Maxwell stress tensor which will be derived in terms of E, D, H and B.

Consider the forces acting across an arbitrary surface s of the volume Ω in a medium for which the electromagnetic fields are defined. If ρ_e is the electrical charge density in Ω, then the Lorentz force acting on an element $d\Omega$ is

$$d\mathbf{F} = (\rho_e \mathbf{E} + \rho_e \mathbf{v} \times \mathbf{B})\, d\Omega \tag{2.123}$$

Defining J to be the current density then

$$\rho_e \mathbf{v} = \mathbf{J} = \nabla \times \mathbf{H} - \frac{\partial \mathbf{D}}{\partial t} \tag{2.124}$$

and

$$\rho_e = \nabla \cdot \mathbf{D} \tag{2.125}$$

from Ampère's law and the charge continuity equation. The force equation is now

$$d\mathbf{F} = \left[\mathbf{E}(\nabla \cdot \mathbf{D}) + \left(\nabla \times \mathbf{H} - \frac{\partial \mathbf{D}}{\partial t}\right) \times \mathbf{B}\right] d\Omega \tag{2.126}$$

and the total force throughout the volume Ω is

$$\mathbf{F} = \int_v \left[\mathbf{E}(\nabla \cdot \mathbf{D}) - \mathbf{B} \times (\nabla \times \mathbf{H}) - \frac{\partial \mathbf{D}}{\partial t} \times \mathbf{B}\right] d\Omega \tag{2.127}$$

or, by using Faraday's law

$$\mathbf{F} = \int_v \left[\mathbf{E}(\nabla \cdot \mathbf{D}) - \mathbf{B} \times (\nabla \times \mathbf{H}) - \frac{\partial (\mathbf{D} \times \mathbf{B})}{\partial t} - \mathbf{D} \times (\nabla \times \mathbf{E})\right] d\Omega \tag{2.128}$$

Recalling the vector expansion of $\nabla(\mathbf{H} \cdot \mathbf{B})$ and since $\mathbf{B} = \mu_0 \mathbf{H}$ then

$$2\mathbf{B} \times (\nabla \times \mathbf{H}) = \nabla(\mathbf{H} \cdot \mathbf{B}) - 2\mathbf{B} \cdot \nabla \mathbf{H} \tag{2.129}$$

and

$$2\mathbf{D} \times (\nabla \times \mathbf{E}) = \nabla(\mathbf{E} \cdot \mathbf{D}) - 2\mathbf{D} \cdot \nabla \mathbf{E} \tag{2.130}$$

since $\mathbf{D} = \epsilon \mathbf{E}$. Substituting these relations in Eq. (2.128) produces

$$\mathbf{F} = \int_v \left\{\mathbf{E}(\nabla \cdot \mathbf{D}) - \frac{\partial (\mathbf{D} \times \mathbf{B})}{\partial t} + (\mathbf{B} \cdot \nabla \mathbf{H} + \mathbf{D} \cdot \nabla \mathbf{E}) - \frac{1}{2}\left[\nabla(\mathbf{H} \cdot \mathbf{B} + \mathbf{E} \cdot \mathbf{D})\right]\right\} d\Omega \tag{2.131}$$

Defining w_{em}, the integrand of Eq. (2.109), as the electromagnetic energy density and

$$\mathcal{T} = \mathbf{ED} + \mathbf{HB} \tag{2.132}$$

which is a tensor of the second rank and is the sum of two dyads[5] (called a *dyadic*), then scaler products of vectors and dyadics can be formed and the force equation becomes

$$\mathbf{F} = \int_v \left[\nabla \cdot \mathcal{T} - \nabla w_{em} - \mathbf{H}\nabla \cdot \mathbf{B} - \frac{\partial(\mathbf{D} \times \mathbf{B})}{\partial t} \right] d\Omega \tag{2.133}$$

Now $\nabla \cdot \mathbf{B}$ is zero, thus

$$\mathbf{F} = \int_v (\nabla \cdot \mathcal{T} - \nabla w_{em}) d\Omega - \int_v \frac{\partial(\mathbf{D} \times \mathbf{B})}{\partial t} d\Omega \tag{2.134}$$

Introducing the unit dyadic $\mathcal{I}$ that is symmetric and does not change a vector, i.e.,

$$\mathbf{r} \cdot \mathcal{I} = \mathcal{I} \cdot \mathbf{r} = \mathbf{r} \tag{2.135}$$

then

$$\nabla w_{em} = \nabla \cdot (\mathcal{I} w_{em}) = \nabla \cdot (w_{em} \mathcal{I}) \tag{2.136}$$

and

$$\mathbf{F} = \int_v \nabla \cdot \{\mathcal{T} - w_{em} \mathcal{I}\} d\Omega - \int_v \frac{\partial(\mathbf{D} \times \mathbf{B})}{\partial t} d\Omega \tag{2.137}$$

since scaler products of vectors and dyadics are distributive. Applying Gauss' divergence theorem for tensors,

$$\mathbf{F} = \int_s (\mathcal{T} - w_{em} \mathcal{I}) \cdot \mathbf{n} \, ds - \int_v \frac{\partial(\mathbf{D} \times \mathbf{B})}{\partial t} d\Omega \tag{2.138}$$

The electromagnetic stress tensor can now be defined to be

$$\zeta = \mathfrak{T} - w_{em}\mathfrak{I} \tag{2.139}$$

and the force per unit area across a plane perpendicular to $\mathbf{n}$ is $\mathbf{n} \cdot \zeta$. Also, define

$$\mathbf{G} = \mathbf{D} \times \mathbf{B} \tag{2.140}$$

which is called the *electromagnetic momentum* since a force is equal to the time rate of change of momentum.

2.12 ELECTROMAGNETIC EQUATIONS SUMMARY

The Maxwell equations presented in previous sections represent mathematical expressions of certain experiments and cannot be proved. However, their applicability can be verified and, as a result of extensive experimental research, these equations are known to apply for almost all macroscopic electromagnetic phenomena. The Maxwell equations are now accepted as guiding principles. However, they were obtained for a medium at rest which could be set in motion. The equations can be reconstructed applying relativistic considerations with the result that they are unaltered except that the constitutive equations for $\mathbf{D}$ and $\mathbf{B}$ contain an additional term. These terms are negligible when the medium's velocity is small compared to the light speed. Thus, the constitutive relations for a medium at rest are approximately correct for a medium in motion at modest speeds. The electromagnetic equations are listed again for convenience and the first five are the Maxwell equations. The constitutive equations are then the assumptions.

Charge continuity $$\nabla \cdot \mathbf{D} = \rho_e \tag{2.141}$$

Current continuity $$\nabla \cdot \mathbf{J} = -\frac{\partial \rho_e}{\partial t} \tag{2.142}$$

Magnetic field continuity $$\nabla \cdot \mathbf{B} = 0 \tag{2.143}$$

Ampère's law $$\nabla \times \mathbf{H} = \mathbf{J} + \frac{\partial \mathbf{D}}{\partial t} \tag{2.144}$$

Faraday's law	$\nabla \times \mathbf{E} = -\dfrac{\partial \mathbf{B}}{\partial t}$	(2.145)
Constitutive	$\mathbf{D} = \epsilon \mathbf{E}$	(2.146)
and	$\mathbf{B} = \mu_0 \mathbf{H}$	(2.147)
Lorentz force on a charge	$\mathbf{F}_p = q(\mathbf{E} + \mathbf{v}_p \times \mathbf{B})$	(2.148)
Electromagnetic energy	$w = \dfrac{1}{2}\displaystyle\int_v (\mathbf{E}\cdot\mathbf{D} + \mathbf{H}\cdot\mathbf{B})\,dv$	(2.149)
Current density	$\mathbf{J} = \sigma(\mathbf{E} + \mathbf{v} \times \mathbf{B}) + \rho_e \mathbf{v}$	(2.150)
Maxwell stress tensor	$\zeta = \mathcal{T} - w_{em}\mathcal{I}$	(2.151)
Electromagnetic momentum	$\mathbf{G} = \mathbf{D} \times \mathbf{B}$	(2.152)
Poynting vector	$\mathbf{P} = \mathbf{E} \times \mathbf{H}$	(2.153)

The first six equations can also be written in terms of the scalar and vector potentials that were introduced earlier and are:

$$\nabla^2 \phi = -\frac{\rho_e}{\epsilon} \tag{2.141a}$$

$$\sigma\{-\nabla^2\phi + \nabla\cdot[\mathbf{v} \times (\nabla \times \mathbf{A})]\} = -\frac{\partial \rho_e}{\partial t} \tag{2.142a}$$

$$\nabla\cdot(\nabla \times \mathbf{A}) = 0 \tag{2.143a}$$

$$\nabla(\nabla\cdot\mathbf{A}) + \frac{1}{c^2}\frac{\partial^2\mathbf{A}}{\partial t^2} - \nabla^2\mathbf{A} + \mu_0\sigma\left[\frac{\partial \mathbf{A}}{\partial t} - \mathbf{v} \times (\nabla \times \mathbf{A})\right] = -\frac{1}{c^2}\left(\frac{\partial \nabla\Phi}{\partial t} + \mathbf{v}\nabla^2\Phi\right) - \mu_0\sigma\nabla\Phi \tag{2.144a}$$

where

$$-\nabla\Phi = \mathbf{E} + \frac{\partial \mathbf{A}}{\partial t}$$

$$\nabla \times \nabla\Phi = 0 \tag{2.145a}$$

$$\mathbf{D} = -\epsilon\nabla\phi \tag{2.146a}$$

$$\nabla \times \mathbf{A} = \mu_0 \mathbf{H} \tag{2.147a}$$

2.13 ELECTROMAGNETIC BOUNDARY CONDITIONS

In an infinite unbounded homogeneous medium, the physical properties vary continuously and solutions to the field equations are relatively easy to obtain. However, in practical cases, solutions are required for fields that penetrate different media through a separating boundary. If the material properties are different in the adjacent media, then components of the fields may be discontinuous and a description of the boundary conditions to be applied to the field vectors is required.

The boundary conditions for the normal field components may be determined by considering Fig. 2.3 which is a differential element of the interface separating two media with different physical properties. A coin-shaped volume element of height h, surface area Δs, with its broad faces parallel to the interface, penetrates the boundary and contains portions of both media. The flux leaving the side surface will be negligible and that leaving the broad surfaces will be unaffected if the height is permitted to shrink to zero. Applying Gauss' flux theorem to this volume and taking the limit

$$\lim_{h \to 0} \int_v \nabla \cdot \mathbf{D}\, dv = \lim_{h \to 0} \oint_s \mathbf{D} \cdot d\mathbf{s} = \lim_{h \to 0} \int_v \rho_e\, dv = \lim_{h \to 0} Q \tag{2.154}$$

or

$$(\mathbf{n} \cdot \mathbf{D}_1)\Delta s - (\mathbf{n} \cdot \mathbf{D}_e)\Delta s = \rho_s \Delta s \tag{2.155}$$

and by setting, for free space media,

$$\mathbf{D}_1 = \epsilon_0 \mathbf{E}_1 \quad \text{and} \quad \mathbf{D}_2 = \epsilon_0 \mathbf{E}_2 \tag{2.156}$$

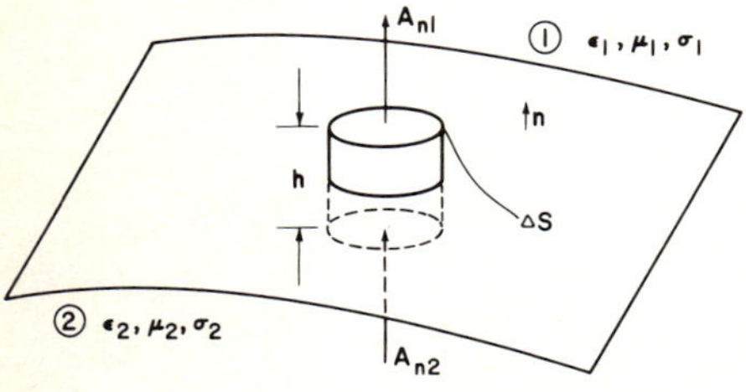

Fig. 2.3 Differential element of an interface between two media.

then

$$\mathbf{n}\cdot(\mathbf{E}_1 - \mathbf{E}_2) = \frac{\rho_s}{\epsilon_0} \tag{2.157}$$

and the normal component of the E field is discontinuous through a surface free-charge. The magnitude of the jump is ρ_s/ϵ_0. But usually there is no surface charge and then

$$\mathbf{n}\cdot\mathbf{D}_1 = \mathbf{n}\cdot\mathbf{D}_2 \tag{2.158}$$

and the normal component of the D field is continuous and since

$$\mathbf{D}_1 = \epsilon_1\mathbf{E}_1 \quad \text{and} \quad \mathbf{D}_2 = \epsilon_2\mathbf{E}_2 \tag{2.159}$$

for two general media, then

$$\frac{\mathbf{n}\cdot\mathbf{E}_1}{\mathbf{n}\cdot\mathbf{E}_2} = \frac{\epsilon_2}{\epsilon_1} \tag{2.160}$$

or

$$\epsilon_1 E_{n1} = \epsilon_2 E_{n2} \tag{2.161}$$

and the normal component of the E field is discontinuous when the permittivities are different. Now

$$\nabla\cdot\mathbf{B} = 0 \tag{2.162}$$

and

$$\lim_{h\to 0}\int_v \nabla\cdot\mathbf{B}\,dv = \lim_{h\to 0}\oint_s \mathbf{B}\cdot d\mathbf{s} \tag{2.163}$$

or

$$(\mathbf{n}\cdot\mathbf{B}_1)\Delta s - (\mathbf{n}\cdot\mathbf{B}_2)\Delta s = 0 \tag{2.164}$$

and

$$\mathbf{n} \cdot \mathbf{B}_1 = \mathbf{n} \cdot \mathbf{B}_2 \tag{2.165}$$

and the normal component of the B field is continuous. Using the definitions

$$\mathbf{B}_1 = \mu_1 \mathbf{H}_1 \quad \text{and} \quad \mathbf{B}_2 = \mu_2 \mathbf{H}_2 \tag{2.166}$$

then

$$\frac{\mathbf{n} \cdot \mathbf{H}_1}{\mathbf{n} \cdot \mathbf{H}_2} = \frac{\mu_2}{\mu_1} \tag{2.167}$$

or

$$\mu_1 H_{n1} = \mu_2 H_{n2} \tag{2.168}$$

and the normal component of the H field is discontinuous when the permeability of the media are different. Also

$$\nabla \cdot \mathbf{J} = 0 \tag{2.169}$$

and

$$\lim_{b \to 0} \int_v (\nabla \cdot \mathbf{J})\, dv = \lim_{b \to 0} \oint_s \mathbf{J} \cdot ds = 0 \tag{2.170}$$

and

$$(\mathbf{n} \cdot \mathbf{J}_1)\Delta s - (\mathbf{n} \cdot \mathbf{J}_2)\Delta s = 0 \tag{2.171}$$

or

$$\mathbf{n} \cdot \mathbf{J}_1 = \mathbf{n} \cdot \mathbf{J}_2 \tag{2.172}$$

and the normal component of the current is continuous. With Ohm's law for the two media, i.e.,

$$\mathbf{J}_1 = \sigma_1 \mathbf{E}_1 \quad \text{and} \quad \mathbf{J}_2 = \sigma_2 \mathbf{E}_2 \tag{2.173}$$

then

$$\frac{\mathbf{n} \cdot \mathbf{E}_1}{\mathbf{n} \cdot \mathbf{E}_2} = \frac{\sigma_2}{\sigma_1} \tag{2.174}$$

or

$$\sigma_1 E_{n1} = \sigma_2 E_{n2} \tag{2.175}$$

and the normal component of the E field is also discontinuous when the electrical conductivities are different.

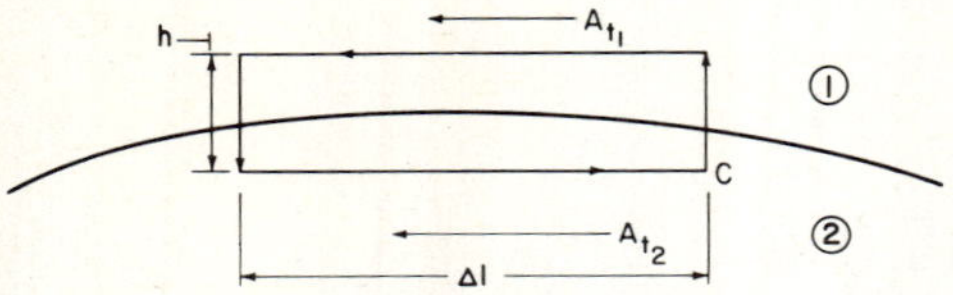

Fig. 2.4 Cross section of the interface.

Figure 2.4 is a cross section of the interface and since

$$\nabla \times \mathbf{E} = 0 \tag{2.176}$$

the E field is conservative and its line integral around the closed path C must be zero. Thus,

$$\lim_{h \to 0} \oint \mathbf{E} \cdot d\mathbf{l} = 0 \tag{2.177}$$

and

$$(\mathbf{n} \times \mathbf{E}_1)\Delta l - (\mathbf{n} \times \mathbf{E}_2)\Delta l = 0 \tag{2.178}$$

or

$$\mathbf{n} \times \mathbf{E}_1 = \mathbf{n} \times \mathbf{E}_2 \tag{2.179}$$

and the tangential component of the E field is continuous. Using the constitutive relations, Eq. (2.159), then

$$\frac{\mathbf{n} \times \mathbf{D}_1}{\mathbf{n} \times \mathbf{D}_2} = \frac{\epsilon_1}{\epsilon_2} \tag{2.180}$$

or

$$\epsilon_2 D_{t_1} = \epsilon_1 D_{t_2} \tag{2.181}$$

and the tangential component of the D field is discontinuous when the permittivities are different. Assuming that the magnetic field is produced by a steady current, then

$$\nabla \times \mathbf{H} = \mathbf{J} \tag{2.182}$$

and

$$\lim_{b \to 0} \int_{\Delta s} (\nabla \times \mathbf{H}) \cdot d\mathbf{s} = \lim_{b \to 0} \oint_{c} \mathbf{H} \cdot d\mathbf{l} = \lim_{b \to 0} \int_{\Delta s} \mathbf{J} \cdot d\mathbf{s} \tag{2.183}$$

consequently,

$$(\mathbf{n} \times \mathbf{H}_1)\Delta l - (\mathbf{n} \times \mathbf{H}_2)\Delta l = \mathbf{J}_s \Delta l \tag{2.184}$$

where $\mathbf{J}_s$ is the surface current density. Thus, the jump in the tangential component is due to the surface current, i.e.,

$$\mathbf{n} \times (\mathbf{H}_1 - \mathbf{H}_2) = \mathbf{J}_s \tag{2.185}$$

In many problems, the surface current is negligibly small, thus,

$$\mathbf{n} \times \mathbf{H}_1 = \mathbf{n} \times \mathbf{H}_2 \tag{2.186}$$

and the tangential component of the H field is continuous. Again, using the relations, Eq. (2.166),

$$\frac{\mathbf{n} \times \mathbf{B}_1}{\mathbf{n} \times \mathbf{B}_2} = \frac{\mu_1}{\mu_2} \tag{2.187}$$

or

$$\mu_2 B_{t_1} = \mu_1 B_{t_2} \tag{2.188}$$

and the tangential component of the B field is discontinuous when the permeabilities are different. Since the tangential component of the E field is continuous, then

$$\frac{\mathbf{J}_{t_1}}{\sigma_1} = \frac{\mathbf{J}_{t_2}}{\sigma_2} \tag{2.189}$$

or

$$\sigma_2 J_{t_1} = \sigma_1 J_{t_2} \tag{2.190}$$

where Ohm's law was used for each medium. Thus, the tangential component of the current is discontinuous when the electrical conductivity of the media are different.

All the preceding results are analogous to the refraction of a light ray passing from one medium to another and obey Snell's law relations. These relations can be derived by observing in Fig. 2.5 that

$$\tan\theta_i = \frac{A_{t_i}}{A_{n_i}} \tag{2.191}$$

and since

$$E_{t_1} = E_{t_2} \tag{2.192}$$

$$\epsilon_1 E_{n_1} = \epsilon_2 E_{n_2} \tag{2.193}$$

$$B_{n_1} = B_{n_2} \tag{2.194}$$

$$\mu_2 B_{t_1} = \mu_1 B_{t_2} \tag{2.195}$$

$$J_{n_1} = J_{n_2} \tag{2.196}$$

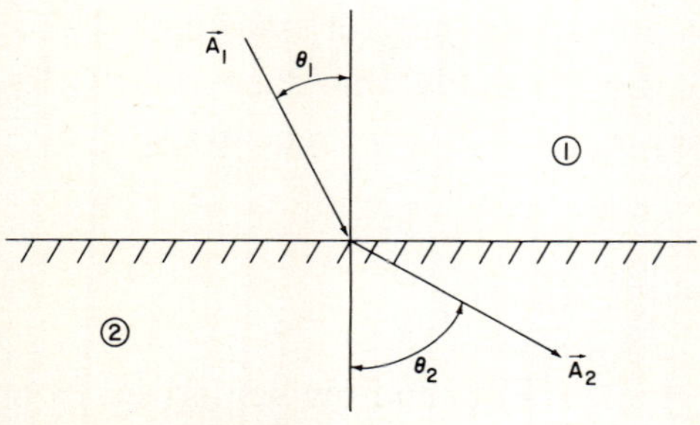

Fig. 2.5 Snell's law.

and

$$\sigma_2 J_{t_1} = \sigma_1 J_{t_2} \tag{2.197}$$

then

$$\frac{\tan\theta_1}{\tan\theta_2} = \frac{\epsilon_1}{\epsilon_2} = \frac{\mu_1}{\mu_2} = \frac{\sigma_1}{\sigma_2} \tag{2.198}$$

These relations demonstrate that the magnetic field lines are bent away from the normal in the highest permeability medium and when

$$\mu_2 \gg \mu_1 \tag{2.199}$$

then

$$\tan\theta_1 \approx 0 \tag{2.200}$$

and the flux lines are perpendicular to the surface in medium 1 if θ_2 is not exactly 90°. Also if region 1 is a good conductor and region 2 an insulator, then

$$\sigma_1 \gg \sigma_2 \tag{2.201}$$

and

$$\tan\theta_2 \approx 0 \tag{2.202}$$

and the current leaves the surface in a normal direction which corresponds to the requirement that the E field be perpendicular to the surface of a good conductor.

These boundary conditions were determined for stationary fields. However, similar arguments may be made for time-dependent fields with the results that the boundary conditions are unchanged.

2.14 ELECTROMAGNETIC WAVES

A simple, but historically important, result can now be derived from the Maxwell equations by asking, "What is the effect of changing the electrical field strength with time and at a given point?" This question can be answered by assuming that the medium has no charges or currents and by eliminating either E or H from the equations. By taking the curl of Faraday's law and using Ampère's law, then

$$\nabla \times (\nabla \times \mathbf{E}) = -\mu_0 \frac{\partial}{\partial t}(\nabla \times \mathbf{H}) = -\mu_0 \epsilon_0 \frac{\partial^2 \mathbf{E}}{\partial t^2} \tag{2.203}$$

and since

$$\nabla \times (\nabla \times \mathbf{E}) = \nabla(\nabla \cdot \mathbf{E}) - \nabla^2 \mathbf{E} = -\nabla^2 \mathbf{E} \tag{2.204}$$

then

$$\nabla^2 \mathbf{E} - \mu_0 \epsilon_0 \frac{\partial^2 \mathbf{E}}{\partial t^2} = 0 \tag{2.205}$$

or

$$\nabla^2 \mathbf{E} - \frac{1}{c^2} \frac{\partial^2 \mathbf{E}}{\partial t^2} = 0 \tag{2.206}$$

A similar equation can be obtained for the magnetic field wave. The parameter c has the dimensions of velocity and numerically is equal to 3×10^8 m/sec. This is exactly the speed of light in free space or in a vacuum. This was the result that Maxwell used to demonstrate that the propagation speed of an electromagnetic disturbance is the speed of light and light is an electromagnetic wave. This important result was the first root of the theory of relativity which was later developed by Einstein and has profoundly influenced recent physics and engineering. The more complex magnetogas-dynamic waves will be studied in Chap. 5.

2.15 MAGNETIC FIELD DIFFUSION

In the introduction to this chapter, the magnetic Reynolds number was introduced with the comment that magnetic fields readily diffuse through a permeable medium with little distortion when R_σ is small and are frozen in the medium when it is very large. These results can be demonstrated simply by deriving a nondimensional diffusion equation for a medium at rest and examining it and the Maxwell equations for very large and small magnetic Reynolds numbers. First, assume that $\rho_e, \partial\rho_e/\partial t$ as well as $\partial \mathbf{D}/\partial t$ are zero according to the MFD assumptions of Sec. 2.24, and then the Ampère, Faraday, and Ohm laws are

$$\nabla \times \mathbf{H} = \mathbf{J} \tag{2.207}$$

$$\nabla \times \mathbf{E} = -\frac{\partial \mathbf{B}}{\partial t} \tag{2.208}$$

$$\mathbf{J} = \sigma \mathbf{E} \tag{2.209}$$

where E is in the moving coordinate system. Taking the curl of Ampère's law, then

$$\nabla \times (\nabla \times \mathbf{H}) = \nabla \times \mathbf{J} \tag{2.210}$$

or

$$-\nabla^2 \mathbf{H} = \nabla \times \mathbf{J} \tag{2.211}$$

since the divergence of the magnetic field is zero. Multiplying Faraday's law by the conductivity of the medium, then

$$-\nabla^2 \mathbf{H} = \nabla \times \mathbf{J} = \sigma \nabla \times \mathbf{E} = -\sigma\mu_e \frac{\partial \mathbf{H}}{\partial t} \tag{2.212}$$

or

$$\nabla^2 \mathbf{H} = \sigma\mu_e \frac{\partial \mathbf{H}}{\partial t} \tag{2.213}$$

This equation is nondimensionalized by defining

$$\mathbf{H}_1 = \frac{\mathbf{H}}{\mathbf{H}_0} \qquad x_{i0} = \frac{x_i}{L} \qquad \text{and} \qquad t_1 = \frac{U}{L}t \tag{2.214}$$

where $\mathbf{H}_0$ is an applied magnetic field, L a characteristic length, and U the velocity of the applied magnetic field relative to the medium.

The diffusion equation is now

$$\nabla^2 \mathbf{H}_1 = R_\sigma \frac{\partial \mathbf{H}_1}{\partial t_1} \tag{2.215}$$

or

$$\frac{1}{R_\sigma} \nabla^2 \mathbf{H}_1 = \frac{\partial \mathbf{H}_1}{\partial t} \tag{2.216}$$

If the magnetic Reynolds number is very large due to an infinite conductivity or a great characteristic length as in astrophysical problems, then

$$\frac{\partial \mathbf{H}_1}{\partial t} \approx 0 \tag{2.217}$$

Thus, both the curl and divergence of the electric field are zero and the electric field is at most a constant. The voltage gradient is negligible when either the conductivity is infinite or the distance is great and then the constant electric field is negligible. The current must now be determined from Ampère's law and the magnetic field effectively controls the current. It is called *frozen-in* since the time to diffuse the characteristic distance is very large. The other limiting case is for very small magnetic Reynolds numbers due to a very small conductivity. The diffusion equation is

$$\nabla^2 \mathbf{H}_1 \approx 0 \tag{2.218}$$

and the magnetic field is a linear function at most. Since the conductivity is very small, then the current is very small and Ampère's law is

$$\nabla \times \mathbf{H}_1 \approx 0 \tag{2.219}$$

Again, the magnetic field must be a constant since both the curl and divergence are zero and the applied magnetic field is undistorted by the current flow. Since ν_H is very large for a small conductivity, then the magnetic field diffuses easily through the medium.

2.16 FLUID KINEMATICS[6]

In this and the next five sections, we will consider the equations that describe the motion of a general continuum fluid in a system of three orthogonal coordinates, that is, x_i $(i = 1,2,3)$. When a repeated subscript occurs in an equation, then the summation convention is to be applied and three terms will appear in the expansion, i.e.,

$$A_i B_i = A_1 B_1 + A_2 B_2 + A_3 B_3 \tag{2.220}$$

When ν appears as a subscript A_ν, it refers to the outward normal component of the vector $\mathbf{A}_\nu$ and $\partial\phi/\partial\nu$ is the component of $\nabla\phi$ in the $\boldsymbol{\nu}$ direction which is the outward normal unit vector.

The motion of the fluid will be observed from a convenient coordinate system that may be in motion relative to another system. It may be desirable to transform the equations of mechanics of the fluid motion from the original system to the other system. If the one system is moving at a constant velocity relative to the other, then the transformation will be a galilean transformation and Newton's second law will be invariant. This result is easily demonstrated by first defining

$$\bar{y}_i = y_i - u_i t \qquad i = 1,2,3 \tag{2.221}$$

and

$$\bar{t} = t \tag{2.222}$$

where u_i is the constant velocity of the origin of $\bar{y}$ relative to y at zero time. These relations define a general galilean transformation. Newton's second law will not change if the accelerations in both systems are equivalent. This requirement is satisfied by all permissible galilean transformations since

$$\frac{d^2\bar{y}_i}{d\bar{t}^2} \stackrel{!}{=} \frac{d^2 y_i}{dt^2} \tag{2.223}$$

There are two systems for specifying the fluid motions which will be individually used to provide the most simple physical description.

These systems are the lagrangian and the eulerian systems which have the following definitions. In the lagrangian system, the material point is located by a position vector a relative to some origin at time t_0, and time. Any fluid quantity is then a function of the material point moving with the fluid and time. The velocity and acceleration are

$$v_i = \frac{\partial x_i}{\partial t} \tag{2.224}$$

and

$$f_i = \frac{\partial v_i}{\partial t} = \frac{\partial^2 x_i}{\partial t^2} \tag{2.225}$$

In the eulerian system, the fluid quantities are functions of a position in space and time, i.e., $\mathbf{x}$ and t. In this system, $\partial \mathbf{v}/\partial t$ is the rate of change of $\mathbf{v}$ at a particular point in space and is not the acceleration. Now if a quantity F is a function of x_i and t and x_i is taken as a function of time so that the same fluid element is considered, then DF/Dt is obtained by expressing x_i as a function of a_i and t_1. Thus,

$$\frac{DF}{Dt} = \frac{\partial F}{\partial t} + \frac{\partial x_j}{\partial t}\frac{\partial F}{\partial x_j} = \frac{\partial F}{\partial t} + v_j \frac{\partial F}{\partial x_j} \tag{2.226}$$

and three terms are required by the repeated index. The total change is equal to the local change plus the convective change. The acceleration is obtained by replacing F with v_i and is

$$f_i = \frac{Dv_i}{Dt} = \frac{\partial v_i}{\partial t} + v_j \frac{\partial v_i}{\partial x_j} . \tag{2.227}$$

When F is a scalar, then

$$\frac{DF}{Dt} = \frac{\partial F}{\partial t} + \mathbf{v} \cdot \nabla F \tag{2.228}$$

is the invariant form. Now

$$\mathbf{f} = \frac{\partial \mathbf{v}}{\partial t} + \nabla\left(\frac{\mathbf{v}^2}{2}\right) + (\nabla \times \mathbf{v}) \times \mathbf{v} \tag{2.229}$$

and the change of any vector $\mathbf{A}$ is

$$\frac{D\mathbf{A}}{Dt} = \frac{\partial \mathbf{A}}{\partial t} + \frac{1}{2}[\nabla(\mathbf{v}\cdot\mathbf{A}) + (\nabla\times\mathbf{v})\times\mathbf{A} + (\nabla\times\mathbf{A})\times\mathbf{v} - \nabla\times(\mathbf{v}\times\mathbf{A}) + \mathbf{v}(\nabla\cdot\mathbf{A}) - \mathbf{A}(\nabla\cdot\mathbf{v})] \quad (2.230)$$

We may also wish to determine the rate of change of the total amount of a general quantity associated with the fluid that is inside of a surface s at any instant. Let $F(\mathbf{x}, t)$ be the quantity per unit mass and then the amount in a volume element $d\tau$ is $\rho F d\tau$. The momentum is obtained when F is the velocity and the energy is determined when F is equal to one-half the velocity squared. The total time rate of change of the quantity in the given volume is

$$\frac{D}{Dt}\int_v \rho F d\tau \quad (2.231)$$

But

$$\frac{D}{Dt}\int_v \rho F d\tau = \frac{D}{Dt}\int_v F dm = \int_v \frac{DF}{Dt} dm = \int_v \rho \frac{DF}{Dt} d\tau \quad (2.232)$$

since dm is the differential mass, which is invariant, and if the integrand is continuous in the volume and on the boundary. Now the change of the quantity F in a small time change is computed with the change of F in the original volume plus the change in the volume which is a layer of the thickness $v_\nu \Delta t$. Thus,

$$\int_v \left(\rho F + \frac{\partial \rho F}{\partial t}\Delta t\right) d\tau + \int_s \rho F (v_\nu \Delta t)\, ds \quad (2.233)$$

and the time rate of increase is

$$\frac{1}{\Delta t}\left[\int_v \frac{\partial \rho F}{\partial t}\Delta t d\tau + \int_s \rho F (v_\nu \Delta t)\, ds\right] \quad (2.234)$$

and then

$$\frac{D}{Dt}\int_v \rho F d\tau = \int_v \frac{\partial \rho F}{\partial t}\, d\tau + \int_s \rho F v_\nu \, ds$$
$$= \int_v \left[\frac{\partial \rho F}{\partial t} + \nabla\cdot(\rho F \mathbf{v})\right] d\tau$$
$$= \int_v \left\{ F\left[\frac{\partial \rho}{\partial t} + \nabla\cdot(\rho \mathbf{v})\right] + \rho\left(\frac{\partial F}{\partial t} + \mathbf{v}\cdot\nabla F\right)\right\} d\tau \qquad (2.235)$$

The quantity in the bracket is the continuity equation which will be derived in the next section and is equal to zero. Equation (2.235) is then

$$\frac{D}{Dt}\int_v \rho F d\tau = \int_v \rho\left(\frac{\partial F}{\partial t} + \mathbf{v}\cdot\nabla F\right) d\tau \qquad (2.236)$$

A similar expression is derived for a general quantity $G(\mathbf{x}, t)$ per unit volume as follows:

$$\frac{D}{Dt}\int_v G d\tau = \int_v \frac{\partial G}{\partial t}\, d\tau + \int_s G v_\nu \, ds$$
$$= \int_v \left[\frac{\partial G}{\partial t} + \nabla\cdot(G\mathbf{v})\right] d\tau$$
$$= \int_v \left(\frac{\partial G}{\partial t} + \mathbf{v}\cdot\nabla G + G\nabla\cdot\mathbf{v}\right) d\tau \qquad (2.237)$$

or

$$\frac{D}{Dt}\int_v G d\tau = \int_v \left(\frac{DG}{Dt} + G\nabla\cdot\mathbf{v}\right) d\tau \qquad (2.238)$$

2.17 CONTINUITY EQUATION

In a fixed volume in space, the rate of increase of the mass must equal the rate of mass flux inward through the surface. Thus

$$\frac{\partial}{\partial t}\int_v \rho\, d\tau = -\int_s \rho v_\nu\, ds \tag{2.239}$$

and since ρ is a continuous function of the coordinates, then

$$\frac{\partial}{\partial t}\int_v \rho\, d\tau = \int_v \frac{\partial \rho}{\partial t}\, d\tau \tag{2.240}$$

Applying Gauss' divergence theorem

$$\int_v \frac{\partial \rho}{\partial t}\, d\tau = -\int_v \nabla \cdot (\rho \mathbf{v})\, d\tau \tag{2.241}$$

or

$$\int_v \left[\frac{\partial \rho}{\partial t} + \nabla \cdot (\rho \mathbf{v})\right] d\tau = 0 \tag{2.242}$$

If the integrand is continuous throughout the volume and on its boundaries, then the equation for the fluid continuity is

$$\frac{\partial \rho}{\partial t} + \nabla \cdot (\rho \mathbf{v}) = 0 \tag{2.243}$$

Expanding the divergence terms, the continuity equation can also be expressed as

$$\frac{1}{\rho}\frac{D\rho}{Dt} + \nabla \cdot \mathbf{v} = 0 \tag{2.244}$$

If $\Delta\tau$ is a volume element that moves with the fluid so that the surface passes through the same particles and remains connected without holes or folds, the fluid must remain inside. Then

$$\lim_{\Delta\tau \to 0} \left[\frac{1}{\Delta\tau} \frac{D(\Delta\tau)}{Dt} \right] = \nabla \cdot \mathbf{v} = -\frac{1}{\rho} \frac{D\rho}{Dt} \tag{2.245}$$

and continuity is a physical requirement. A fluid is then incompressible if the density of any fluid element remains constant as time changes during the fluid motion being considered. Thus

$$\frac{D\rho}{Dt} = 0 \tag{2.246}$$

and

$$\nabla \cdot \mathbf{v} = 0 \tag{2.247}$$

which is true whether ρ is constant throughout the fluid or not.

2.18 THE STRESS TENSOR[7,9]

A surface s in a fluid in equilibrium experiences forces due to the molecular action of the fluid. The forces on both sides of the surface are equal and opposite and the forces on one side are equivalent to a force and a couple. We assume that, as the surface element shrinks to zero, the ratio of the force to the area has a finite, nonzero limit and the couple goes to zero. The intensity of the force at a point on the surface is this limit and is called a *stress* which, by assumption, is continuous up to and including the limit. The stress can be exerted in any direction and when the fluid is at rest the stresses are normal and result in pressure on the fluid. Tangential stresses can also occur in fluids in motion as a result of the viscosity or internal friction. We will consider problems in which the only other forces are body forces that act on the fluid volume. Defining $\mathbf{F}$ to be these forces per unit mass, then $\rho\mathbf{F}$ will be the forces per unit volume. An example of such a force is the buoyancy force on a unit mass due to temperature differences in a fluid in a gravitational field; this will be discussed in Sec. 4.10.

An element of fluid flowing under the action of forces can be studied in static equilibrium by the use of D'Alembert's principle which associates with each mass element an opposite force. These forces are equal to the mass times the acceleration and are called *inertia forces*. When the inertia forces are included with the other forces, the resultant force and moment are zero.

The specification of a stress requires the designation of a magnitude and two directions and is a tensor quantity that is invariant under all allowable coordinate transformations. The directions are the surface normal and the stress direction. Let $\boldsymbol{\nu}$ be a unit vector normal to the surface and $\mathbf{P}_\nu$ be the stress in any direction with the components $P_{\nu i}$ $(i = 1,2,3)$ in the coordinate direction (x_1, x_2, x_3). A rectangular volume will be in equilibrium if the inertia forces are included and the resultant moment about the centroid will be zero. If we let the sides shrink to zero, then $P_{ij} = P_{ji}$ and the stresses on the sides are symmetric. Now consider a surface of a triangular volume element with a $\boldsymbol{\nu}$ normal to the surface and a stress $\mathbf{P}_\nu$ in an arbitrary direction. $\mathbf{P}_j$ are the stresses on the other surfaces and have arbitrary directions. Figure 2.6 is due to Cauchy and is called a *stress quadic*. By shrinking the volume element to zero and maintaining $\boldsymbol{\nu}$ in the same direction, the surface stress $\mathbf{P}_\nu$ can be computed from the other stresses if the volume element is in rotational equilibrium. The body forces will go to zero as the volume goes to zero and $\mathbf{P}_\nu$ will balance the $\mathbf{P}_j$ stresses. Thus, the components of $\mathbf{P}_\nu$ along the three coordinate axes x_i are

$$P_{\nu i} = \nu_j P_{ji} \tag{2.248}$$

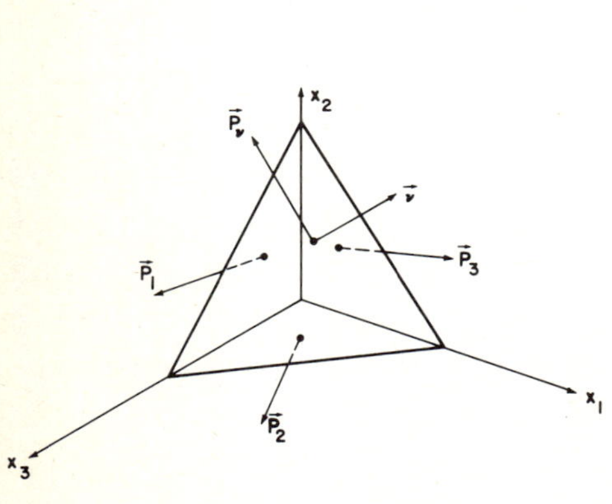

Fig. 2.6 Stress quadic.

and ν_j are the j direction cosines. There are then three equations with three terms that are required to compute P_ν. When $i = j$ the stresses are normal and are pressures when directed inward and tensions when directed outward.

2.19 EQUATIONS OF EQUILIBRIUM OF LINEAR AND ANGULAR MOMENTUM

The fluid inside a surface s that is fixed in space will be in equilibrium when the forces and moments are equated to zero by including the inertia forces. The force along each x_i axis is then the integral of the stress over each surface, i.e.,

$$\int_s P_{\nu i}\, ds = \int_s \nu_j P_{ji}\, ds = \int_v \frac{\partial P_{ji}}{\partial x_j}\, d\tau \tag{2.249}$$

The requirement for equilibrium along the x_i direction is

$$0 = \int_v \rho(f_i - F_i)\, d\tau - \int_s P_{\nu i}\, ds \tag{2.250}$$

or

$$0 = \int_v \left(\rho f_i - \rho F_i - \frac{\partial P_{ji}}{\partial x_j}\right) d\tau\, . \tag{2.251}$$

If the integrand is continuous, then

$$\rho f_i = \rho F_i + \frac{\partial P_{ji}}{\partial x_j} \tag{2.252}$$

and is called the *equation of linear momentum*. By expressing the stress tensor as

$$P_{ij} = -p\delta_{ij} + \pi_{ij} \tag{2.253}$$

where δ_{ij} is equal to unity if $i = j$ and is equal to zero if $i \neq j$ and p is the thermodynamic pressure due to molecular action, then

$$\rho f_i = \rho F_i - \frac{\partial p}{\partial x_i} + \frac{\partial \pi_{ji}}{\partial x_j} \tag{2.254}$$

The requirement for a zero resultant moment about the x_k axis is:

$$\begin{aligned} 0 &= \int_s (x_i P_{\nu j} - x_j P_{\nu i})\, ds = \int_s (\nu_l x_i P_{lj} - \nu_l x_j P_{li})\, ds \\ &= \int_v \frac{\partial}{\partial x_l}(x_i P_{lj} - x_j P_{li})\, d\tau \\ &= \int_v \left(x_i \frac{\partial P_{lj}}{\partial x_l} - x_j \frac{\partial P_{li}}{\partial x_l} + P_{ij} - P_{ji} \right) d\tau \end{aligned} \tag{2.255}$$

where l are the coordinate directions. When the integrand is continuous, then

$$x_i \frac{\partial P_{lj}}{\partial x_l} - x_j \frac{\partial P_{li}}{\partial x_l} + P_{ij} - P_{ji} = 0 \tag{2.256}$$

is the equation for equilibrium of angular momentum about the x_k axis. Two other equations can be obtained in a similar manner or by changing i, j, k in cyclic order. Since,

$$\frac{\partial P_{li}}{\partial x_l} = \rho(f_i - F_i) \tag{2.257}$$

$$\frac{\partial P_{lj}}{\partial x_l} = \rho(f_j - F_j) \tag{2.258}$$

and

$$\frac{\partial P_{lk}}{\partial x_l} = \rho(f_k - F_k) \tag{2.259}$$

from the linear momentum equations, then

$$x_i \rho(f_j - F_j) - x_j \rho(f_i - F_i) + P_{ij} - P_{ji} = 0 \tag{2.260}$$

$$x_j \rho (f_k - F_k) - x_k \rho (f_j - F_j) + P_{jk} - P_{kj} = 0 \quad (2.261)$$

and

$$x_k \rho (f_i - F_i) - x_i \rho (f_k - F_k) + P_{ki} - P_{ik} = 0 \quad (2.262)$$

But $f_i - F_i$, $f_j - F_j$ and $f_k - F_k$ act through the origin and then the moments

$$x_i(f_j - F_j) = x_j(f_i - F_i) = x_j(f_k - F_k) = x_k(f_j - F_j)$$
$$= x_k(f_i - F_i) = x_i(f_k - F_k) = 0 \quad (2.263)$$

This result reduces the equations to

$$P_{ij} = P_{ji} \quad (2.264)$$

$$P_{jk} = P_{kj} \quad (2.265)$$

$$P_{ki} = P_{ik} \quad (2.266)$$

which demonstrates that the stresses on the volume sides are symmetric. Thus, the requirement of local equilibrium of moments results in the equilibrium of the volume element if the equations of linear momentum are satisfied and the stress tensor is symmetric.

2.20 THE NAVIER-STROKE EQUATIONS

A specific form can be given for π_{ij} by recalling that it is zero for fluids at rest and in rigid body motion, there are no viscous stresses. Thus it must be related to quantities associated with the fluid motion and depends on the rate of strain tensor which is determined as follows. Consider two points P and Q at x_i and $x_i + \delta x_i$ then, for small deformations, the velocity of Q relative to P has the components

$$\delta v_i = \left(\frac{\partial v_i}{\partial x_j}\right)\delta x_j \quad (2.267)$$

By setting

$$e_{ij} = \frac{\partial v_j}{\partial x_i} + \frac{\partial v_i}{\partial x_j} \quad (2.268)$$

and

$$\xi_{ij} = \frac{\partial v_j}{\partial x_i} - \frac{\partial v_i}{\partial x_j} \tag{2.269}$$

so that e_{ij} is symmetric and ξ_{ij} is antisymmetric, i.e., $e_{ij} = e_{ji}$ and $\xi_{ij} = -\xi_{ji}$, then

$$e_{ij} - \xi_{ij} = 2\frac{\partial v_i}{\partial x_j} \tag{2.270}$$

and

$$\delta v_i = \frac{e_{ij}}{2}\delta x_j - \frac{\xi_{ij}}{2}\delta x_j \tag{2.271}$$

The term ξ_{ij} represents a rotation about the k axis and the second term in Eq. (2.271) is the velocity of Q due to a rigid body rotation. The first term describes the distortion of the fluid element in the neighborhood of P which is more readily understood by setting ξ_{ij} equal to zero and all e_{ij} equal to zero except for the one being considered. Therefore, e_{11} is twice the rate of extensional strain of a line element in the x_1 direction and the results are similar for e_{22} and e_{33}. The rate of shearing strain is e_{23} and represents the rate of change of the angle between two lines of particles that lie on x_2 and x_3 at that instant. The same is true for e_{31} and e_{12}.

The principal axes of the strain tensor e_{ij} are e_1, e_2, and e_3 and are chosen so that the strains are only in normal directions. If the coordinate axes are selected to coincide with the principal axes, then the rates of shear are zero and e_1, e_2, and e_3 are equal to twice the rate of extension along the axes. Since π_{ij} and e_{ij} are symmetrical and ξ_{ij} is antisymmetrical, then an external moment would be required to maintain the rigid body motion if π_{ij} depended on ξ_{ij}. This is known to be physically incorrect. Thus we assume that π_{ij} are linear functions of the strain components according to Hooke's law for small deformations. This relation is a generalization of Newton's concept of simple shearing motion and is called a *newtonian fluid*. We also assume that the fluid properties are the same in all directions or isotropic. With this assumption, the principal axes of stress and rate of strain coincide and if we consider a volume element with its edges parallel to these axes, then there is no shear. These π_1, π_2, and π_3 and e_1, e_2, and e_3 are called the principal viscous stresses and principal rates of extension.

According to our assumptions, we write

$$\pi_i = A \frac{e_1 + e_2 + e_3}{2} + Be_i \tag{2.272}$$

or

$$\pi_i = A \frac{e_{ii}}{2} + Be_i \tag{2.273}$$

Stokes concluded from Hooke's law and symmetry that in isotropic bodies, the pressure, equal in all directions around a point, is attended by a proportional compression without shear, and a shearing stress is accompanied by a proportional shearing strain without a compression. Thus,

$$e_{ii} = 3e_i \tag{2.274}$$

and

$$\pi_i = \mu e_i \tag{2.275}$$

for these two simple cases. By using Eqs. (2.274) and (2.273), the definition for the bulk viscosity β is obtained:

$$\frac{\pi_i}{3} = \beta\left(\frac{e_i}{2}\right) \tag{2.276}$$

The coefficients of shear viscosity μ and bulk viscosity β vary with the temperature and pressure in the fluid and are nonnegative physical quantities. Negative quantities would violate the Second Law of Thermodynamics. The coefficients of shear viscosity and bulk viscosity generally depend on the thermodynamic state and may vary with position in the fluid. Inserting these results in Eq. (2.273), it becomes

$$\pi_i = \left(\beta - \frac{2\mu}{3}\right)\frac{e_{ii}}{2} + \mu e_i \tag{2.277}$$

which can be transformed from the principal axes to any axes by the tensor transformation formula

$$\pi_{ij} = \alpha_{j(i)}\pi_{(i)} \tag{2.278}$$

where α_{ji} are the direction cosines between the sets of axes. Therefore,

$$\pi_{ij} = \beta\left(\frac{e_{ii}}{2}\right)\alpha_{j(i)} + \mu\left[e_i\alpha_{ji} - \frac{2}{3}\left(\frac{e_{ii}}{2}\right)\alpha_{j(i)}\right] \tag{2.279}$$

and by setting

$$\alpha_{ji} = \delta_{ij} \tag{2.280}$$

to preserve Stokes' conclusions, then

$$\pi_{ij} = \beta\left(\frac{e_{ii}}{2}\right)\delta_{ij} + \mu\left[e_{ij} - \frac{2}{3}\left(\frac{e_{ii}}{2}\right)\delta_{ij}\right] \tag{2.281}$$

Substituting this expression for π_{ij} in the ith equation for the linear momentum and noting that

$$\frac{e_{ii}}{2} = \nabla \cdot \mathbf{v} \tag{2.282}$$

then

$$\rho f_i = \rho F_i - \frac{\partial p}{\partial x_i} + \mu\nabla^2 v_i + \frac{\mu}{3}\frac{\partial(e_{ii}/2)}{\partial x_i} + e_{ij}\frac{\partial\mu}{\partial x_j} - \frac{2}{3}\left(\frac{e_{ii}}{2}\right)\frac{\partial\mu}{\partial x_i} + \frac{\partial(\beta e_{ii}/2)}{\partial x_i} \tag{2.283}$$

The bulk viscosity is important only if the dilation $e_{ii}/2$ is large and occurs in problems of sound absorption, acoustic streaming, and shock waves. Thus, for $\beta = 0$ and for an incompressible fluid with a constant viscosity, μ,

$$f_i = F_i - \frac{1}{\rho}\frac{\partial p}{\partial x_i} + \frac{\mu}{\rho}\nabla^2 v_i \tag{2.284}$$

is the equation for the linear momentum of a fluid with small deformations. Equation (2.284) is the incompressible form of the Navier-Stokes equations named in honor of Navier, who was the first to correctly present the inviscid equations of motion in 1826 and Stokes, who provided the viscous equations of motion in 1847 by decomposing the most general motion of a fluid element into one of pure translation, pure rotation, and pure strain.

2.21 THE ENERGY EQUATION[6,8]

We assume the existence of an internal energy function per unit mass and the usual thermodynamic variables which are related by reversible thermodynamics. Since the energy balance will require inclusion of the rate at which work is done by body forces and stresses over the surface, it is best to examine a surface S that encloses a portion of the fluid and moves with the fluid. The general energy balance can be stated as:

Rate at which surface stresses do work
+ Rate at which body forces do work
+ Rate at which heat energy is conducted inward through S
+ Rate at which heat is added by chemical reaction, radiation, electromagnetic action, etc.

is equal to

Rate of increase of kinetic energy
+ Rate of increase of internal energy.

The energy balance for the volume element is then

$$\int_S P_{\nu i}\, v_i\, ds + \int_V \rho F_i\, v_i\, d\tau + \int_S q_\nu\, ds + \int_V \rho R\, d\tau = \frac{D}{Dt}\int_V \rho\left(\frac{v_i v_i}{2} + \xi\right) d\tau \tag{2.285}$$

$$= \int_V \rho \frac{D}{Dt}\left(\frac{v_i v_i}{2} + \xi\right) d\tau$$

$$= \int_V \rho \left(v_i f_i + \frac{D\xi}{Dt}\right) d\tau \qquad (2.285)\ \text{(cont'd)}$$

where q_ν is the normal component of the general heat flux vector $\mathbf{q}$ and R is the rate of heat addition per unit mass. The surface stress term can be expanded with our tensor formula and Gauss' divergence law to give

$$\int_S P_{\nu i} v_i ds = \int_S \nu_j P_{ji} v_i ds = \int_V \left(v_i \frac{\partial P_{ji}}{\partial x_j} + P_{ji} \frac{\partial v_i}{\partial x_j}\right) d\tau$$
$$= \int_V v_i(\rho f_i - \rho F_i) d\tau + \int_V \pi_{ji} \frac{\partial v_i}{\partial x_j} d\tau$$
$$- \int_V p \frac{\partial v_i}{\partial x_i} d\tau \qquad (2.286)$$

Recalling that $\pi_{ij} = \pi_{ji}$, then

$$\pi_{ij} \frac{\partial v_i}{\partial x_j} = \pi_{ij} \frac{\partial v_j}{\partial x_i} \qquad (2.287)$$

and let

$$\Phi = \pi_{ij} \frac{\partial v_i}{\partial x_j} = \pi_{ij} \frac{\partial v_j}{\partial x_i} = \tfrac{1}{2} \pi_{ij} e_{ij} \qquad (2.288)$$

which can be related to the viscosities and dilation by

$$\Phi = \tfrac{1}{2} \pi_{ij} e_{ij} = \tfrac{1}{2} \mu e_{ij} e_{ij} - \tfrac{2}{3} \mu \left(\frac{e_{ii}}{2}\right)^2 + \beta \left(\frac{e_{ii}}{2}\right)^2 \qquad (2.289)$$

The dissipation function Φ is always nonnegative. The last term of Eq. (2.286) can be rewritten with the continuity equation since

$$\frac{\partial v_i}{\partial x_i} = \nabla \cdot \mathbf{v} = -\frac{1}{\rho}\frac{D\rho}{Dt} = \rho\frac{D(1/\rho)}{Dt} \tag{2.290}$$

The energy equation is now

$$\int_V \left\{ \rho \left[\frac{D\xi}{Dt} + p\frac{D(1/\rho)}{Dt} \right] - \Phi - \nabla \cdot \mathbf{q} - \rho R \right\} d\tau = 0 \tag{2.291}$$

and if the integrand is continuous, then

$$\rho \left[\frac{D\xi}{Dt} + p\frac{D(1/\rho)}{Dt} \right] = \Phi + \nabla \cdot \mathbf{q} + \rho R = \rho T\frac{Ds}{Dt} \tag{2.292}$$

A specific relation can be given for the heat flux vector $\mathbf{q}$ by assuming the form

$$\mathbf{q} = k\nabla T \tag{2.293}$$

where k is the coefficient of thermal conductivity of the fluid which can depend on the temperature. This relation is consistent with the linear relation that we assumed for π_{ij}.

The energy equation can be stated in many forms and some of the most important are derived below. Let

$$I = \xi + \frac{p}{\rho} \tag{2.294}$$

be the enthalpy per unit mass and S be the entropy per unit mass and then

$$\frac{DI}{Dt} = \frac{D\xi}{Dt} + p\frac{D(1/\rho)}{Dt} + \frac{1}{\rho}\frac{Dp}{Dt} \tag{2.295}$$

The First Law of Thermodynamics is

$$\delta Q = D\xi + pD\frac{1}{\rho} \tag{2.296}$$

and

$$DS = \frac{\delta Q}{T} \tag{2.297}$$

thus

$$TDS = T\frac{\delta Q}{T} = D\xi + pD\frac{1}{\rho} \tag{2.298}$$

or

$$T\frac{DS}{Dt} = \frac{D\xi}{Dt} + p\frac{D(1/\rho)}{Dt} \tag{2.299}$$

Equation (2.295) becomes

$$\frac{DI}{Dt} = T\frac{DS}{Dt} + \frac{1}{\rho}\frac{Dp}{Dt} \tag{2.300}$$

and when the entropy is taken as a function of pressure and temperature, then

$$\frac{DS}{Dt} = \left(\frac{\partial S}{\partial T}\right)_p \frac{DT}{Dt} + \left(\frac{\partial S}{\partial p}\right)_T \frac{Dp}{Dt} \tag{2.301}$$

If $S(1/\rho, T)$, Eq. (2.301) in a different form is

$$\frac{DS}{Dt} = \left(\frac{\partial S}{\partial T}\right)_\rho \frac{DT}{Dt} + \left(\frac{\partial S}{\partial(1/\rho)}\right)_T \frac{D(1/\rho)}{Dt} \tag{2.302}$$

The specific heats at constant pressure and constant volume are defined as

$$\left(\frac{\partial S}{\partial T}\right)_\rho = \frac{c_V}{T} \tag{2.303}$$

and

$$\left(\frac{\partial S}{\partial T}\right)_p = \frac{c_p}{T} \tag{2.304}$$

and the particular one chosen will be determined by the problem which is being studied. By using two of Maxwell's thermodynamic relations, the two equations for the total entropy change are

$$\frac{DS}{Dt} = \frac{c_p}{T}\frac{DT}{Dt} - \left[\frac{\partial(1/\rho)}{\partial T}\right]_p \frac{Dp}{Dt} \tag{2.305}$$

and

$$\frac{DS}{Dt} = \frac{c_V}{T}\frac{DT}{Dt} + \left(\frac{\partial p}{\partial T}\right)_\rho \frac{D(1/\rho)}{Dt} \tag{2.306}$$

since

$$\left(\frac{\partial S}{\partial T}\right)_\rho = -\left[\frac{\partial(1/\rho)}{\partial T}\right]_p \qquad \left[\frac{\partial(1/\rho)}{\partial T}\right]_\rho = -\left(\frac{\partial S}{\partial p}\right)_T \tag{2.307}$$

and

$$\left[\frac{\partial S}{\partial(1/\rho)}\right]_T = \left(\frac{\partial P}{\partial T}\right)_\rho \tag{2.308}$$

are the Maxwell relations. The various forms of the energy equation are obtained by observing that the bracketed term in Eq. (2.292) can be related to the total entropy change, i.e.,

$$\rho\left[\frac{D\xi}{Dt} + p\,\frac{D(1/\rho)}{Dt}\right] = \rho\,\frac{DI}{Dt} - \frac{Dp}{Dt} = \rho T\,\frac{DS}{Dt} \tag{2.309}$$

The equations are now

$$\rho c_p \frac{DT}{Dt} + \frac{T}{\rho}\left(\frac{\partial\rho}{\partial T}\right)_p \frac{Dp}{Dt} = \Phi + \nabla\cdot\mathbf{q} + \rho R \tag{2.310}$$

$$\rho c_v \frac{DT}{Dt} + T\nabla \cdot \mathbf{v}\left(\frac{\partial P}{\partial T}\right)_\rho = \Phi + \nabla \cdot \mathbf{q} + \rho R \tag{2.311}$$

and

$$\rho c_p \frac{DT}{Dt} - \frac{Dp}{Dt} = \rho c_v \frac{DT}{Dt} + p\nabla \cdot \mathbf{v} = \Phi + \nabla \cdot \mathbf{q} + \rho R \tag{2.312}$$

Equation (2.311) was obtained with the continuity equation and the two forms in Eq. (2.312) are for a perfect gas where

$$p = R\rho T \tag{2.313}$$

Notice that when an incompressible fluid is being considered, expressions for the entropy change that contain c_v should be used and not those employing c_p.

2.22 FLUID-DYNAMIC BOUNDARY CONDITIONS

The boundary conditions for the fluid at a surface is the no-slip condition; i.e., the tangential velocity component is zero. Since most surfaces are not penetrated by the fluid, then the normal velocity component is also zero. The thermal boundary condition is either that the fluid has the same temperature as the wall, which is prescribed, or the heat flow through the surface is prescribed. The flow and thermal boundary conditions will be discussed in detail for individual problems in the succeeding chapters.

2.23 THE MAGNETOFLUID-DYNAMIC EQUATIONS

We can generalize the previously derived momentum and energy equations to describe the flow of an electrically conducting fluid through electric and magnetic fields. By neglecting relativistic terms in the constitutive relations for the electric displacement and the magnetic induction, the Maxwell stress tensor is symmetric and the derivation of the MFD momentum equation will be the same as in Secs. 2.18 and 2.19 except the stress tensor will be replaced by the sum of p_{ij} and $\mathcal{T}_{ij}$ which will remain symmetric. The result will be

$$\rho f_i = \rho F_i + \frac{\partial p_{ji}}{\partial x_j} + \frac{\partial \mathcal{T}_{ji}}{\partial x_j} - \frac{\partial G_i}{\partial t} \tag{2.314}$$

and the last term is the rate of change of the electromagnetic momentum. The electromagnetic stress tensor term can be expanded as

$$\frac{\partial \mathcal{T}_{ji}}{\partial x_j} = \frac{\partial E_i}{\partial x_j} D_j + E_i \frac{\partial D_j}{\partial x_j} + \frac{\partial H_i}{\partial x_j} B_j + H_i \frac{\partial B_j}{\partial x_j} - D_k \frac{\partial E_k}{\partial x_i} - B_k \frac{\partial H_k}{\partial x_i} \tag{2.315}$$

and since the summation convention is used

$$\begin{aligned} \frac{\partial \mathcal{T}_{ji}}{\partial x_j} &= \rho_e E_i + [(\nabla \times \mathbf{E}) \times \mathbf{D}]_i + [(\nabla \times \mathbf{H}) \times \mathbf{B}]_i \\ &= \rho_e E_i - \left(\frac{\partial \mathbf{B}}{\partial t} \times \mathbf{D}\right)_i + \left(\frac{\partial \mathbf{D}}{\partial t} \times \mathbf{B}\right)_i + (\mathbf{J} \times \mathbf{B})_i \end{aligned} \tag{2.316}$$

or

$$\frac{\partial \mathcal{T}_{ji}}{\partial x_j} = \rho_e E_i + (\mathbf{J} \times \mathbf{B})_i + \frac{\partial G_i}{\partial t} \tag{2.317}$$

The MFD momentum equations are then

$$\rho f_i = \rho F_i - \frac{\partial p}{\partial x_i} + \rho_e E_i + (\mathbf{J} \times \mathbf{B})_i + \frac{\partial \pi_{ji}}{\partial x_j} \tag{2.318}$$

The new terms are (1) the force of the electric field due to the presence of the electric charges in the fluid and (2) the force on the fluid produced by the interaction of the current and the magnetic field (called a *J cross B* force or the *Lorentz force*).

The derivation of the energy equation will be repeated with the addition of the work performed by the Maxwell stress, the rate of increase of the electromagnetic energy and the rate of flux of the electromagnetic energy across the surface. But, first, it is useful to obtain an energy balance with only these additional quantities, i.e.,

Rate of work done by Maxwell stresses = Rate of increase of energy in the volume + Rate of flux out of volume.

In analytical form, this balance is

$$\int_S \zeta_{\nu i} v_i \, ds = \left[\int_V \frac{\partial w_{em}}{\partial t} \, d\tau + \int_S w_{em} v_\nu \, ds \right] + \int_S N'_\nu \, ds \qquad (2.319)$$

where N' is the Poynting vector in the local moving coordinates. Adding these terms to Eq. (2.285) of Sec. 2.21

$$\int_S \zeta_{\nu i} v_i \, ds + \int_S P_{\nu i} v_i \, ds + \int_V \rho F_i v_i \, d\tau + \int_V \nabla \cdot \mathbf{q} \, d\tau + \int_V \rho R \, d\tau$$

$$= \frac{D}{Dt} \int_V \rho \left(\frac{v_i v_i}{2} + \xi \right) d\tau + \left(\int_V \frac{\partial w_{em}}{\partial t} \, d\tau + \int_S w_{em} v_\nu \, ds \right) + \int_S N'_\nu \, ds \qquad (2.320)$$

Now

$$\int_S P_{\nu i} v_i \, ds = \int_S \nu_j P_{ji} v_i \, ds = \int_V v_i \frac{\partial P_{ji}}{\partial x_j} \, d\tau + \int_V P_{ji} \frac{\partial v_i}{\partial x_j} \, d\tau$$

$$= \int_V v_i [\rho f_i - \rho F_i - \rho_e E_i - (\mathbf{J} \times \mathbf{B})_i] \, d\tau$$

$$- \int_V p \frac{\partial v_i}{\partial x_i} \, d\tau + \int_V \Phi \, d\tau \qquad (2.321)$$

and

$$\zeta_{\nu i} v_i = \nu_j \zeta_{ji} v_i = \nu_j (E_i D_j v_i + H_i B_j v_i - w_{em} \nu_j)$$

$$= (\mathbf{E} \cdot \mathbf{v}) D_\nu + (\mathbf{H} \cdot \mathbf{v}) B_\nu - w_{em} v_\nu \qquad (2.322)$$

or

$$\zeta_{\nu i} v_i = N'_\nu - N_\nu + w_{em} v_\nu \qquad (2.323)$$

also

$$\int_V \frac{\partial w_{em}}{\partial t} d\tau = \int_V (\mathbf{E} \cdot \dot{\mathbf{D}} + \mathbf{H} \cdot \dot{\mathbf{B}}) d\tau$$

$$= \int_V [\mathbf{E} \cdot (\nabla \times \mathbf{H} - \mathbf{J}) - \mathbf{H} \cdot (\nabla \times \mathbf{E})] d\tau$$

$$= -\int_V \mathbf{E} \cdot \mathbf{J} d\tau - \int_V \nabla \cdot \mathbf{N} d\tau \qquad (3.324)$$

or

$$-\int_V \frac{\partial w_{em}}{\partial t} d\tau = \int_V \mathbf{E} \cdot \mathbf{J} d\tau + \int_S N_\nu ds \qquad (2.325)$$

Substituting these in Eq. (2.320) an energy equation similar to Eq. (2.291) is obtained,

$$\int_V \left\{ \rho \left[\frac{D\xi}{Dt} + p \frac{D(1/\rho)}{Dt} \right] - \Phi - \nabla \cdot \mathbf{q} - \rho R - \mathbf{E} \cdot \mathbf{J} + \rho_e \mathbf{v} \cdot \mathbf{E} + \mathbf{v} \cdot (\mathbf{J} \times \mathbf{B}) \right\} d\tau = 0 \qquad (2.326)$$

The term $\mathbf{E} \cdot \mathbf{J}$ is called the *Joule heating* and is due to the resistance of the fluid to the flow of current. This form of the energy equation can still be refined by noting that

$$\mathbf{v} \cdot (\mathbf{J} \times \mathbf{B}) = -\mathbf{J} \cdot (\mathbf{v} \times \mathbf{B}) \qquad (2.327)$$

and then the group of new terms becomes

$$\mathbf{E} \cdot \mathbf{J} - \rho_e \mathbf{v} \cdot \mathbf{E} - \mathbf{v} \cdot (\mathbf{J} \times \mathbf{B}) = (\mathbf{E} + \mathbf{v} \times \mathbf{B}) \cdot (\mathbf{J} - \rho_e \mathbf{v}) = \overline{\mathbf{E}} \cdot \overline{\mathbf{J}} = \frac{\overline{J}^2}{\sigma} \qquad (2.328)$$

since $\mathbf{v} \cdot (\mathbf{v} \times \mathbf{B})$ is zero and

$$\overline{\mathbf{E}} = \mathbf{E} + \mathbf{v} \times \mathbf{B} \qquad (2.329)$$

and

$$\bar{\mathbf{J}} = \mathbf{J} - \rho_e \mathbf{v} \tag{2.330}$$

for the local moving coordinate system. With this simplification, and if the integrand of Eq. (2.279) is continuous, the MFD energy equation is

$$\rho \left[\frac{D\xi}{Dt} + p \frac{D(1/\rho)}{Dt} \right] = \Phi + \frac{J^2}{\sigma} + \nabla \cdot \mathbf{q} + \rho R \tag{2.331}$$

The left side can also be expressed in the other forms given in Sec. 2.20.

Equations (2.318) and (2.331) are then the momentum and energy equations for MFD which will be used in the forms presented or in forms convenient for the problem that is being considered.

2.24 THE MAGNETOFLUID-DYNAMIC APPROXIMATIONS

Equations (2.318) and (2.331) are the momentum and energy equations which, with the Maxwell equations, form the basis for studying magnetofluid dynamics. However, the equations are not usually applied in their present form and require interpretation and several assumptions to provide the set to be used in all but the last chapter. In magnetofluid dynamics, we consider a conducting fluid that is approximately grossly neutral; the charge density in the Maxwell equations must then be interpreted as an excess charge density which is generally not large. If we disregard the excess charge density, then we must disregard the displacement current according to the discussion in Sec. 2.5. In most problems, the displacement current, excess charge density, excess charge body force, and the current due to convection of the excess charge are small. The electrodynamic equations to be used are then the pre-Maxwell equations and the complete set becomes

$$\nabla \cdot \mathbf{D} = 0 \tag{2.332}$$

$$\nabla \cdot \mathbf{J} = 0 \tag{2.333}$$

$$\nabla \cdot \mathbf{B} = 0 \tag{2.334}$$

$$\nabla \times \mathbf{H} = \mathbf{J} \tag{2.335}$$

$$\nabla \times \mathbf{E} = -\frac{\partial \mathbf{B}}{\partial t} \tag{2.336}$$

$$\mathbf{D} = \epsilon \mathbf{E} \tag{2.337}$$

$$\mathbf{B} = \mu_e \mathbf{H} \tag{2.338}$$

$$\mathbf{J} = \sigma(\mathbf{E} + \mathbf{v} \times \mathbf{B}) \tag{2.339}$$

$$\frac{\partial \rho}{\partial t} + \frac{\partial \rho v_i}{\partial x_i} = 0 \tag{2.340}$$

$$\rho \frac{Dv_i}{Dt} = \rho F_i - \frac{\partial p}{\partial x_i} + (\mathbf{J} \times \mathbf{B})_i + \frac{\partial \pi_{ji}}{\partial x_j} \tag{2.341}$$

$$\rho T \frac{DS}{Dt} = \Phi + \frac{J^2}{\sigma} + \nabla \cdot \mathbf{q} + \rho R \tag{2.342}$$

Notice that the number of equations can now be reduced by solving for E in Eq. (2.339), substituting for J from Eq. (2.335) and then combining with Eq. (2.336) to yield one equation for the magnetic field, i.e.,

$$\frac{\partial \mathbf{H}}{\partial t} = \nabla \times (\mathbf{v} \times \mathbf{H}) - \nabla \times [\nu_H(\nabla \times \mathbf{H})] \tag{2.343}$$

Also eliminating J in Eqs. (2.341) and (2.342) gives

$$\rho \frac{D\mathbf{v}}{Dt} - (\mathbf{B} \cdot \nabla)\mathbf{H} = \rho F_i - \nabla\left(p + \frac{1}{2}\mu_e H^2\right) + \nabla \cdot \pi_i \tag{2.344}$$

and

$$\rho T \frac{DS}{Dt} = \Phi + (\nabla \times \mathbf{H}) \cdot [\nu_H(\nabla \times \mathbf{B}) - \mathbf{v} \times \mathbf{B}] + \nabla \cdot \mathbf{q} + \rho R \tag{2.345}$$

and we only need to consider the interaction of the fluid-dynamic variables with the magnetic field strength H. The set of equations reduces to Eqs. (2.333), (2.340), (2.343), (2.344), and (2.345).

Additional assumptions are revealed by nondimensionalizing the set of equations, Eqs. (2.332) through (2.342), with the aid of the following:

$$x_i^* = x_i/L,\ v_i^* = v_i/U,\ t^* = t/t_0,\ p^* = p/\rho_0 U^2,\ F_i^* = F_i L/U^2,$$
$$\Phi^* = \Phi L/\rho_0 U^3,\ \mathbf{E}^* = \mathbf{E}/E_0,\ \mathbf{B}^* = \mathbf{B}/B_0,\ \mathbf{H}^* = \mathbf{H}/H_0,\ \mathbf{J}^* = \mathbf{J}/\sigma_0 U B_0,$$
$$\pi_{ji}^* = \pi_{ji} L/\mu_0 U,\ \mu_e^* = \mu_e/\mu_{e0},\ \sigma^* = \sigma/\sigma_0,\ \epsilon^* = \epsilon/\epsilon_0,\ \nu_H^* = \nu_H/\nu_{H0},$$
$$k^* = k/k_0,\ \rho^* = \rho/\rho_0,\ \mu^* = \mu/\mu_0,\ c_p^* = c_p/c_{p0},\ S^* = S T_0/U^2,$$
$$R^* = RL/U^3,\ R_t = t_0 U/L,\ R_E = E_0/UB_0,\ R_\sigma = \sigma_0 \mu_{e0} UL,$$
$$R_H = \mu_0 H_0^2/\rho_0 U^2,\ R_l = UL/\nu_0,\ P_r = c_{p0}\mu_0/k_0 \tag{2.346}$$

The equations become

$$\nabla^* \cdot \mathbf{D}^* = 0 \tag{2.347}$$

$$\nabla^* \cdot \mathbf{J}^* = 0 \tag{2.348}$$

$$\nabla^* \cdot \mathbf{B}^* = 0 \tag{2.349}$$

$$\nabla^* \times \mathbf{H}^* = R_\sigma \mathbf{J}^* \tag{2.350}$$

$$R_E (\nabla^* \times \mathbf{E}^*) = -\frac{1}{R_t}\frac{\partial \mathbf{B}^*}{\partial t^*} \tag{2.351}$$

$$\mathbf{D}^* = \epsilon^* \mathbf{E}^* \tag{2.352}$$

$$\mathbf{B}^* = \mu_e^* \mathbf{H}^* \tag{2.353}$$

$$\mathbf{J}^* = \sigma^* (R_E \mathbf{E}^* + \mathbf{v}^* \times \mathbf{B}^*) \tag{2.354}$$

$$\frac{1}{R_t}\frac{\partial \rho^*}{\partial t^*} + \frac{\partial \rho^* v_i^*}{\partial x_i^*} = 0 \tag{2.355}$$

$$\frac{1}{R_t}\frac{\partial v_i^*}{\partial t^*} + v_j^* \frac{\partial v_i^*}{\partial x_j^*} = F_i^* - \frac{1}{\rho^*}\frac{\partial p^*}{\partial x_i^*} + R_\sigma R_H \frac{(\mathbf{J}^* \times \mathbf{B}^*)_i}{\rho^*} + \frac{1}{R_e}\frac{\partial \pi_{ji}^*}{\partial x_j^*} \tag{2.356}$$

$$\rho^* T^* \left(\frac{1}{R_t}\frac{\partial S^*}{\partial t^*} + v_j^* \frac{\partial S^*}{\partial x_j^*}\right) = \frac{1}{R_e}\Phi^* + \frac{1}{P_r R_e}\nabla^* \cdot (k^* \nabla^* T^*) + \rho^* R^* + R_\sigma R_H \frac{\mathbf{J}^{*2}}{\sigma^*} \tag{2.357}$$

All the terms that contain the nondimensional parameters will be the same magnitude as the remaining terms if the multipliers are the order of unity. Thus, the time scale of problems to be studied is the same order as L/U and high frequency problems are not considered. Also the velocities are small compared to the speed of light since the relativistic terms in the constitutive relations, Eqs. (2.352) and (2.353) were neglected. The electric field strength is then at most the same order as UB_0 and the magnetic field will influence the flow and fluid temperature when the product $R_\sigma R_H$, or M_m^2 is the order of unity. The Reynolds and Peclet numbers' influence is the same as in ordinary fluid dynamics.

With the assumption of negligible excess charge and negligible surface charges, the electromagnetic boundary conditions of Sec. 2.13 remain unchanged and are

	Continuous	*Discontinuous*
	$\mathbf{n}\cdot\mathbf{D}_1^* = \mathbf{n}\cdot\mathbf{D}_2^*$	$\epsilon_1^* E_{n1}^* = \epsilon_2^* E_{n2}^*$
Normal	$\mathbf{n}\cdot\mathbf{B}_1^* = \mathbf{n}\cdot\mathbf{B}_2^*$	$\mu_{e1}^* H_{n1}^* = \mu_{e2}^* H_{n2}^*$
components	$\mathbf{n}\cdot\mathbf{J}_1^* = \mathbf{n}\cdot\mathbf{J}_2^*$	$\sigma_1^* E_{n1}^* = \sigma_2^* E_{n2}^*$
		$\epsilon_2^* D_{t1}^* = \epsilon_1^* D_{t2}^*$
Tangential	$\mathbf{n}\times\mathbf{E}_1^* = n\times\mathbf{E}_2^*$	$\mu_{e2}^* B_{t1}^* = \mu_{e1}^* B_{t2}^*$
components	$\mathbf{n}\times\mathbf{H}_1^* = \mathbf{n}\times\mathbf{H}_2^*$	$\sigma_2^* J_{t1}^* = \sigma_1^* J_{t2}^*$

(2.358)

The no-slip condition and the prescribed temperatures or heat flux at the wall are again the fluid boundary conditions.

2.25 PROPERTIES OF MAGNETOFLUID-DYNAMIC EQUATIONS

The MFD equations in the previous section are applicable when the electric and magnetic fields are equally important. However, magnetofluid dynamics is concerned with problems in which the magnetic field is of prime importance, and electrofluid dynamics considers problems in which the electric field is dominant and the excess charge may be important. Magnetogas dynamics and magnetohydrodynamics are also concerned primarily with the magnetic field since they are subtopics of magnetofluid dynamics.

Even though the MFD equations are more complicated than are the Navier-Stokes equations, there are similarities between them and we may use the results of fluid dynamics to study MFD.

2.25.1 Analogy between MFD Magneto-Field Equation and Fluid-Dynamic Vorticity. Assuming that the electrical conductivity is a constant, the magnetic field equations, Eqs. (2.349) and (2.343), become

$$\nabla^* \cdot \mathbf{B}^* = 0 \tag{2.359}$$

$$\frac{1}{R_t}\frac{\partial \mathbf{H}^*}{\partial t^*} + \mathbf{H}^*(\nabla^* \cdot \mathbf{v}^*) + (\mathbf{v}^* \cdot \nabla^*)\mathbf{H}^* - (\mathbf{H}^* \cdot \nabla^*)\mathbf{v}^* = \frac{\nu_H^*}{R_\sigma}\nabla^{*2}\mathbf{H}^* \tag{2.360}$$

By defining the fluid vorticity vector as

$$\boldsymbol{\omega}^* = \nabla^* \times \mathbf{v}^* \tag{2.361}$$

and assuming that the viscosity is constant and the fluid is baratropic, i.e., $p = f(\rho)$ only, the continuity and momentum equations become

$$\nabla^* \cdot \boldsymbol{\omega}^* = 0 \tag{2.362}$$

$$\frac{1}{R_t}\frac{\partial \omega^*}{\partial t^*} + \omega^*(\nabla^* \cdot \mathbf{v}^*) + (\mathbf{v}^* \cdot \nabla^*)\boldsymbol{\omega}^* - (\boldsymbol{\omega}^* \cdot \nabla^*)\mathbf{v}^* = \frac{\nu^*}{R_e}\nabla^{*2}\omega^* \tag{2.363}$$

and both sets of equations have the same form. We may now apply all the theorems of vorticity in a baratropic fluid to the magnetic field of magnetofluid dynamics. By Helmholtz's theory, we know that the vortex lines move with the fluid if the viscosity is zero and analogously, the magnetic lines move with the fluid when ν_H^* is zero which is called the *frozen-field case* since $\sigma^* \to \infty$.

2.25.2 Diffusion Phenomena in MFD. There are three diffusion coefficients in MFD, i.e., the coefficient of viscosity μ^*, which represents the diffusion of momentum; the coefficient of thermal conductivity k^*, and ν_H^* which represents the diffusion of the magnetic field. These diffusion phenomena may be represented by the three nondimensional parameters, i.e., the Reynolds number R_e, the Peclet number $P_e = P_r R_e$, and the magnetic Reynolds number R_σ. When these nondimensional parameters are large, we have the boundary layer phenomena, i.e., the velocity boundary layer, thermal boundary layer, and magnetic boundary layer. Since we have one more diffusion coefficient in MGD than in ordinary gas dynamics, we have some new phenomena in MGD that will not occur in ordinary gas dynamics as we shall demonstrate in Chap. 4.

2.25.3 Ideal Plasma. When the parameters R_e, P_e, and R_σ are very large, outside the boundary layer region, the fluid may be considered as inviscid, non-heat-conducting, and infinitely electrically conducting. Such a fluid may be referred to as *ideal plasma*. The system of the fundamental equations of MGD for an ideal plasma is hyperbolic and it is similar to that of ordinary gas dynamics for inviscid fluid. For hyperbolic equations, disturbances are propagated with finite speeds. For an inviscid fluid in ordinary gas dynamics, disturbances travel with the speed of sound relative to the motion of the fluid. In MGD of an ideal plasma, there is more than one mode of disturbances at each point. We shall discuss these waves in detail in Chap. 5.

2.26 MAGNETOFLUID-DYNAMIC EQUATIONS–SPECIAL CASES

Most problems employ only special forms of the MFD equations for restricted conditions. The following are the more common special cases.

2.26.1 Small R_σ. The left side of Eq. (2.360) will now be small and the magnetic field will be independent of the flow. As a consequence, we can assume that the magnetic field is given and we neglect the induced magnetic field. The equations become

$$R_E\,(\nabla^* \times \mathbf{E}^*) = -\frac{1}{R_t}\frac{\partial \mathbf{B}^*}{\partial t^*} \tag{2.364}$$

$$\mathbf{J}^* = \sigma^*(R_E\,\mathbf{E}^* + \mathbf{v}^* \times \mathbf{B}^*) \tag{2.365}$$

$$\frac{1}{R_t}\frac{\partial \rho^*}{\partial t^*} + \frac{\partial \rho^* v_i^*}{\partial x_i^*} = 0 \tag{2.366}$$

$$\frac{1}{R_t}\frac{\partial v_i^*}{\partial t^*} + v_j^*\frac{\partial v_i^*}{\partial x_j^*} = F_i^* - \frac{1}{\rho^*}\frac{\partial p^*}{\partial x_i^*} + \frac{R_\sigma R_H}{\rho^*}(\mathbf{J}^* \times \mathbf{B}^*)_i + \frac{1}{R_e}\frac{\partial \pi_{ji}^*}{\partial x_j^*} \tag{2.367}$$

$$\rho^* T^*\left(\frac{1}{R_t}\frac{\partial S^*}{\partial t^*} + v_j^*\frac{\partial S^*}{\partial x_j^*}\right) = \frac{1}{R_e}\Phi^* + \frac{1}{P_e}\nabla^* \cdot (k^*\nabla^* T^*) + \rho^* R^* + R_\sigma R_H \frac{J^{*2}}{\sigma^*} \tag{2.368}$$

These equations will predict the first-order effects of the magnetic field on the flow.

2.26.2 Small M_m. The fluid-dynamic equations are now independent of the magnetic field and are

$$\frac{1}{R_t}\frac{\partial \rho^*}{\partial t^*} + \frac{\partial \rho^* v_i^*}{\partial x_i^*} = 0 \tag{2.369}$$

$$\frac{1}{R_t}\frac{\partial v_i^*}{\partial t^*} + v_j^*\frac{\partial v_i^*}{\partial x_j^*} = F_i^* - \frac{1}{\rho^*}\frac{\partial p^*}{\partial x_i^*} + \frac{1}{R_e}\frac{\partial \pi_{ji}^*}{\partial x_j^*} \tag{2.370}$$

$$\rho^* T^*\left(\frac{1}{R_t}\frac{\partial S^*}{\partial t^*} + v_j^*\frac{\partial S^*}{\partial x_j^*}\right) = \frac{1}{R_e}\Phi^* + \frac{1}{R_e}\nabla^*\cdot(k^*\nabla^* T^*) + \rho^* R^* \tag{2.371}$$

These are the equations of ordinary fluid dynamics. However, since ν_H^* is generally a function of temperature, these equations determine the effect of the flow on the magnetic field.

2.26.3 Constant Material Properties and Low Speed. With these assumptions, we obtain the set of equations most commonly applied in engineering problems,

$$\nabla^* \times \mathbf{H}^* = R_\sigma \mathbf{J}^* \tag{2.372}$$

$$R_E(\nabla^* \times \mathbf{E}^*) = -\frac{1}{R_t}\frac{\partial \mathbf{B}^*}{\partial t^*} \tag{2.373}$$

$$\mathbf{J}^* = \sigma^*(R_E \mathbf{E}^* + \mathbf{v}^* \times \mathbf{B}^*) \tag{2.374}$$

$$\frac{\partial v_i^*}{\partial x_i^*} = 0 \tag{2.375}$$

$$\frac{1}{R_t}\frac{\partial v_i^*}{\partial t^*} + v_j^*\frac{\partial v_i^*}{\partial x_j^*} = F_i^* - \frac{1}{\rho^*}\frac{\partial p^*}{\partial x_i^*} + \frac{R_\sigma R_H}{\rho^*}(\mathbf{J}^* \times \mathbf{B}^*)_i + \frac{\nu^*}{R_e}\nabla^{*2} v_i^* \tag{2.376}$$

$$\rho^* T^*\left(\frac{1}{R_t}\frac{\partial S^*}{\partial t^*} + v_j^*\frac{\partial S^*}{\partial x_j^*}\right) = \frac{k^*}{P_e}\nabla^{*2} T^* + \rho^* R^* + R_\sigma R_H \frac{J^{*2}}{\sigma^*} \tag{2.377}$$

2.26.4 Ideal Plasma. An inviscid, non-heat-conducting, and infinitely electrically conducting fluid is called an *ideal plasma* and the equations are

$$\nabla^* \times (\mathbf{v}^* \times \mathbf{H}^*) = \frac{1}{R_t} \frac{\partial \mathbf{B}^*}{\partial t^*} \tag{2.378}$$

$$\frac{1}{R_t} \frac{\partial \rho^*}{\partial t} + \frac{\partial \rho^* v^*}{\partial x_i^*} = 0 \tag{2.379}$$

$$\frac{1}{R_t} \frac{\partial v_i^*}{\partial t^*} + v_j^* \frac{\partial v_i^*}{\partial x_j^*} = F_i - \frac{1}{\rho^*} \frac{\partial [p^* + (\mu_e^* H^{*2}/2) R_H]}{\partial x_i^*} + \mu_e^* (\mathbf{H}^* \cdot \nabla^*) \mathbf{H}^* R_H \tag{2.380}$$

$$\frac{1}{R_t} \frac{\partial S^*}{\partial t^*} + v_j^* \frac{\partial S^*}{\partial x_j^*} = \frac{R^*}{T^*} \tag{2.381}$$

where the electric field is only the induced field $\mathbf{v}^* \times \mathbf{B}^*$.

REFERENCES

1. L. B. Loeb, "Fundamentals of Electricity and Magnetism," John Wiley & Sons, Inc., New York, 1947.
2. R. Plonsey and R. E. Collin, "Principles and Applications of Electromagnetic Fields," McGraw-Hill Book Company, New York, 1961.
3. R. K. Wangsness, "Introduction to Theoretical Physics—Classical Mechanics and Electrodynamics," John Wiley & Sons, Inc., New York, 1963.
4. R. H. Atkin, "Theoretical Electromagnetics," John Wiley & Sons, Inc., New York, 1962.
5. L. Page, "Introduction to Theoretical Physics," 2d ed., D. Van Nostrand Company, Inc., New York.
6. S. Goldstein, Lectures on Fluid Mechanics, "Lectures in Applied Mathematics," vol. II, Interscience Publishers, Ltd., London, 1960.
7. A. E. Love, "A Treatise on the Mathematical Theory of Elasticity," 4th ed., Dover Publications, Inc., New York, 1944.
8. S. O. Soo, "Analytical Thermodynamics," Prentice-Hall, Inc., Englewood Cliffs, N.J., 1962.
9. I. S. Sokohnikoff, "Tensor Analyses Theory and Application to Geometry and Mechanics of Continua," 2d ed., John Wiley & Sons, Inc., New York, 1964.

Chapter 3

MAGNETOFLUID-DYNAMIC ONE-DIMENSIONAL FLOWS

3.1 INTRODUCTION

Magnetofluid-dynamic generation of electric power may be accomplished by extracting electric energy directly from a moving conducting fluid. Even though a great many possible flow configurations of an MFD generator have been suggested, the simplest and the most promising one is the Faraday generator which consists of a pipe or a channel of slowly varying cross-section area in which the conducting fluid flows under transverse electromagnetic fields and an axial pressure gradient. In engineering practice, we may study the performance of MFD generators by analyzing the quasi-one-dimensional MFD flow[1,2] in a channel. Quasi-one-dimensional flow analysis is also useful for other engineering problems of MFD such as propulsion devices.

There are two distinct types of analyses of one-dimensional-flow problems of MFD.[3] One is referred to as "approximately" one-dimensional flow and the other as "strictly" one-dimensional flow. The main difference in these analyses is in the expression of the electrical current density. In this chapter, the equations for quasi-one-dimensional flow will be derived from the MFD equations and the other two cases will be described by appropriate assumptions.

In the quasi-one-dimensional flow analysis, we use the average value of the MFD variables such as the x-wise velocity u, temperature T, y-wise magnetic field Hy, and so forth. These values are averaged over the cross-section area $A(x)$, that is,

$$A(x) = \iint dydz \tag{3.1}$$

at a fixed distance x along the axis of the channel, that is,

$$\overline{Q}(x, t) = \frac{1}{A(x)} \iint_{A(x)} Q(x, y, z, t)\, dydz \tag{3.2}$$

where $Q(x, y, z, t)$ may be any one of the MFD variables and $\overline{Q}(x, t)$ is the corresponding average value used in the quasi-one-dimensional analysis. The average value $\overline{Q}(x, t)$ is, in general, a function of the axial coordinate x and the time t only.

We will derive the equations for all the average quantities $\overline{Q}(x, t)$ of MFD so that the accurate expression for the average electrical current density can be obtained. We shall also show the effects of nonuniform profiles in the channel which may be important. For instance, one should not simply replace the velocity u by its average or its mass average [see Eq. (3.13)] whenever it occurs in the equations.

In Secs. 3.2 to 3.9, we derive the exact quasi-one-dimensional MFD flow equations and in Secs. 3.11 and 3.12, we discuss various types of quasi-one-dimensional flows of MFD.

3.2 QUASI-ONE-DIMENSIONAL ASSUMPTIONS

We consider the flow of an electrically conducting fluid in a channel of slowly varying cross section with transverse externally applied electromagnetic fields. For simplicity, we assume that the axis of the channel is straight and is denoted as the x axis. The cross section of the channel is assumed to be rectangular. The height of the channel in the y direction is $b(x)$ which is a given, slowly varying function of x; the width of the channel in the z direction is b_z which is assumed to be a constant. We assume that b_z is always larger than $b(x)$ so that the variation of the MFD variables in the z direction is negligibly small compared to those in the x or y direction. There is an externally applied magnetic field in the y direction, $H_y = H_0(x)$, on the boundary of the channel which is a slowly varying function of x and an externally applied electric field in the z direction, $E_z = E_0(x)$, which is also a slowly varying function of x. In general, we should solve the fundamental equations of MFD with all four independent variables x, y, z, and t. Such an integration is usually too complicated to be useful in practical problems. Hence, for engineering analyses, we consider only the average variables $\overline{Q}(x, t)$ in the channel defined by Eq. (3.2) or similar expressions such as Eq. (3.14). From Eq. (3.2) we may define the average x-, y-, and z-velocity components by the following relations:

$$A(x)\,\overline{u}(x, t) = \iint_A u(x, y, z, t)\,dydz \tag{3.3}$$

$$A(x)\,\overline{v}(x, t) = \iint_A v(x, y, z, t)\,dydz \tag{3.4}$$

$$A(x)\,\overline{w}(x, t) = \iint_A w(x, y, z, t)\,dydz \tag{3.5}$$

We also have the expression for the average density $\overline{\rho}$:

$$\overline{\rho} = \frac{1}{A(x)} \iint_A \rho(x, y, z, t)\,dydz \tag{3.6}$$

Similar expressions may be obtained for other MFD variables such as average temperature $\overline{T}$ or average magnetic field strength $\overline{\mathbf{H}}$.

The average value of the product of two quantities f and g is in general not equal to the product of the average values of the two quantities. The difference between these two averages is known as the *covariance* of f and g, i.e.,

$$\overline{fg} - \overline{f}\,\overline{g} = \operatorname{cov}(f, g) \tag{3.7}$$

where the bar refers to the average value of a quantity defined by Eq. (3.2).

Now we will derive the basic equations of quasi-one-dimensional flow by integrating the fundamental equations of MFD over a cross section $A(x) = b_z b(x)$ at a fixed value of x.

3.3 EQUATION OF CONTINUITY

The equation of continuity in cartesian coordinates is

$$\frac{\partial \rho}{\partial t} + \frac{\partial \rho u}{\partial x} + \frac{\partial \rho v}{\partial y} + \frac{\partial \rho w}{\partial z} = 0 \tag{3.8}$$

If we integrate Eq. (3.8) over the cross section $A(x)$ of Eq. (3.1) and use the definition of the average value, Eq. (3.2) and the boundary conditions that $v = 0$ and $w = 0$ on the surface of the channel, we have

$$\frac{\partial A\bar{\rho}}{\partial t} + \frac{\partial A\overline{\rho u}}{\partial x} = 0 \tag{3.9}$$

or

$$\frac{\partial \bar{\rho}}{\partial t} + \bar{u}\frac{\partial \bar{\rho}}{\partial x} + \bar{\rho}\frac{\partial \bar{u}}{\partial x} + \frac{\partial}{\partial x}[\mathrm{cov}(\rho, u)] + [\bar{\rho}\,\bar{u} + \mathrm{cov}(\rho, u)]\frac{1}{A}\frac{dA}{dx} = 0 \tag{3.10}$$

In ordinary quasi-one-dimensional analysis, the $\mathrm{cov}(\rho, u)$ is usually neglected.

For steady flow, Eq. (3.9) gives

$$A\overline{\rho u} = A\bar{\rho}\,\bar{u} + A\,\mathrm{cov}(\rho, u) = \mathrm{const} = m \tag{3.11}$$

By means of the following definition of the mean value of density, the covariance term in Eq. (3.11) may be absorbed into the mean value of the density. If we define a mass average value of the density ρ_m by the relation

$$\frac{1}{\rho_m} = \frac{\int (1/\rho)dm}{m} = \frac{\iint u\,dy\,dz}{m} = \frac{A\bar{u}}{m} \tag{3.12}$$

where $dm = \rho u dA$, Eq. (3.11) becomes

$$A\rho_m \bar{u} = \text{const} = m \tag{3.13}$$

The mean density ρ_m is the quotient of the total rate of mass flow m and the total rate of the volume flow of the fluid $A\bar{u}$. The mean value of density ρ_m is usually referred to as the *mass average value* and is different from the average value of density $\bar{\rho}$ over a section $A(x)$. We may generalize the definition, Eq. (3.12), by considering the mass average value of any variable Q with the definition

$$mQ_m = \int Q dm \tag{3.14}$$

However, we shall not use the mean values defined by this equation because it is not applicable for unsteady flow where m is not a constant and it is not suitable for electromagnetic variables. Furthermore, it will not give any simplification for the equation of state or equation of energy and when two different averaging processes $\bar{u}$ and ρ_m are used in one analysis, it may cause confusion.

Since it is inconvenient to use Eq. (3.10) or (3.11) because we do not know the functional variation of cov (ρ, u), a nondimensional covariance coefficient k_ρ may be introduced in the expression $\overline{\rho u}$ such that

$$\overline{\rho u} = k_\rho \bar{\rho} \bar{u} \tag{3.15}$$

The factor k_ρ depends on the shape of the channel and its operating conditions and should be determined experimentally. However, for a given channel at a given operating condition, we may assume that k_ρ is a constant. If one of the variables ρ or u is a constant, k_ρ is unity. Hence we would expect that the value of k_ρ is of the order of unity. Figure 3.1 shows a rough estimate of k_ρ and cov (ρ, u) by assuming $u^* = 1 - y^{*2}$ and $\rho^* = [1 - 0.2M^2(1 - y^{*2})^2]^{2.5}$ where u^* and ρ^* are respectively the nondimensional velocity and density with respect to their values at the center of the channel and y^* is the nondimensional y-distance. Figure 3.1 shows that k_ρ may differ significantly from

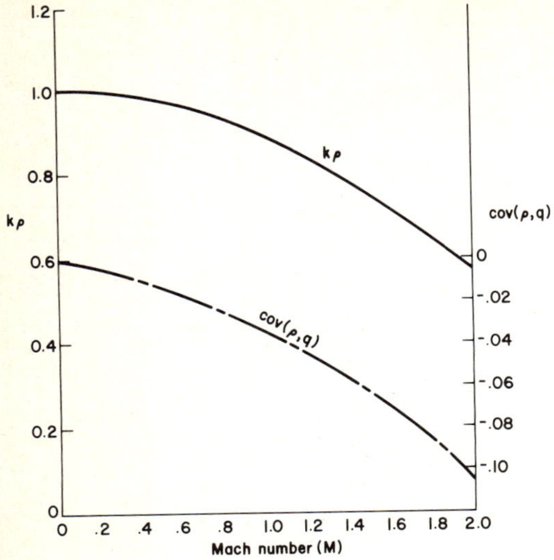

Fig. 3.1 Covariance of density and velocity cov (ρ, q) and its corresponding factor k_ρ.

unity and that the value of k_ρ depends not only on the velocity profile, but also on Mach number M which represents the operating conditions.

3.4 EQUATION OF STATE

Applying the averaging process Eq. (3.2) to the equation of state of a perfect gas, we have

$$\bar{p} = R\overline{\rho T} = R[\bar{\rho}\,\bar{T} + \mathrm{cov}(\rho, T)] = k_p R\bar{\rho}\bar{T} \tag{3.16}$$

Similar to Eq. (3.15), we introduced a nondimensional factor k_p in Eq. (3.16). If any one of the variables p, ρ, or T is a constant over the section at a given station x, the cov (ρ, T) is zero and k_p is unity. Thus, in quasi-one-dimensional flow of ordinary gas dynamics, one usually assumes that p is constant at any given x station, and cov $(\rho, T) = 0$. Even though such an assumption is a good approximation for ordinary gas dynamics, it is not a good one for MFD. For instance, in the Poiseuille flow of an incompressible fluid, the pressure is independent of the transverse coordinates in ordinary gas dynamics. However, in MFD flow, the pressure depends on the transverse coordinates in MFD flow because of the variation of the axial magnetic field strength H_x over the cross section. Since the variation of pressure over the cross section is proportional to the square of the magnetic Reynolds number R_σ, the covariance coefficient k_p depends on R_σ. When R_σ is very small, we may assume that k_p is equal to unity.

3.5 EQUATION FOR AVERAGE ELECTRIC CURRENT DENSITY

We consider only the case of scalar electrical conductivity. The Hall current effects will be discussed in a later chapter.

For simplicity, we shall consider essentially a two-dimensional analysis by assuming

$$w = \overline{w} \simeq 0 \quad B_z = \overline{B}_z \simeq 0 \tag{3.17}$$

and all variables are independent of the z coordinate. The assumptions will not seriously restrict the final results. They are consistent with our configuration of the channel and with disregarding the Hall current.

Under these assumptions, we have

$$J_x \simeq \overline{J}_x \simeq 0 \quad J_y \simeq \overline{J}_y \simeq 0 \tag{3.18}$$

The z component of the electrical current density is

$$J_z = \frac{\partial H_y}{\partial x} - \frac{\partial H_x}{\partial y} \tag{3.19}$$

Applying the averaging process in Eq. (3.2) to Eq. (3.19) we have

$$A\overline{J}_z = \frac{\partial \overline{H}_y A}{\partial x} - b_z[H_x\,(y = b) - H_x\,(y = 0)] \tag{3.20}$$

Equation (3.20) shows that there are two parts for the average $\overline{J}_z$, which are

$$\overline{J}_{z1}(x,t) = -\frac{1}{b}[H_x(y=b) - H_x(y=0)] \tag{3.21}$$

and

$$\overline{J}_{z2}(x,t) = \frac{1}{A}\frac{\partial A\overline{H}_y}{\partial x} \tag{3.22}$$

In approximately one-dimensional MFD flow, one assumes $\overline{J}_z = \overline{J}_{z1}$, while in the strictly one-dimensional MFD flow, one assumes $\overline{J}_z = \overline{J}_{z2}$. In general, the two terms are of the same order of magnitude and we should not neglect either of them.

Since the exact expression of $\overline{J}_{z1}$ depends on the solution of a two-dimensional problem, Eq. (3.21) is not very useful in the quasi-one-dimensional analysis. We should express $\overline{J}_{z1}$ in terms of the variables in the quasi-one-dimensional analysis. From the exact solution of the plane Poiseuille flow, we have [see Chap. 4, Eq. (4.18)]

$$H_x\left(y=\frac{b}{2}\right) - H_x\left(y=-\frac{b}{2}\right) = -\sigma b(\mu_e \overline{u}\overline{H}_y + \overline{E}_z) \tag{3.23}$$

With the help of Eq. (3.23), we may write

$$\overline{J}_{z1} = k_{\sigma 1}\overline{\sigma}(\overline{E}_z + k_{j1}\overline{u}\overline{B}_y) \tag{3.24}$$

where $k_{\sigma 1}$ and k_{j1} are factors that account for the actual averaging process. Equation (3.24) agrees with the expression of $\overline{J}_z$ derived from the generalized Ohm's law.

According to the generalized Ohm's law

$$J_z = \sigma E_{zu} = \sigma(E_z + uB_y - vB_x) \tag{3.25}$$

The averaging process of Eq. (3.25) gives

$$\overline{J}_z = \overline{\sigma E_{zu}} = \overline{\sigma}(\overline{E}_z + \overline{uB_y} - \overline{vB_x}) + \operatorname{cov}(\sigma, E_{zu}) = k_\sigma\overline{\sigma}(\overline{E}_z + k_j\overline{u}\overline{B}_y) \tag{3.26}$$

The factors k_σ and k_j take the corresponding covariance into consideration and k_j includes the effect of $\overline{vB_x}$ which is usually small in our problem. Because of the similarity of Eqs. (3.24) and (3.26), many research workers thought that Eq. (3.24) was the correct one; in other words, they neglected $\overline{J}_{z2}$. The differences between the coefficients k_σ and $k_{\sigma 1}$, as well as k_j and k_{j1} should be noted and we shall show later that Eq. (3.26) is identical to Eq. (3.20) which contains both $\overline{J}_{z1}$ and $\overline{J}_{z2}$.

3.6 EQUATION FOR AVERAGE ELECTRIC FIELD

Because of the assumptions made in Eq. (3.17), we have $E_x \simeq \overline{E} = 0$ and $E_y \simeq \overline{E}_y \simeq 0$. The z component of the electrical field strength is

$$E_z = \nu_H \left(\frac{\partial B_y}{\partial x} - \frac{\partial B_x}{\partial y} \right) - (uB_y - vB_x) \tag{3.27}$$

The average value of E_z is then

$$\overline{E}_z = \frac{1}{k_\sigma \overline{\sigma} A} \frac{\partial A\overline{H}_y}{\partial x} - \frac{1}{k_\sigma \overline{\sigma} b} [H_x (y = b) - H_x (y = 0)] - k_j \overline{u}\overline{B}_y \tag{3.28}$$

3.7 EQUATION FOR AVERAGE MAGNETIC FIELD

The equation for the magnetic field for the case of scalar electrical conductivity is

$$\frac{\partial \mathbf{B}}{\partial t} = \nabla \times (\mathbf{q} \times \mathbf{B}) - \nabla \times [\nu_H (\nabla \times \mathbf{B})] \tag{3.29}$$

The x component of the equation of magnetic induction, Eq. (3.29), is

$$\frac{\partial B_x}{\partial t} = \frac{\partial}{\partial y} \left[(uB_y - vB_x) - \nu_H \left(\frac{\partial B_y}{\partial x} - \frac{\partial B_x}{\partial y} \right) \right] - \frac{\partial}{\partial z} \left[(wB_x - uB_z) - \nu_H \left(\frac{\partial B_x}{\partial z} - \frac{\partial B_z}{\partial x} \right) \right] \tag{3.30}$$

Under the two-dimensional assumption made in Eq. (3.17), the averaging process [Eq. (3.2)] gives as the average value of $\partial \overline{B}_x/\partial t$,

$$\frac{\partial \overline{B}_x}{\partial t} = -\frac{1}{b}\left[\left(\nu_H \frac{\partial B_y}{\partial x} - \nu_H \frac{\partial B_x}{\partial y}\right)_{y=+b/2} - \left(\nu_H \frac{\partial B_y}{\partial x} - \nu_H \frac{\partial B_x}{\partial y}\right)_{y=-b/2}\right] \tag{3.31}$$

For steady flow, we have

$$\left[\frac{1}{\sigma}\left(\frac{\partial H_x}{\partial y} - \frac{\partial H_y}{\partial x}\right)\right]_{y=b/2} = \left[\frac{1}{\sigma}\left(\frac{\partial H_x}{\partial y} - \frac{\partial H_y}{\partial x}\right)\right]_{y=-b/2} \tag{3.32}$$

Equation (3.32) shows that the local z component of the electric field E_z at $y = b$ is equal to that at $y = 0$ in steady flow. But in unsteady flow, these two values are not equal.

The y component of Eq. (3.29) is

$$\frac{\partial B_y}{\partial t} = \frac{\partial}{\partial z}\left[(vB_z - wB_y) - \nu_H\left(\frac{\partial B_z}{\partial y} - \frac{\partial B_y}{\partial z}\right)\right] - \frac{\partial}{\partial x}\left[(uB_y - vB_x) - \nu_H\left(\frac{\partial B_y}{\partial x} - \frac{\partial B_x}{\partial y}\right)\right] \tag{3.33}$$

With the help of Eqs. (3.2), (3.17), and (3.28), we have the average value of $\partial \overline{B}_y/\partial t$ as follows:

$$A\frac{\partial \overline{B}_y}{\partial t} = \frac{\partial}{\partial x}(A\overline{E}_z) = \frac{\partial}{\partial x}\left\{\frac{1}{k_\sigma \overline{\sigma}}\frac{\partial A\overline{H}_y}{\partial x} - \frac{b_z}{k_\sigma \overline{\sigma}}\left[H_x\left(y = \frac{b}{2}\right) - H_x\left(y = -\frac{b}{2}\right)\right] - k_j \overline{u}\overline{B}_y A\right\} \tag{3.34}$$

For steady flow, Eq. (3.34) gives

$$A\overline{E}_z = \frac{1}{k_\sigma \overline{\sigma}}\frac{dA\overline{H}_y}{dx} - \frac{b_z}{k_\sigma \overline{\sigma}}\left[H_x\left(y = \frac{b}{2}\right) - H_x\left(y = -\frac{b}{2}\right)\right] - Ak_j \overline{u}\overline{B}_y$$
$$= \text{const} = K \tag{3.35}$$

Notice that only when A is a constant, is the average value of E_z a constant. In general, we have

$$A\overline{E}_z = \text{const} = K \tag{3.36}$$

Equation (3.36) is different from the approximation $E\sqrt{A}$ is a constant which has been used by some authors.[4,5] Equation (3.36) is the direct result obtained by the averaging process of Eq. (3.2).

Combining Eqs. (3.21), (3.24), (3.26), and (3.30), we have the differential equation for $\overline{H}_y$:

$$\overline{J}_{z2} = \frac{1}{A}\frac{\partial A\overline{H}_y}{\partial x} = \sigma_0(\overline{E}_z + k_J\overline{u}\overline{B}_y) \tag{3.37}$$

where

and

$$\left.\begin{aligned} \sigma_0 &= (k_\sigma - k_{\sigma 1})\overline{\sigma} = k_0\overline{\sigma} \\ k_J &= \frac{k_\sigma k_j - k_{\sigma_1}k_{j1}}{k_\sigma - k_{\sigma_1}} \end{aligned}\right\} \tag{3.38}$$

If

$$\overline{J}_{z2} \equiv 0 \quad k_\sigma = k_{\sigma 1} \text{ and } k_0 = \sigma_0 = 0 \text{ then } A\overline{H}_y = \text{constant} \tag{3.39}$$

If $\overline{\sigma} = 0$, Eq. (3.39) also holds true.

Equation (3.37) should be considered as one of the basic equations in the quasi-one-dimensional MFD flow. If A is a constant, Eq. (3.37) reduces to the same form as the equation used in a strictly one-dimensional analysis except for the factors k_0 and k_J. $\overline{H}_y$ is not an arbitrary function of x in quasi-one-dimensional flow as it is in approximately one-dimensional analysis of MFD. The value of k_0 should lie between zero and unity and depends on the ratio of $\overline{J}_{z1}/\overline{J}_z$.

3.8 EQUATION OF MOTION

The correct way to average the momentum equation of MFD flow in a channel is over an elementary volume between two sections of the

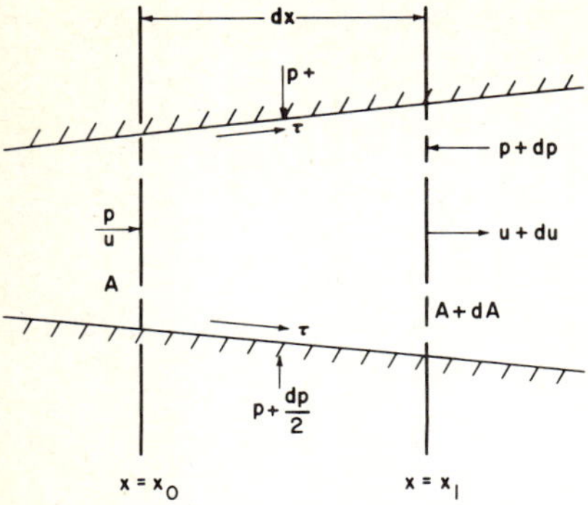

Fig. 3.2 Control surface in an MFD channel flow.

channel with a distance dx apart (Fig. 3.2). We integrate the x-wise equation of motion over this volume and then take the limit $dx \to 0$. We have

$$\frac{\partial A\overline{\rho u}}{\partial t} + \frac{\partial (A\overline{\rho u^2})}{\partial x} = -A\frac{\partial \overline{p}}{\partial x} + A\overline{F}_b - A\overline{J_z B_y} \tag{3.40}$$

Now we write

$$\overline{\rho u^2} = \overline{\rho}\,\overline{u}^2 + \overline{u}\,\mathrm{cov}(\rho, u) + \mathrm{cov}(\rho u, u) = k_u\overline{\rho}\,\overline{u}^2 \tag{3.41}$$

where k_u is a nondimensional covariance coefficient and

$$A\overline{F}_b = b_z[\tau_{xy}(y=b) - \tau_{xy}(y=0)] = C_f\frac{\overline{\rho}\,\overline{u}^2}{2}b_z \tag{3.42}$$

where $\overline{F}_b$ is the wall friction term and $\tau_{xy}(\ )$ is the shearing stress on the wall. We neglect the small difference of the direction of the actual wall friction and the x direction. For engineering problems, the wall friction is usually expressed in terms of the skin friction coefficient C_f, defined in Eq. (3.42). Now

$$\overline{J_z B_y} = \overline{\sigma(E_z + uB_y - vB_x)B_y} = k_\sigma\overline{\sigma}(k_E\overline{E}_z + k_B\overline{u}\overline{B}_y)\overline{B}_y \tag{3.43}$$

where k_E and k_B are new covariance factors.

Substituting Eqs. (3.15) and (3.41)-(3.43) into Eq. (3.40), we have the equation of motion:

$$\bar{\rho}\,\frac{\partial k_\rho \bar{u}}{\partial t} + \bar{\rho}\,\bar{u}\,\frac{\partial k_u \bar{u}}{\partial x} = -\frac{\partial \bar{p}}{\partial x} + \frac{C_f}{2b}\,\bar{\rho}\,\bar{u}^2 - k_\sigma \bar{\sigma}\bar{B}_y(k_E \bar{E}_z + k_B\,\bar{u}\bar{B}_y)$$

$$- \bar{u}\left(k_\rho\,\frac{\partial \bar{\rho}}{\partial t} + k_u\,\frac{\partial \bar{\rho}\,\bar{u}}{\partial x} + k_u\,\frac{\bar{\rho}\,\bar{u}}{A}\,\frac{dA}{dx}\right) \qquad (3.44)$$

If we take all the covariance coefficients, k_ρ, k_u, k_σ, k_E, and k_B to be unity, Eq. (3.44) becomes the customary equation of motion in the quasi-one-dimensional flow:

$$\bar{\rho}\,\frac{\partial \bar{u}}{\partial t} + \bar{\rho}\,\bar{u}\,\frac{\partial \bar{u}}{\partial x} = -\frac{\partial \bar{p}}{\partial x} - \bar{\sigma}\bar{B}_y(\bar{E}_z + \bar{u}\bar{B}_y) + \frac{C_f}{2b}\,\bar{\rho}\,\bar{u}^2 \qquad (3.45)$$

3.9 EQUATION OF ENERGY

If we average the energy equation of MFD with the help of Eq. (3.2) and use the boundary condition $u = v = w = 0$ on the wall and the assumption of Eq. (3.17), we have the following equation of energy of quasi-one-dimensional MFD flow:

$$\frac{\partial A\bar{\rho}\,\bar{e}_m}{\partial t} + \frac{\partial}{\partial x}(A\overline{\rho u e_m}) = -\frac{\partial A\overline{pu}}{\partial x} + \frac{\partial A\overline{u\tau}_{xx}}{\partial x} + \frac{\partial A\overline{v\tau}_{xy}}{\partial x} + \frac{\partial \bar{Q}_x A}{\partial x}$$

$$+ b_z\left[Q_y\left(y = \frac{b}{2}\right) - Q_y\left(y = -\frac{b}{2}\right)\right] + A\overline{E_z J_z} + A\bar{Q}_0 \quad (3.46)$$

The meanings and further simplifications of various terms in Eq. (3.46) follow:

The total energy per unit mass e_m is

$$e_m = U_m + \tfrac{1}{2}(u^2 + v^2)$$

then

$$\bar{e}_m = \bar{U}_m + \tfrac{1}{2}(\overline{u^2} + \overline{v^2}) = \bar{U}_m + \tfrac{1}{2}k_v\bar{u}^2 \qquad (3.47)$$

Where U_m is the internal energy per unit mass and the factor k_v takes the effect of $\overline{v^2}$ as well as the covariance term into consideration,

$$\overline{\rho e_m} = \bar{\rho}\bar{e}_m + \text{cov}(\rho, e_m) = k_e \bar{\rho}\bar{e}_m \tag{3.48}$$

$$\overline{\rho u e_m} = \overline{\rho u}\bar{e}_m + \text{cov}(\rho u, e_m) = k_\rho \bar{\rho}\bar{u}\bar{e}_m + \text{cov}(\rho u, e_m) = k_{\rho u}\bar{\rho}\bar{u}\bar{e}_m \tag{3.49}$$

The new factors k_e and $k_{\rho u}$ are the corresponding covariance coefficients.

The flow work term is

$$\overline{pu} = \bar{p}\bar{u} + \text{cov}(p, u) = k_{pu}\bar{u}\bar{p} \tag{3.50}$$

Where k_{pu} is the covariance coefficient for the flow work.

In engineering problems when the Reynolds number is large and the boundary-layer approximations hold, the viscous dissipations $\partial A\overline{u\tau}_{xx}/\partial x$ and $\partial A\overline{v\tau}_{xy}/\partial x$ are small. We may consider the viscous dissipation as a part of the heat addition Q_0 and need not consider them separately.

The terms of heat loss by conduction are $\partial A\bar{Q}_x/\partial x$ and

$$b_z\left[Q_y\left(y = \frac{b}{2}\right) - Q_y\left(y = -\frac{b}{2}\right)\right] = b_z(Q_c - b\bar{Q}_0) \tag{3.51}$$

Since the major portion of heat conduction occurs at the surface of the channel, we may neglect $\partial A\bar{Q}_x/\partial x$ and consider only the y-wise heat conduction of Eq. (3.51). In fact, the term $b_z Q_c$ may be considered as the resultant heat-transfer term, that is, heat loss by conduction, and heat addition by chemical reaction, viscous dissipation, or other external agency of heat source.

The energy source from the electromagnetic field is

$$\overline{E_z J_z} = \bar{E}_z\bar{J}_z + \text{cov}(E_z, J_z) = k_{EJ}\bar{E}_z\bar{J}_z \tag{3.52}$$

where k_{EJ} is the covariance coefficient.

With the help of Eqs. (3.47)–(3.52), the energy equation of quasi-one-dimensional MFD flow, Eq. (3.46), becomes

$$\bar{\rho}\frac{\partial k_e \bar{e}_m}{\partial t} + \bar{\rho}\bar{u}\frac{\partial k_{\rho u}\bar{e}_m}{\partial x} = -\bar{u}k_{pu}\frac{\partial \bar{p}}{\partial x} - \frac{\bar{p}}{A}\frac{\partial k_{pu}\bar{u}A}{\partial x} + \frac{1}{b}Q_c + k_{EJ}\bar{E}_z\bar{J}_z - \bar{e}_m\left(k_e\frac{\partial \bar{\rho}}{\partial t} + k_{\rho u}\frac{\partial \bar{\rho}\bar{u}}{\partial x} + k_{\rho u}\frac{\bar{\rho}\bar{u}}{A}\frac{dA}{dx}\right) \tag{3.53}$$

If the covariance coefficients k_e and $k_{\rho u}$ are unity, the last term of the parentheses in Eq. (3.53) is zero. Since we expect that these covariance coefficients do not differ greatly from unity, we may neglect these terms. Similarly, we may also neglect the terms in the parentheses of equation of motion, Eq. (3.44).

We may derive an average equation of kinetic energy from the equation of motion and we have approximately the following equation:

$$\bar{\rho}\,\bar{u}\,\frac{\partial k_e k_u \bar{u}}{\partial t} + \bar{\rho}\,\bar{u}^2\,\frac{\partial (k_{\rho u} k_e k_u \bar{u})}{\partial x} + k_{pu}\,\bar{u}\,\frac{\partial \bar{p}}{\partial x}$$

$$= k_T \bar{u} \bar{F}_b - k_{uB} \bar{\sigma} \bar{B}_y \bar{u} (k_E \bar{E}_z + k_B \bar{u} \bar{B}_y) \qquad (3.54)$$

where k_T and k_{uB} are the corresponding new covariance coefficients.

Subtracting Eq. (3.54) from Eq. (3.53) and neglecting the terms in parentheses of Eq. (3.53), we have another form of the energy equation of quasi-one-dimensional MFD flow,

$$\bar{\rho}\,\frac{\partial k_e \bar{U}_m}{\partial t} + \bar{\rho}\,\bar{u}\,\frac{\partial k_e \bar{U}_m}{\partial x} = -\frac{p}{A}\,\frac{\partial A \bar{u} k_{pu}}{\partial x} - k_T \bar{u} \bar{F}_b + \frac{1}{b} Q_c$$

$$+ k_{uB}\,\bar{\sigma}\,\bar{u}\,\bar{B}_y (k_E \bar{E}_z + k_B \bar{u} \bar{B}_y) + k_{EJ}\, k_\sigma \bar{E}_z \bar{\sigma} (\bar{E}_z + k_j \bar{u} \bar{B}_y) \qquad (3.55)$$

Equation (3.55) reduces to the customary energy equation in quasi-one-dimensional MFD flow when we assume that all the covariance coefficients are unity:

$$\bar{\rho}\,\frac{\partial \bar{U}_m}{\partial t} + \bar{\rho}\,\bar{u}\,\frac{\partial \bar{U}_m}{\partial x} = -\frac{\bar{p}}{A}\,\frac{\partial A \bar{u}}{\partial x} + \bar{\sigma}(\bar{E}_z + \bar{u} \bar{B}_y) + \frac{1}{b} Q_c - \frac{C_f}{2b}\,\bar{\rho}\,\bar{u}^3 \qquad (3.56)$$

3.10 SUMMARY AND COMPARISONS

The accurate quasi-one-dimensional equations of MFD should be:

$$\frac{\partial \rho A}{\partial t} + \frac{\partial \rho u k_\rho A}{\partial x} = 0 \qquad \text{(Ia)}$$

$$p = k_p \rho R T \qquad \text{(Ib)}$$

$$\rho \frac{\partial k_\rho u}{\partial t} + \rho u \frac{\partial k_u u}{\partial x} = -\frac{\partial p}{\partial x} - k_\sigma \sigma B_y(k_E E_z + k_B u B_y) + \frac{C_f}{2b} \rho u^2 \quad \text{(Ic)}$$

$$\rho \frac{\partial k_e U_m}{\partial t} + \rho u \frac{\partial k_e U_m}{\partial x} + \frac{p}{A} \frac{\partial k_{pu} A u}{\partial x} = \frac{1}{b} Q_c - k_T \frac{C_f}{2b} \rho u^3$$

$$+ k_{uB} \sigma u B_y (k_E E_z + k_B u B_y) + k_{EJ} k_\sigma \sigma (E_z + k_j u B_y) \quad \text{(Id)}$$

$$\frac{1}{A} \frac{\partial A H_y}{\partial x} = k_0 \sigma (E_z + k_J u B_y) \quad \text{(Ie)}$$

$$A \frac{\partial B_y}{\partial t} = \frac{\partial}{\partial x} (A E_z) \quad \text{(If)}$$

where all the variables refer to their mean values and we drop the bar for the mean value from now on.

The system I equations become system II when we assume that all the covariance coefficients except k_0 are unity. The system II equations are

$$\frac{\partial \rho A}{\partial t} + \frac{\partial \rho u A}{\partial x} = 0 \quad \text{(IIa)}$$

$$p = \rho R T \quad \text{(IIb)}$$

$$\rho \frac{\partial u}{\partial t} + \rho u \frac{\partial u}{\partial x} = -\frac{\partial p}{\partial x} - \sigma B_y (E_z + u B_y) + \frac{C_f}{2b} \rho u^2 \quad \text{(IIc)}$$

$$\rho \frac{\partial U_m}{\partial t} + \rho u \frac{\partial U_m}{\partial x} + \frac{p}{A} \frac{\partial A u}{\partial x} = \sigma (E_z + u B_y)^2 + \frac{1}{b} Q_c - \frac{Q_f}{2b} \rho u^3 \quad \text{(IId)}$$

$$\frac{1}{A} \frac{\partial A H_y}{\partial x} = k_0 \sigma (E_z + u B_y) \quad \text{(IIe)}$$

$$A \frac{\partial B_y}{\partial t} = \frac{\partial A E_z}{\partial x} \quad \text{(IIf)}$$

It is interesting to compare the system of equations (IIa)–(IIf) with those equations of approximately one-dimensional and strictly one-dimensional MFD flow. We consider the two cases that follow.

3.10.1 Constant Area Channel. For a constant area channel, Eqs. (IIa)–(IIf) become exactly the same as those for strictly one-dimensional MFD flow except for the factor k_0. From Eq. (3.38) we see that if $\bar{J}_{z2} = 0$ then $k_0 = 0$ and if $\bar{J}_{z1} = 0$ then $k_0 = 1$. In approximately one-dimensional MFD flow, we take $k_0 = 0$ while in strictly one-dimensional flow, we take $k_0 = 1$. In general, the value of k_0 should lie between zero and unity. As long as k_0 is different from zero, the system of equations (II) behaves the same as those for strictly one-dimensional MFD. Particularly when $\sigma \to \infty$ and $k_0 \neq 0$, we have the frozen flow relation H_y/ρ = constant and an effective sound speed

$$a_{ek} = \sqrt{a^2 + \frac{\mu_e H_y^2}{k_0 \rho}} \tag{3.57}$$

is the characteristic speed of the flow. If $k_0 = 1$, Eq. (3.57) is reduced to the results of strictly one-dimensional flow. If $k_0 = 0$, we do not have the frozen flow relation because AH_y = constant instead of H_y/ρ = constant and the ordinary sound speed $a = \sqrt{\gamma RT}$ is the characteristic speed of the flow. Furthermore, for finite electrical conductivity, the ordinary sound speed $a = \sqrt{\gamma RT}$ is the characteristic speed of the flow. Since we have little knowledge of the covariance coefficients, we suggest that $k_0 = 1/2$ be used in our quasi-one-dimensional analysis. It corresponds to the assumption that $\bar{J}_{z2} = \bar{J}_{z1}$.

3.10.2 Gas-Dynamic Approximation. In many engineering problems, the electrical conductivity is small, that is, $\sigma \to 0$. When $\sigma \to 0$, we have

$$\left.\begin{aligned} AH_y &\simeq \text{const} \\ AE_z &\simeq \text{const} \end{aligned}\right\} \tag{3.58}$$

The two parts of Eq. (3.58) show that both the electric field E_z and the magnetic field H_y are independent of the flow field and they are given functions of x because A is a given function of x. However, both H_y and E_z are not arbitrarily given functions of x but depend on $A(x)$. Since in quasi-one-dimensional analysis, $A(x)$ is a slowly varying function of x, both H_y and E_z are slowly varying functions of x too. Within the limit of slow variations of x, the arbitrary functions of H_y and E_z may not introduce large errors. Hence, the approximate one-dimensional analysis may be considered as a good approximation only in the case of a small magnetic Reynolds number. We shall call it the *gas-dynamic approximation* of quasi-one-dimensional MFD flow in which the influence on electromagnetic fields due to gas-dynamic variables are neglected. We need to consider only Eqs. (IIa)–(IId), for the gas-dynamic variables u, p, ρ, and T only.

3.11 SOLUTIONS FOR STEADY FLOW OF INVISCID FLUIDS

In many MFD engineering problems, we consider the case of a very high Reynolds number. The effects of viscosity and heat conduction may be neglected in the first approximation. Hence, we shall consider the steady flow of an inviscid and non-heat-conducting but electrically conducting fluid in this section. The effects of viscosity and heat conduction will be discussed in Sec. 3.12.

From our preliminary study of k_ρ, covariance factors such as k_ρ and k_u depend not only on the variations of the variables across the channel, but also on some important nondimensional parameters such as Mach number, magnetic Reynolds number, and the like. We have not studied any of the covariance factors.

Since we do not have any information about these covariance factors, we shall assume that all covariance factors are unity. From equations in system II, the basic equations of quasi-one-dimensional steady flow of an inviscid and non-heat-conducting fluid are:

$$A\rho u = \text{const} \tag{3.59}$$

$$p = \rho R T \tag{3.60}$$

$$\rho u \frac{du}{dx} = -\frac{dp}{dx} - \sigma B_y(E_z + uB_y) \tag{3.61}$$

$$\rho u \frac{dU_m}{dx} + \frac{p}{A}\frac{duA}{dx} = \sigma(E_z + uB_y)^2 \tag{3.62}$$

$$\frac{1}{A}\frac{dAH_y}{dx} = k_0\sigma(E_z + uB_y) \tag{3.63}$$

$$AE_z = \text{const} = K \tag{3.64}$$

We shall consider two special cases of these equations.

3.11.1 Case of Infinite Electrical Conductivity. In this case, Eq. (3.63) gives

$$AuB_y = -AE_z = \text{const} \tag{3.65}$$

From Eqs. (3.59) and (3.65), we have

$$\frac{B_y}{\rho} = \text{const} = \frac{B_{y0}}{\rho_0} \tag{3.66}$$

Equation (3.66) is the frozen field relation of MFD for the case $\sigma \to \infty$.

From Eqs. (3.59), (3.62), and (3.65) we have

$$\frac{p}{p_0} = \left(\frac{\rho}{\rho_0}\right)^\gamma \tag{3.67}$$

In the present case, we have isentropic flow. Equations (3.59)-(3.64) give the relation

$$\frac{du}{dx} = \frac{ua^2}{u^2 - a_{ek}^2} \frac{1}{A} \frac{dA}{dx} \tag{3.68}$$

If $k_0 = 1$, Eq. (3.68) becomes the result of strictly one-dimensional MFD analysis. Equation (3.68) shows that the nozzle will choke at $u = a_{ek}$. The approximate analysis is definitely wrong in this case because the induced electromagnetic fields are neglected.

3.11.2 Gas-Dynamic Approximation. When the magnetic Reynolds number is small, we may apply the gas-dynamic approximation and consider only the gas-dynamic variables u, p, ρ, and T by Eqs. (3.59) to (3.62). This system of equations is the same as those of approximately one-dimensional analysis. Since we have two algebraic relations, Eqs. (3.59) and (3.60), the resultant equations consist of two first-order, nonlinear, differential equations. We may use any two of the gas-dynamic variables for these two equations. If we use the velocity u and the local Mach number M as the variables, the corresponding differential equations are as follows:

$$\frac{du}{dx} = \frac{1}{M^2 - 1}\left[\frac{u}{A}\frac{dA}{dx} - \frac{\sigma B_y{}^2}{p}\left(u + \frac{E_z}{B_y}\right)\left(u + \frac{\gamma - 1}{\gamma}\frac{E_z}{B_y}\right)\right]$$

$$= \frac{1}{M^2 - 1} F_1(u, M, x) \tag{3.69}$$

$$\frac{dM}{dx} = \frac{1}{M^2 - 1}\left\{\left(1 + \frac{\gamma - 1}{2}M^2\right)\frac{M}{A}\frac{dA}{dx} - \left(1 + \frac{\gamma - 1}{2}M^2\right)\frac{\sigma B_y^2 M}{up}\left(u + \frac{E_z}{B_y}\right)\right.$$

$$\left.\times\left[u + \frac{(1 + \gamma M^2)(\gamma - 1)E_z}{[2 + (\gamma - 1)M^2]\gamma B_y}\right]\right\} = \frac{1}{M^2 - 1} F_2(u, M, x) \tag{3.70}$$

Where the functions $F_1(u, M, x)$ and $F_2(u, M, x)$ are the abbreviation of the corresponding expressions in the curl brackets. Equations (3.69) and (3.70) illustrate that there are three characteristic velocities which show the effects of electromagnetic fields which are

$$u_1 = -\frac{\gamma - 1}{\gamma}\frac{E_z}{B_y} \qquad u_2 = \frac{1 + \gamma M^2}{2 + (\gamma - 1)M^2}u_1 \qquad u_3 = \frac{\gamma}{\gamma - 1}u_1 \tag{3.71}$$

It is not easy to integrate Eqs. (3.69) and (3.70) from a given initial condition for given values of $A(x)$, E_z, and B_y. In general, numerical methods are required. However, different restrictions may be applied so that these equations are easily integrated. For instance, the constant area channel, constant velocity channel, constant temperature channel, and constant density have been investigated. We may show the effects of electromagnetic fields by considering some of these simple cases.

Constant area channel. If A = constant, Eqs. (3.69) and (3.70) become respectively

$$\frac{du}{dx} = -\frac{\sigma B_y{}^2}{(M^2 - 1)p}(u - u_1)(u - u_3) \tag{3.72}$$

$$\frac{dM}{dx} = -\frac{1 + [(\gamma - 1)/2]M^2}{M^2 - 1}\frac{\sigma B_y{}^2 M}{up}(u - u_2)(u - u_3) \tag{3.73}$$

These equations show that the characteristics of the flow depend on the initial value of the velocity u_0 relative to the characteristic velocities u_1, u_2, and u_3. For instance, if we take $M_0 < 1$ $u_0 < u_2 < u_1 < u_3$, both du/dx and dM/dx are positive initially. This is the case of magnetofluid-dynamic acceleration. If the initial Mach number M_0 is sufficiently close to unity, the acceleration will soon become infinite and the channel is choked. For other initial conditions, the velocity u may increase monotonically toward u_1, while the Mach number M first increases then decreases $(u > u_2)$ and finally approaches an asymptotic value $M_a < 1$. The third possibility is the case of a particular set of initial conditions that M reaches unity and u reaches u_1 simultaneously. We have a smooth acceleration from a subsonic to a supersonic flow in a nozzle of constant cross-sectional area.

A summary of the various possibilities for constant-area flow, provided by Sears and Resler,[6] is presented in Fig. 3.3, where the signs of du/dx and dM/dx in a u, M plane are indicated. There are eight distinct regions in the u, M plane, in which different behaviors of $u(x)$ and $M(x)$ can be predicted from Eqs. (3.72) and (3.73). These are

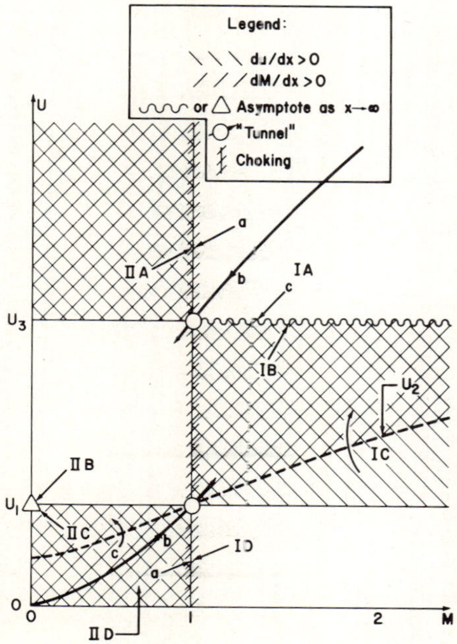

Fig. 3.3 Magnetogas-dynamic acceleration of one-dimensional flow in a constant-area channel (E_z/B_y = const).

identified in Fig. 3.3 as follows:

(I)–$M > 1$	(II)–$M < 1$
(A) $u_3 < u$	(A) $u_3 < u$
(B) $u_2 < u < u_3$	(B) $u_1 < u < u_3$
(C) $u_1 < u < u_2$	(C) $u_2 < u < u_1$
(D) $u < u_1$	(D) $u < u_2$

In general, a "barrier" exists at $M = 1$, where the phenomenon of choking occurs when the denominators of both Eqs. (3.72) and (3.73) disappear. This represents choking since the curve of u versus x turns back upon itself and becomes double valued after reaching the barrier. Thus the sonic speed can never actually be attained in such a channel, and the phenomenon would have to occur, if at all, at the open end of a channel.

There are two exceptional cases, denoted "tunnels" in Fig. 3.3. These are two one-way tunnels through the barrier at sonic speed. For these special combinations of parameters (IAb) and (IIDb) the numerators and denominators of both Eqs. (3.72) and (3.73) disappear simultaneously, and finite and continuous deceleration and acceleration, respectively, occur. The curves that lead to these tunnels have been plotted in Fig. 3.3 as they always separate choking from nonchoking channels and represent the special combinations that permit smooth passage through the sonic speed.

Constant velocity channel. If the velocity u is a constant, then du/dx is zero and then Eqs. (3.69) and (3.70) become

$$\frac{dA}{dx} = \frac{\sigma B_y{}^2 A}{pu}(u - u_1)(u - u_3) \tag{3.74}$$

and

$$\frac{dM}{dx} = \frac{\gamma - 1}{2\gamma}\frac{\sigma B_y{}^2}{pu} M u_3(u - u_3) \tag{3.75}$$

and the factor $1/(M^2 - 1)$ has been eliminated. Thus, a constant velocity channel doesn't choke.

Constant temperature channel. When the temperature T is constant along the channel, Eq. (3.69) becomes

$$\frac{du}{dx} = \frac{u}{\gamma M^2 - 1}\left[\frac{1}{A}\frac{dA}{dx} - \frac{\gamma\sigma B_y{}^2 M^2}{u^2\rho}(u - u_3)\right] \tag{3.76}$$

The critical Mach number is $M = 1/\sqrt{\gamma}$ since the isothermal sound speed is the characteristic speed in this case. If the numerator of Eq. (3.76) is not zero, the acceleration becomes infinite at $M = 1/\sqrt{\gamma}$. However, smooth transition through this critical Mach number may be obtained if the electromagnetic field is so adjusted that the numerator vanishes when $M = 1/\sqrt{\gamma}$.

Constant density channel. We may design the channel so that the cross-sectional area $A(x)$ varies exactly at a rate that is required to keep the density of the fluid constant. Equation (3.69) becomes

$$dx = -\frac{\rho}{\sigma\gamma E_z B_y{}^2}\frac{u^2 du}{(u - u_1)(u - u_3)} \tag{3.77}$$

For the case of constant σ, E_z and B_y, Eq. (3.77) can be integrated from the initial values $x = 0$, $u = u_0$, that is,

$$x = \frac{\rho}{\sigma E_z B_y{}^2}\left\{(u_3 - u_0)(u_0 - u) + u_3{}^2\left[\frac{\gamma - 1}{\gamma}\ln\left(\frac{u - u_1}{u_0 - u_1}\right) - \ln\left(\frac{u - u_3}{u_0 - u_3}\right)\right]\right\} \tag{3.78}$$

and the pressure p and the temperature T always increase with x. The energy is extracted from the flow when $u > u_3$ and added to the flow when $u < u_3$. Since the density is constant in this channel, the flow does not choke.

Nozzle flow. In engineering problems, we study the flow in a channel of a given shape but under various operating conditions such as different axial pressure gradient and different imposed electromagnetic fields. The analyses of flows in a channel of given shapes have not been extensively studied except for the case of constant-area channel. For those solutions of channel flow restricted to special properties of the flow variables such as constant velocity, constant temperature, or constant density, if we change the initial and final conditions of the flow, the shape of the channel will change accordingly. When we compare two solutions under the assumption of a certain constant flow property but different initial and final conditions, we compare the flow field in one channel at one operating condition with that of another channel at another operating condition. Such constant property solutions would be useless to engineers who operate a given MFD channel at various operating conditions. It is much more useful to study the flow in a channel of a given shape under various operating conditions, as we have usually done in ordinary gas dynamics, than to study channel flows with constant flow properties.

Hence, we will now assume that $A(x)$, B_y, and E_z are given and discuss the flow field from Eqs. (3.69) and (3.70).

In the phase plane M versus u, we have from Eqs. (3.69) and (3.70)

$$\frac{dM}{du} = \frac{F_2(u, M, x)}{F_1(u, M, x)} \tag{3.79}$$

We shall examine only a special case of zero electric field which will give all the essential features of the general case.

In this case, Eq. (3.79) becomes

$$\frac{dM}{du} = \frac{M}{u}\left(1 + \frac{\gamma - 1}{2} M^2\right) \tag{3.80}$$

Integration of this equation gives

$$1 + \frac{\gamma - 1}{2} \frac{u^2}{a^2} = \frac{a_0^2}{a^2} \tag{3.81}$$

where a_0 is the stagnation sound speed which is a constant. These relations are the same as those in ordinary gas dynamics. Thus, we need to consider Eq. (3.69) alone which becomes

$$\frac{du}{dx} = \frac{u}{M^2 - 1}\left(\frac{1}{A}\frac{dA}{dx} - \frac{\sigma B_y{}^2 u}{p}\right) \tag{3.82}$$

Equation (3.82) shows that the ponderomotive force in the present case retards the velocity of the flow in the channel if the flow is supersonic ($M > 1$) and accelerates the flow if it is subsonic, ($M < 1$). We consider a channel with one minimum section, de Laval nozzle, as shown in Fig. 3.4 with B_y = constant. For $\sigma = 0$ or $B_y = 0$, the flow field for a given entrance pressure and various exit pressures is identical to that of ordinary gas dynamics whose velocity distributions are shown in the dotted lines. For the pure subsonic cases, the curves (1′) and (1) are the corresponding velocity distributions with and without magnetic field, respectively. The general shapes of these two curves are similar but the velocity on curve (1′) is always larger than on curve (1) the maximum velocity on curve (1) occurs at $x = x_m$ where $dA/dx = 0$ while the maximum velocity on curve (1′) occurs at $x = x_{01}$ where

$$\frac{L}{A}\frac{dA}{dx} = \gamma R_\sigma \frac{M^2}{M_m{}^2} \tag{3.83}$$

In the above equation L is the reference length, $R_\sigma = \sigma\mu_e uL$ is the local magnetic Reynolds number, $M = u/a$ is the local Mach number, and $M_m = u/V_H$ is the local magnetic Mach number when

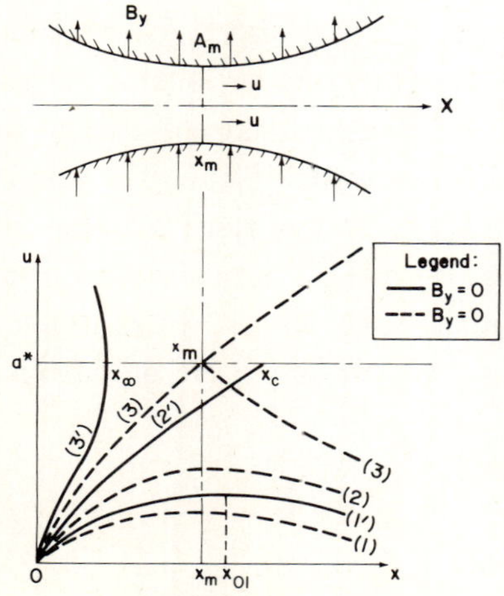

Fig. 3.4 Velocity distribution in a de Laval nozzle with and without magnetic field but $E_z = 0$.

$V_H = H_y\sqrt{\mu_e/\rho}$ is the local Alfvén's wave speed. For given values of A and B_y, the exact location x_{01} depends on the velocity or the axial pressure gradient in the channel.

For the subsonic-supersonic cases, e.g., curves (3) and (3′), the velocity on curve (3′) with B_y is always larger than that on curve (3) without B_y on the subsonic side. Hence the curve (3′) reaches the local Mach number unity ($u = a^*$) before the minimum section (at $x = x_\infty < x_m$). At $x = x_\infty$, the slope du/dx is infinite and our assumption of inviscid fluid breaks down.

There is a critical case (2′) where the conditions $M = 1$ and Eq. (3.83) occur simultaneously. Let us call this location $x = x_c$. Hence $x = x_c$ is a singular point of Eq. (3.82). The nature of this singular point is one of the essential features of MFD nozzle flow. Depending on the value of the critical sound speed and the quantities in the numerator of Eq. (3.82), the singularity may be a saddle-point, nodal-point, vortex-point, or a spiral-point type. By comparison, in ordinary fluid-dynamic nozzle flow, only the saddle-point type occurs. The nature of the flow in the region of x_c will be determined by the type of singularity. However, this feature of MFD nozzle flow has not been studied in detail, although Pai,[2] discusses the integral curves for the various types.

3.12 SOLUTIONS FOR STEADY FLOW OF VISCOUS AND HEAT-CONDUCTING FLUID

We now consider the case when the wall friction and heat transfer and heat addition are not negligible. In engineering problems, these terms are usually small; hence, it is sufficiently accurate to use a constant skin coefficient C_f to represent the wall friction effect and a simple expression for Q_c to represent the effect of heat transfer or heat addition, as we have used in Eqs. (I) or (II). Since we do not have any information for these covariance coefficients, we shall use our basic equations from Eq. (II). Thus our present basic equations are:

$$A\rho u = \text{const} = \dot{m} \tag{3.84}$$

$$p = \rho RT \tag{3.85}$$

$$\rho u \frac{du}{dx} + \frac{dp}{dx} + \sigma B_y (E_z + uB_y) = \frac{C_f}{2b} \rho u^2 \tag{3.86}$$

$$\rho u \frac{dU_m}{dx} + \frac{p}{A} \frac{duA}{dx} = \sigma (E_z + uB_y)^2 + \frac{1}{b} Q_c - \frac{C_f}{2b} \rho u^3 \tag{3.87}$$

$$\frac{1}{A} \frac{dAH_y}{dx} = k_0 \sigma (E_z + uB_y) \tag{3.88}$$

$$AE_z = \text{const} \tag{3.89}$$

A consideration of the two special cases of these equations follows.

3.12.1 Case of Infinite Electrical Conductivity. Since the wall friction and heat-transfer terms do not affect Eqs. (3.84), (3.88), and (3.89), the frozen field relation of Eq. (3.66) holds for viscous and heat-conducting fluids.

The isentropic relation [Eq. (3.67)] does not hold true here. The energy equation (3.87) may be written as follows:

$$\frac{\dot{m}p}{(\gamma - 1)\rho} \frac{d}{dx}\left[\log\left(\frac{p}{\rho^{\gamma}}\right)\right] = b_z\left(Q_c - \frac{C_f}{2} \rho u^3\right) \tag{3.90}$$

or

$$\dot{m}T \frac{dS}{dx} = b_z\left(Q_c - \frac{C_f}{2} \rho u^3\right) \tag{3.91}$$

Only when both C_f and Q_c are zero, may we have isentropic flow, i.e., S = constant. Another special case of possible isentropic flow is $Q_c = (C_f/2)\rho u^3$. In other words, the heat addition just balances the heat loss due to wall friction.

Equation of motion (3.86) may be transformed into

$$\frac{du}{dx} = \frac{u}{u^2 - a_{ek}^2}\left[\frac{a^2}{A}\frac{dA}{dx} + \frac{(\gamma - 1) b_z}{m}\left(-Q_c + \frac{\gamma}{\gamma - 1}\frac{C_f}{2}\rho u^3\right)\right] \tag{3.92}$$

Again we have the effective sound speed (a_{ek}) [cf. Eq. (3.57)] as the critical speed for the frozen field case. If $k_0 = 1$, we have the results of strictly one-dimensional analysis. For effective subsonic flow $u < a_{ek}$, the heat addition Q_c increases the velocity gradient but the wall friction decreases the velocity gradient. For effective supersonic flow, $u > a_{ek}$, the heat addition decreases the velocity gradient but the wall friction increases the velocity gradient. With heat transfer or wall friction or both, the singular point will, in general, not occur at the neck except when $Q_c = [\gamma/(\gamma - 1)](C_f/2)\rho u^3$.

Let us consider the constant area channel flow of fluid with infinite electrical conductivity. This case has been studied by many authors[7-9] by assuming $k_0 = 1$. We shall retain the factor k_0 in our analysis so that k_0 is a constant, but lies between 0 and 1.

Rayleigh flow of MFD. In the ordinary gas dynamics of constant area channel flow with heat addition but without wall friction, the flow is referred to as *Rayleigh flow*. In the present case, we have similar results. From Eqs. (3.86) and (3.88), we have

$$F = \rho u^2 + p + \frac{\mu_e H_y^2}{2k_0} = \text{const} \tag{3.93}$$

If $H_y = 0$, Eq. (3.93) reduces to the results of Rayleigh flow of ordinary gas dynamics. In the present case, if we define a total pressure $p_t = p + \mu_e H_y^2/2k_0$, p_t replaces the gas pressure in ordinary Rayleigh flow. If $k_0 = 1$, we obtain the strictly one-dimensional result. The line of F = constant in the T–S plane is known as the *Rayleigh line*.

From the energy equation (3.87), we have

$$C_p T + \frac{u^2}{2} + \frac{\mu_e H_y^2}{k_0 \rho} = \int^x \frac{Q_c}{b\rho u}\, dx + \text{const} = C_p T_0 \tag{3.94}$$

Without the heat-transfer term, ($Q_c = 0$), the effective stagnation temperature T_0 is a constant, but with heat transfer, the effective stagnation temperature T_0 is not a constant, but a function of the heat transfer Q_c.

The similarity of Eq. (3.94) with that of ordinary Rayleigh flow is evident. The only difference is the effective magnetic energy term

$\mu_e H_y^2/k_0\rho$. If $k_0 = 1$, we have the strictly one-dimensional MFD result.

Fanno flow of MFD. In ordinary gas dynamics, constant area flow with wall friction but without heat transfer, is referred to as *Fanno flow.* We have similar results for MFD. When $Q_c = 0$, but $C_f \neq 0$, from Eq. (3.87) with the help of Eqs. (3.86) and (3.88), we have

$$H = C_p T + \frac{u^2}{2} + \frac{\mu_e H_y^2}{k_0 \rho} = \text{const} \tag{3.95}$$

Equation (3.95) differs from the equation of ordinary gas dynamics in the addition of magnetic energy term $\mu_e H_y^2/k_0\rho$. The line H = constant in the T–S plane is known as the *magnetic Fanno line.*

From Eqs. (3.91), (3.93), and (3.95), we have the well-known thermodynamic relation $dF + \rho T ds = \rho dH$.

3.12.2 Gas-Dynamic Approximation. The complete system of Eqs. (3.84)–(3.89) has not been studied for finite electrical conductivity. When the magnetic Reynolds number is small, we may apply the gas-dynamic approximations and consider only the gas-dynamic variables u, p, ρ, and T. Similar to the results for inviscid fluid of Sec. 3.11, we also have a system of two first-order differential equations as follows:

$$\frac{du}{dx} = \frac{1}{M^2 - 1}\left[\frac{u}{A}\frac{dA}{dx} - \frac{\sigma B_y^2}{p}\left(u + \frac{E_z}{B_y}\right)\left(u + \frac{\gamma - 1}{\gamma}\frac{E_z}{B_y}\right) - \frac{\gamma - 1}{\gamma p}\frac{Q_c}{b} + \frac{\gamma C_f}{2b} u M^2\right] = \frac{F_{1v}}{M^2 - 1} \tag{3.97}$$

$$\begin{aligned}\frac{dM}{dx} = \frac{1}{M^2 - 1}\Bigg\{&\left(1 + \frac{\gamma - 1}{2}M^2\right)\frac{M}{A}\frac{dA}{dx} - \left(1 + \frac{\gamma - 1}{2}M^2\right)\frac{\sigma B_y^2 M}{ap} \\ &\times\left(u + \frac{E_z}{B_y}\right)\left(u + \frac{(1 + \gamma M^2)(\gamma - 1)E_z}{[2 + (\gamma - 1)M^2]\gamma B_y}\right) - \frac{\gamma - 1}{\gamma}\frac{1 + \gamma M^2}{2ap}\frac{Q_c}{b} \\ &+ \left(1 + \frac{\gamma - 1}{2}M^2\right)\frac{\gamma C_f}{2b}M^3\Bigg\} = \frac{F_{2v}}{M^2 - 1}\end{aligned} \tag{3.98}$$

The general properties of Eqs. (3.97) and (3.98) are the same as those of Eqs. (3.69) and (3.70). The effects of wall friction C_f and heat-transfer terms Q_c follow.

First, the location of the singular point will be changed by the wall friction C_f and the heat-transfer terms Q_c, because F_{1v} gives a different location of singular point $x = x_{cv}$ from that given by $F_1 = 0$.

Second, the behavior of the singularity may also be changed by C_f and Q_c. The integral curves depend on both C_f and Q_c also.

In the simple case of zero electrical field, if there is wall friction $C_f \neq 0$ but no heat addition $Q_c = 0$, the relation in Eq. (3.80) holds i.e., there is no singularity in the M-u plane for finite u and M and M is a function of u.

If there is a heat-transfer term $Q_c \neq 0$, Eq. (3.80) does not hold for $E_z = 0$. When $Q_c \neq 0$, the stagnation temperature T_0 will not be a constant, and we do not have a simple relation for M and u such as that given by Eq. (3.80).

In Ref. 4, the case of $Q_c = 0$ with $C_f \neq 0$ has been extensively discussed.

From Eq. (3.97), some general conclusions may be drawn about the effect of wall friction C_f and heat-transfer terms.

1. For subsonic flow $M < 1$, the wall friction reduces the velocity of the flow and heat addition ($Q_c > 0$) increases the velocity of the flow.

2. For supersonic flow, $M > 1$, the wall friction increases the velocity of the flow and heat addition ($Q_c > 0$) reduces the velocity of the flow.

3.13 GENERAL DISCUSSIONS

In this chapter, we derived the exact quasi-one-dimensional equations of magnetofluid dynamics for the case of a fluid with scalar electrical conductivity. Some essential features of these exact quasi-one-dimensional MFD equations follow.

1. The different covariance terms are important in the quasi-one-dimensional flow equations. These covariance terms,

within the treatment of quasi-one-dimensional analysis, may be expressed in terms of covariance coefficients which have not been investigated in the literature, and they depend not only on the configuration of the channel, but also on some nondimensional parameters such as the Mach number and magnetic Reynolds number. The variation of each covariance coefficient with the nondimensional parameters is different from that of other covariance coefficients. For a good understanding of quasi-one-dimensional MFD flow, an extensive investigation on these covariance coefficients is essential.

2. As a first approximation, we may assume that all the covariance coefficients are unity except k_0 in the equation of magnetic field [Eq. (Ie) or (IIe)]. The value of k_0 lies between 0 and 1. If we take $k_0 = 1$, our basic equations (IIa)–(IIf) reduce to those of strictly one-dimensional MFD flow. If we take $k_0 = 0$, our equations (II) reduce to those of approximate one-dimensional MFD flow. We suggest that the value of k_0 should be taken as 1/2 for a first approximation which means that the two parts of electrical current density J_{z1} and J_{z2} are of the same importance.

3. With $k_0 \neq 0$, the properties of our exact basic differential equations (II) are similar to those of strictly one-dimensional flow, except that the effective sound speed is a_{ek} which reduces to the conventional value a_e is we take $k_0 = 1$.

4. The most extensive investigation of quasi-one-dimensional MFD flow in the literature is the approximate one-dimensional analysis. According to our study, the approximate one-dimensional equations may be considered as a gas-dynamic approximation in which the effects of gas-dynamic variables on the electromagnetic fields are negligible. Hence we need to consider the gas-dynamic variables u, T, ρ, and p only by assuming that the electrical and magnetic fields are given. The gas-dynamic approximation is a good approximation only when k_0 is small, i.e., either for a small magnetic Reynolds number or for a very small k_0.

5. In many investigations of quási-one-dimensional MFD flow, the solutions are restricted to such special properties of the flow variables as constant velocity flow, constant temperature flow, and constant density flow. Even though these solutions may be of some

theoretical interest, they are useless to engineers who operate a given MFD channel under various conditions, because when we compare two solutions of certain constant-flow properties but of different initial and final conditions, we compare the flow field of one channel at one operating condition to another channel at another operating condition. Little correlation of practical interest can be obtained.

6. Within the limitation of quasi-one-dimensional analysis, the general properties of the flow field with wall friction and heat-transfer terms are similar to those of inviscid fluid. However, the wall friction and heat-transfer terms may affect the detailed flow field as well as the behavior of the singular point. The similarity of MFD channel flow, wall friction and heat-transfer terms to ordinary gas-dynamic flow is amazing.

REFERENCES

1. E. I. Andriankin, Generation of Electric Current by a Plasma Flowing in a Tube, *Russ. J. Appl. Mech. Tech. Phys.* (*RJAMTP*) no. 2, pp. 20–31, 1962.
2. S. I. Pai, A Critical Survey of Magnetofluid Dynamics, part III, Quasi-One Dimensional Analysis of Magnetogas Dynamics, *Aerospace Res. Lab. Rep.* ARL 63-175, Office of Aerospace Research, USAF, 1965.
3. S. I. Pai, A Critical Survey of Magnetofluid Dynamics, part I, General Discussion (October, 1963); part II, Magnetohydrodynamic Channel Flow (May, 1964), *Aerospace Res. Lab. Rept.* ARL 63-175, Office of Aerospace Research, USAF.
4. E. Dahlberg, Quasi-One Dimensional Analysis of Magnetohydrodynamic Channel Flows, *Trans. Roy. Inst. Tech.* (*Stockholm*), no. 209, pp. 1–16, 1963.
5. J. Naze, The Effects of Magnetic Fields on the Flow of a Conducting Fluid Through a Duct, *Astro. Acta*, vol. 7, no. 4, pp. 261-275, March 1961.
6. E. L. Resler, Jr., and W. R. Sears, The Prospect for Magnetoaerodynamics, *J. Aeron. Sci.*, vol. 25, no. 4, pp. 235–245, April 1958.
7. C. S. Liu and A. B. Cambel, One Dimensional Magnetogasdynamic Flow with Wall Friction, *Phys. Fluids*, vol. 6, no. 6, pp. 792–796, June 1963.
8. C. S. Liu and A. B. Cambel, One Dimensional Magnetogasdynamic Flow with Heat Transfer, *Phys. Fluids*, vol. 7, no. 4, pp. 564–567, April 1964.
9. J. A. Shercliff, Some Generalization in One-Dimensional Magnetogasdynamics, *Rev. Mod. Phys.*, vol. 32, no. 4, pp. 980–986, October 1960.

Chapter 4

MAGNETOFLUID-DYNAMIC VISCOUS FLOWS

4.1 INTRODUCTION

In all devices that are designed to exploit the many MFD phenomena, the conducting fluid flows over surfaces. Fluids of engineering interest with viscosities and thermal conductivities that are not negligible in the vicinity of walls may produce pressure and heat losses that degrade the device's function. Calculating these losses for selected simple geometries and highlighting special attributes of MFD are the primary purposes of this chapter.

The first three sections of this chapter consider MFD flows inside channels and pipes which represent simplifications of configurations that are used in heat exchangers, flow meters, and pipes that connect system components. The MFD boundary-layer assumptions are discussed in Sec. 4.5 and are required for the analysis of the flows in Secs. 4.6 to 4.10. In Sec. 4.6, the MFD flow in the region of a stagnation point is described and is applicable to bodies reentering the atmosphere. Flat plate solutions discussed in succeeding sections are used to determine the losses in large volume to surface area devices such as those considered in Chap. 3. Unsteady flows have been included in Secs. 4.3 and 4.7 to provide an understanding of MFD flows in the starting phase of a device. In Sec. 4.9, the completely new phenomenon of upstream wakes is presented to illustrate another unique attribute of MFD. The final section discusses the impact of several assumptions that were used in earlier sections and includes a stern warning against misapplying MFD solutions that contain restrictive assumptions.

4.2 GENERALIZED HARTMANN-COUETTE FLOW

The simplest channel flow is the fully developed flow between two parallel plates under transverse electromagnetic fields. By fully developed flow, we mean that the velocity profile in the channel is independent of the axial coordinate of the channel. Such an MHD channel flow was first investigated in 1930 by Hartmann, who considered the plane Poiseuille flow with only a transverse magnetic field, but without a transverse electric field. We shall generalize Hartmann's solution[1] by including (1) an arbitrary transverse electric field and (2) Couette flow, i.e., a relative motion between the two parallel infinite plates. Hartmann's solution will be a special case of this generalized solution.

4.2.1 Hartmann-Couette Flow. We consider a two-dimensional steady laminar flow of an incompressible and electrically conducting fluid of constant viscosity and constant electrical conductivity between two parallel walls with the applied magnetic field H_{y0} in the y direction. We assume that the two walls are at $y = \pm L$. One of the walls, at $y = -L$, is at rest and insulated, while the other wall is moving with a speed u_w in the x direction or at rest (Fig. 4.1). There is a constant electric field E_{z0} in the z direction. Except for the pressure p, all the variables are functions of y only.

The parameters and conditions of this problem permit defining the following nondimensional variables.

$$\begin{aligned}
&y^* = y/L,\ u^*(y^*) = u(y)/U_0,\ v^* = w^* = 0,\ E_z^* = E_z/E_{z0} = 1,\\
&R_E = E_{z0}/U_0 B_{y0},\ E_x^* = E_y^* = 0,\ H_y^* = H_y/H_{y0} = 1,\\
&H_x^*(y^*) = H_x(y)/H_{y0},\ H_z^* = 0,\ H_a^2 = \sigma B_0^2 L^2/\mu,\\
&p^*(x^*, y^*) = p/\rho U_0^2 = x^* p_x^* + p_1^*(y^*)
\end{aligned} \tag{4.1}$$

where we take the channel half-width L as the characteristic length and the pressure relation represents the experimental observation that the pressure falls linearly along the channel in fully developed flow. There are several ways to choose the characteristic velocity U_0. We may choose either the velocity of the upper wall u_w or the mean flow velocity across the channel as the reference velocity. The

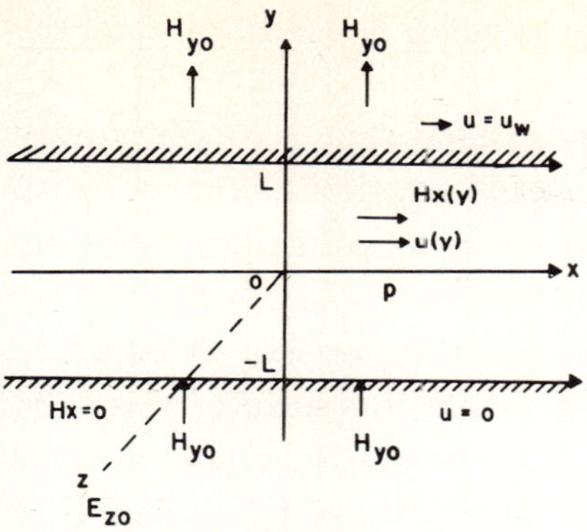

Fig. 4.1 Generalized Hartmann channel flow.

expression for the nondimensional velocity u^* will be different for different choices of the reference velocity which will be specified later. With these nondimensional variables, assumptions and Ohm's law, Eq. (3.25), the MFD flow equations of Chap. 2 become

$$\frac{1}{H_a^2}\frac{d^2u^*}{dy^2} - u^* = \frac{R_e}{H_a^2}p_x^* + R_E = -C \text{ (const)} \tag{4.2}$$

$$\frac{dp_1^*}{dy^*} = R_\sigma R_H H_x^*(u^* + R_E) \tag{4.3}$$

$$\frac{1}{R_\sigma}\frac{d^2H_x^*}{dy^{*2}} + \frac{du^*}{dy^*} = 0 \tag{4.4}$$

Equation (4.3) demonstrates that the pressure variation in the transverse direction is produced by the magnetic interaction and is small for small magnetic Reynolds numbers since it is proportional to R_σ^2 because H_x^* is proportional to R_σ. [See Eq. (4.4).] In Sec. 3.4 (Chap. 3), the average pressure across the channel is used in determining the covariance coefficient k_p. The required pressure variation across the channel can be obtained from the solution of Eq. (4.3) after H_x^* and u^* are determined with Eqs. (4.4) and (4.2) for the specific problem being solved.

For a given type of channel flow, we have to solve the fundamental equations of MHD with given initial and boundary conditions. Customarily, we do not give the spatial distribution of the initial values of all our variables, but we require that the initial condition be consistent with the initial boundary conditions and the fundamental equations.

For the boundary conditions, we assume that across a surface separating a body and the fluid or two fluids, the velocity components, the stresses and the electromagnetic field strengths are all continuous.

These boundary conditions for velocity components are based on the experimental fact that no slip occurs at the surface of a solid body for ordinary fluids. Hence, on the surface of a channel, the velocity of the fluid must be equal to the velocity of the surface. For any particular channel, the velocity of the surface is known. Hence, the boundary conditions of velocity are well defined.

The boundary conditions of the electromagnetic fields are based on the assumption that the surface current and surface charge are zero. It is a good assumption for the case of finite electrical conductivity, particularly under the magnetofluid-dynamic approximation that the excess electrical charge is negligible. However, the application of these boundary conditions for electromagnetic fields is not so well defined as in the case of velocity.

Let us consider a cross section of a straight channel shown in Fig. 4.2. For simplicity, we assume that the channel is at rest. Hence, on the surface, the velocity components of the fluid vanish.

As to the electromagnetic properties, we should assume that the electrical conductivity of the wall σ_w may be different from the

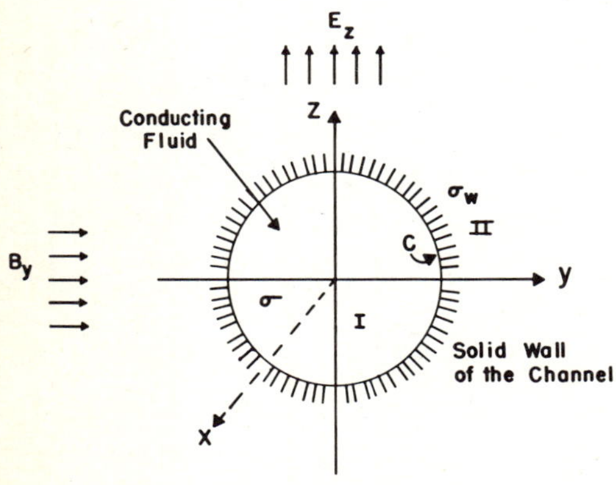

Fig. 4.2 A cross section of an MHD channel.

electrical conductivity of the fluid σ. Furthermore, the electrical conductivity of the wall may not be the same for the whole channel. For instance, some part of the channel may be an insulated wall with $\sigma_w = 0$ while another part of the channel may be a good conductor with $\sigma_w \gg 0$. Another complication may result from some externally applied magnetic field, that is, B_y, and electric field E_z. We do not know exactly the value of the electromagnetic field on the boundary of the channel. In general, we have to solve simultaneously an electrodynamical problem in region II where $u_i = 0$ and the MHD flow problem inside the channel. These two solutions are matched by the electromagnetic boundary conditions on the boundary. No such general solution by matching these two problems has been reported in the literature.

In most of the investigations of MHD channel flow, simplified assumptions have been made for the electrodynamic problems. The following two cases have been extensively discussed.

Insulated channel walls. In this case, the electrical conductivity of the wall σ_w is zero. There is no electrical current in the wall, i.e., in region II of Fig. 4.2. Hence, we may assume that in region II

$$H_y = H_0 = \text{const} \qquad H_z = J_z = E_z = E_x = E_y = 0 \tag{4.5}$$

and

$$H_x = 0 \tag{4.6}$$

For a straight channel and in the fully developed flow region, all the variables (except pressure) are independent of x, the axial distance along the channel. Finally, $H_x = 0$ on the boundary is the boundary condition for the magnetic field.

Thin pipe with an electrically conducting wall.[2] In this case, we consider a straight pipe with walls of a constant thickness h and a constant electrical conductivity σ_w. Outside the pipe, the electrical conductivity of the medium is zero. We have three regions now, i.e., between the region of the fluid (region I) and the region of the insulator (region II), we have a region III of an electrically conducting wall. (See Fig. 4.3.)

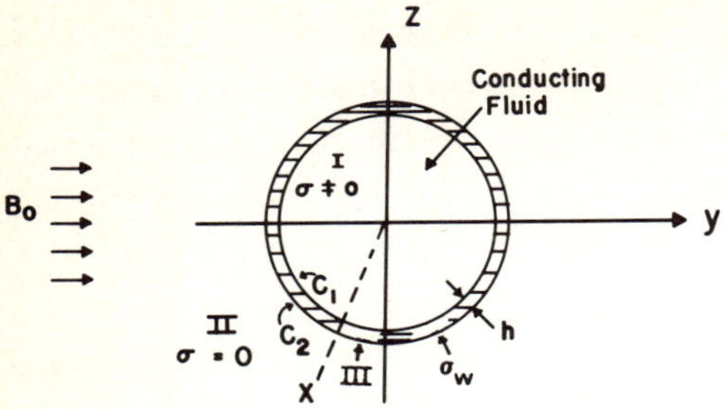

Fig. 4.3 A cross section of a thin pipe in MHD channel flow.

In region II, we have the same situation as in the case of insulated channel walls. Hence, on the outer boundary, we have $H_x = 0$.

On the inner boundary, we have

$$q = 0 \quad H_{x_1} = H_{x_3} \quad \sigma_w \frac{\partial H_{x_1}}{\partial n} = \sigma \frac{\partial H_{x_3}}{\partial n} \tag{4.7}$$

where the last condition represents the continuity of the tangential electric field across the inner boundary between the fluid and the wall, and n is the normal coordinate of the wall. For a thin wall, we may assume that the magnetic field H_{x_3} varies linearly from zero at the outer boundary to the value H_{x_1} at the inner boundary, i.e.,

$$\frac{\partial H_{x_3}}{\partial n} = -\frac{H_{x_1}}{h} \tag{4.8}$$

Hence, the boundary condition of Eq. (4.7) becomes on the inner boundary

$$q = 0 \quad \frac{\partial H_x}{\partial n^*} + \frac{H_x}{\phi} = 0 \quad H_{x_1} = H_{x_3} \tag{4.9}$$

where n^* is n/L, L is a characteristic length of the flow,

$$\phi = \frac{\sigma_w h}{\sigma L}$$

We have dropped subscript 1 on H_x in Eq. (4.9). Equation (4.9) becomes $H_x = 0$ when $\sigma_w = 0$; and when $\sigma_w \to \infty$, $(\phi \to \infty)$, it becomes

$\partial H_x/\partial n^* = 0$. We use the boundary condition [Eq. (4.9)] when we wish to consider the effect of electrical conductivity σ_w of the wall on the MHD channel flow. The boundary conditions for the problem being considered are:

$$
\begin{aligned}
&\text{at } y^* = -1 \qquad u^* = 0 \\
&\text{at } y^* = +1 \qquad u^* = \frac{u_w}{U_0} = u_w^* \\
&\text{at } y^* = -1 \qquad H_x^* = 0
\end{aligned}
\tag{4.10}
$$

The first two boundary conditions are due to the no-slip condition of the fluid and the third one is due to the continuity of the x-wise magnetic field. We assume that the lower plate is an insulated wall at which the x-wise magnetic field is zero. Since there may be a net current in the z direction in the channel, the value of H_x on the upper wall depends on the value of R_E as we shall compute below.

The general solution of Eq. (4.2) for u^* is

$$
u^* = A \cosh H_a y^* + B \sinh H_a y^* + C \tag{4.11}
$$

where A and B are arbitrary constants to be determined by the boundary conditions. The constant C is the effective electrical field parameter or the effective x-wise pressure gradient which is given in our problem.

The first two boundary conditions of Eq. (4.10) are used to determine the arbitrary constants A and B which are:

$$
A = \frac{u_w^* - 2C}{2 \cosh H_a} \qquad B = \frac{u_w^*}{2 \sinh H_a} \tag{4.12}
$$

and then

$$
u^* = \frac{1}{2} u_w^* \left(\frac{\cosh H_a y^*}{\cosh H_a} + \frac{\sinh H_a y^*}{\sinh H_a} \right) + C\left(1 - \frac{\cosh H_a y^*}{\cosh H_a}\right) \tag{4.13}
$$

Equation (4.13) shows that the nondimensional velocity distribution u^* depends on the Hartmann number H_a, effective x-wise pressure gradient $R_e p_x^*$, electric field parameter R_E and the velocity of the upper plate u_w^*. When we plot u^* versus y^*, we have to specify these parameters. Furthermore, the shape of curve u^* versus y^* depends also on the choice of the reference velocity U_0. It is usually convenient to use u_w as the reference velocity for plane Couette flow and the mean flow velocity for plane Poiseuille flow.

If we use the velocity of the upper wall as the reference velocity,[3] we have $u_w^* = 1$ and $u^*(y^*)$ is a function of H_a and C as shown in Fig. 4.4. If the mean flow velocity is used as the reference velocity, we have

$$\int_{-1}^{1} u^* dy^* = 2 \tag{4.14}$$

and when we substitute Eq. (4.13) for the velocity in this relation, we obtain

$$C = \frac{H_a - \frac{1}{2} u_w^* \tanh H_a}{H_a - \tanh H_a} \tag{4.15}$$

Then the velocity distribution u^* depends on H_a and u_w^*. Figure 4.5 shows the case of $u_w^* = 0$.[4] Even though the velocity distribution u^*

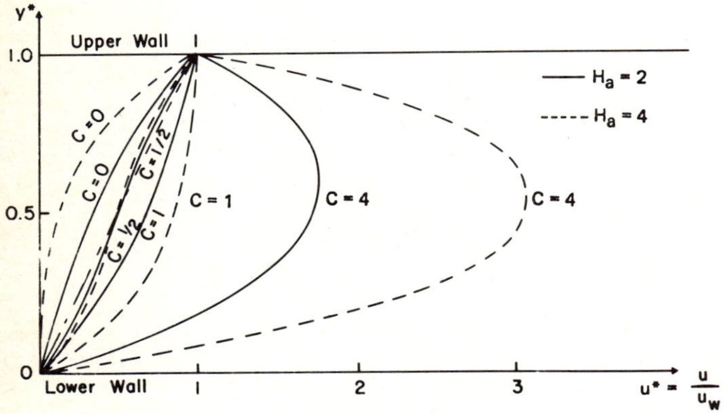

Fig. 4.4 Velocity distributions of generalized Hartmann channel flow (MHD Couette flow).

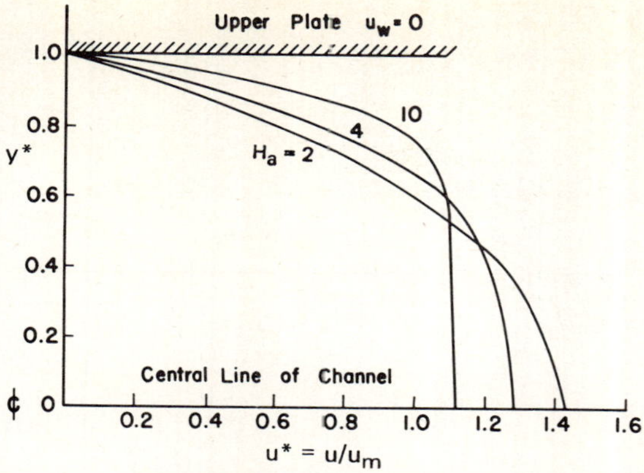

Fig. 4.5 Velocity distributions of Hartmann channel flow ($u_w = 0$).

of Fig. 4.5 is independent of C explicitly, the actual mass flow rate depends on both H_a and C because the characteristic velocity is

$$U_0 = -\frac{\dfrac{E_z}{B_0} + \dfrac{p_x}{\sigma B_0^2}}{\dfrac{H_a - \frac{1}{2}u_w^* \tanh H_a}{H_a - \tanh H_a}} \tag{4.16}$$

For given values of p_x and E_z, the mass flow $\dot{m}$ decreases as the strength of magnetic field H_0 increases, since $\dot{m} = \rho U_0 A$.

The z-wise current density is

$$J_z = \sigma(E_z + uB_y) = \sigma U_0 B_0 (R_E + u^*) \tag{4.17}$$

The x-wise magnetic field may be calculated from the following equation

$$\frac{dH_x}{dy} = -J_z = -\sigma U_0 B_0 (R_E + u^*) \tag{4.18}$$

With the boundary condition [Eq. (4.10)], Eq. (4.18) gives

$$H_x^* = -R_\sigma\left[\int_{-1}^{y^*} u^* dy^* + R_E(y^* + 1)\right] \tag{4.19}$$

Substituting Eq. (4.13) into Eq. (4.19), we have

$$H_x^* = -R_\sigma\left\{\frac{1}{2}u_w^*\left(\frac{\sinh H_a y^* + \sinh H_a}{H_a \cosh H_a} + \frac{\cosh H_a y^* - \cosh H_a}{H_a \sinh H_a}\right)\right.$$
$$\left. + C\left[(y^* + 1) - \frac{\sinh H_a y^* + \sinh H_a}{H_a \cosh H_a}\right] + R_E(y^* + 1)\right\} \tag{4.20}$$

Figure 4.6 shows the x-wise magnetic field distributions[3] in Couette flow of MHD, that is, $p_x = 0, C = -R_E$; while Fig. 4.7 shows the corresponding case for Hartmann channel flow[4] $(u_w^* = 0)$.

The electric field E_z may be considered as a free parameter and may be chosen by the designer. Some values of R_E have special significance.

1. Without an electric field, we have $R_E = 0$. In this case, the total current through the channel is not zero, and the x-wise magnetic field on the upper wall is also not zero.

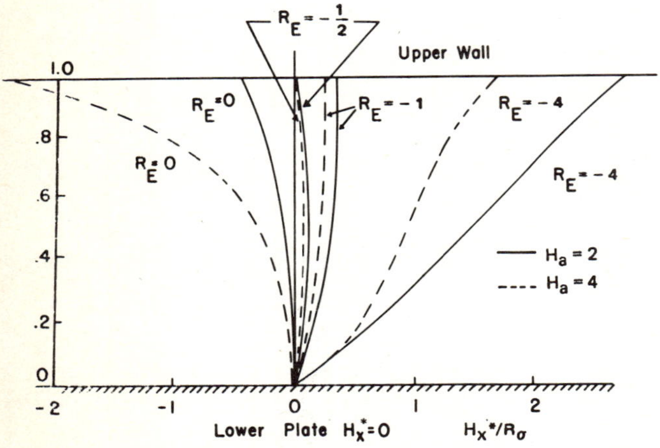

Fig. 4.6 x-wise Magnetic field distribution in Couette flow of MHD with $P_x = 0, C = -R_E$.

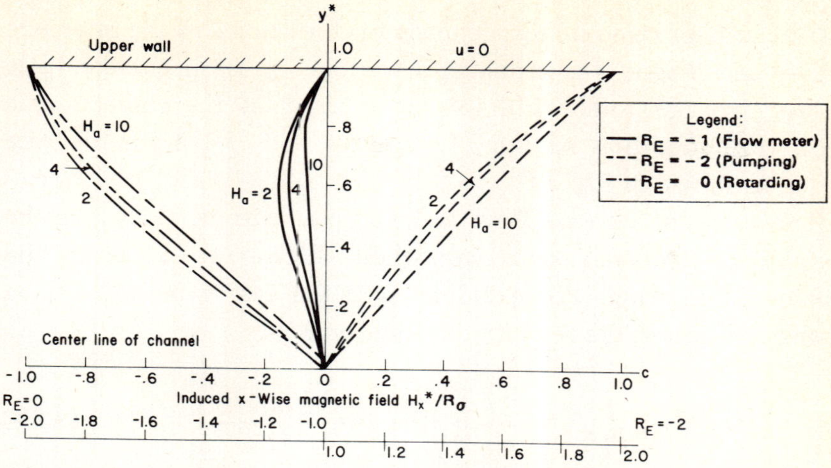

Fig. 4.7 x-wise Magnetic field distributions in Hartmann channel flow ($u_w = 0$).

2. If the total current flowing through the channel is zero, we have

$$I_t = \int_{-1}^{1} J_z \, dy^* = 0 \tag{4.21}$$

or

$$\left(\frac{u_w^*}{2} - C\right) \frac{\sinh H_a}{H_a \cosh H_a} + (C + R_E) = 0 \tag{4.22}$$

and

$$R_E = -C - \frac{1}{H_a}\left(\frac{u_w^*}{2} - C\right) \tanh H_a \tag{4.23}$$

If we consider the case of $u_w^* = 0$ and take the mean flow velocity as the reference velocity, we have from Eqs. (4.15) and (4.22),

$$R_E = -1 \tag{4.24}$$

This critical value of R_E depends on the choice of the reference velocity and length. For instance, if we take u_w as the characteristic velocity and the total width of the channel as the characteristic length, the condition of Eq. (4.23) gives $R_E = -1/2$ for the case $p_x = 0$, as shown in Fig. 4.6.

The cases of current flow and no total current flow must be carefully remembered since important differences occur in the velocity distribution. For example, take the case when $u_w^* = 0$; in dimensional terms, the velocity distribution is

$$u = -\left(\frac{E_{z0}}{B_{y0}} + \frac{p_x}{\sigma B_{y0}^2}\right)\left[1 - \frac{\cosh(H_a y/L)}{\cosh H_a}\right] \tag{4.25}$$

for any E_{z0}. The wall velocity gradients are then

$$\left.\frac{\partial u}{\partial y}\right|_{y=\pm L} = \pm\frac{H_a}{L}\left(\frac{E_{z0}}{B_{y0}} + \frac{p_x}{\sigma B_{y0}^2}\right)\tanh H_a \tag{4.26}$$

and they are functions of the magnetic field strength in dimensional terms for an arbitrary E_{z0}. Now if we require the total current to be zero, then

$$\int_{-L}^{L} (\sigma E_{z0} + \sigma B_{y0}\, u)\, dy = 0 \tag{4.27}$$

which requires that

$$E_{z0} = \frac{p_x}{\sigma B_{y0}}\left(\frac{H_a - \tanh H_a}{\tanh H_a}\right) \tag{4.28}$$

The velocity distribution for this case is

$$u = -\frac{p_x H_a}{\sigma B_{y0}^2}\left[\frac{\cosh H_a - \cosh(H_a y/L)}{\sinh H_a}\right] \tag{4.29}$$

and the wall velocity gradients are

$$\left.\frac{du}{dy}\right|_{y=\pm L} = \pm\frac{p_x}{\mu} \tag{4.30}$$

which are not functions of the magnetic field strength in dimensional terms. However, if the mean velocity computed for each case is used to nondimensionalize the velocity distributions, then the velocity distributions and velocity gradients will have the same form. The mean velocity for an arbitrary E_{z0} was previously computed and the following is the calculation for the zero total current case:

$$U_0 = \frac{1}{2L}\int_{-L}^{L} u\,dy = -\frac{1}{2L}\frac{p_x H_a}{\sigma B_{y0}^2}\int_{-L}^{L}\left[\coth H_a - \frac{\cosh(H_a y/L)}{\sinh H_a}\right]dy \tag{4.31}$$

or

$$U_0 = -\frac{p_x}{\sigma B_{y0}^2}\left(\frac{H_a - \tanh H_a}{\tanh H_a}\right) \tag{4.32}$$

Both velocity distributions and velocity gradients are now

$$u^* = H_a\left[\frac{\cosh H_a - \cosh(H_a y/L)}{H_a \cosh H_a - \sinh H_a}\right] \tag{4.33}$$

and

$$\left.\frac{du^*}{dy^*}\right|_{y^*=\pm 1} = \mp\frac{H_a^2 \tanh H_a}{H_a - \tanh H_a} \tag{4.34}$$

Since both sets of results can be presented in the same form, they appear to be similar. However, remember that the mean velocity is different for each case. Compare Eqs. (4.16) and (4.32).

The influence of the magnetic field on the flow can best be seen in the case of Couette flow in Fig. (4.4). When $C = 0$, the motion is produced only by the moving upper wall and the fluid's viscosity. Increasing the magnetic field retards the flow and tends to separate the flow at the fixed surface. However, when $C > 1/2$, the magnetic field can produce pumping if the increase in C is due to an increase in the current flow by an increase in E_{z0}. A stronger magnetic field provides more pumping and increases the velocity gradient at the fixed surface. The increased pumping can also be produced by a larger pressure gradient. For a fixed value of $C > 1/2$, a higher Hartmann number then corresponds to a larger pressure gradient and greater pumping. The special value of $C = 1/2$ corresponds to the case of maximum power generated by the channel when the internal resistance and external load are equal and the pressure drop along the channel is negligible. The Hartmann flow results in Fig. (4.5) have been non-dimensionalized by the mean flow and an increased magnetic field strength flattens the velocity distribution and increases the surface velocity gradient. However, the latter result is somewhat deceiving because the nondimensional presentation contains all values of R_E. When $R_E = 0$ (short circuit), the magnetic field decreases the wall velocity gradient. [See Eq. (4.26).] The wall velocity gradient [Eq. (4.30)] is independent of the magnetic field when the total current is zero. However, other values of R_E can produce pumping which will increase the wall velocity gradient. The deception is due to nondimensionalizing the velocity distribution by the mean flow which is also changed by the magnetic field.

The induced magnetic field component H_x^* is shown in Figs. (4.6) and (4.7) for the Couette and Hartmann flows. An increase in the Hartmann number reduces the velocity at each point and consequently, the current produced by the $\mathbf{u}^* \times \mathbf{B}^*$ voltage. The value of H_x^* at a given point is then less negative and more positive as the Hartmann number increases. The variation of H_x^* across the complete channel in both Hartmann and Couette flow is best understood by simply sketching the flow and current fields and remembering that at the lower wall, H_x^* is required to be zero. See the problems for Chap. 4.

4.2.2 Hartmann Flow Heat Transfer. The effect of the magnetic field on the fluid's temperature and the heat transferred between the fluid and walls can be easily demonstrated by considering two simple cases of fully developed constant property flow. The assumption of constant properties uncouples the energy and momentum equations which permits use of the Hartmann solution of the momentum equation that did not consider a heat-transfer problem. The previous fully developed assumption is also expanded to include thermally fully developed which means that the thermal boundary conditions apply everywhere on the infinite channel walls. With these assumptions and by neglecting the thermal conduction in the flow direction, the energy equation of Chap. 2 becomes

$$\rho u c_p \frac{\partial T}{\partial x} = \frac{c_p \mu}{\mathrm{P_r}} \frac{\partial^2 T}{\partial y^2} + \frac{J^2}{\sigma} + \mu \left(\frac{\partial u}{\partial y}\right)^2 \tag{4.35}$$

where Pr is the Prandtl number.

The first case to be considered, MFD flow between constant-temperature walls,[4] would produce the elementary linear temperature profile in ordinary fluid dynamics with negligible viscous heating. However, in MFD, the current produces Joule heating in the fluid, which depends on the magnetic field strength. The boundary conditions for this example are now

$$T(\pm L) = T_w = \text{const} \tag{4.36}$$

Since T_w is a constant everywhere, then

$$\frac{\partial T}{\partial x} = 0 \tag{4.37}$$

everywhere in the fluid. Thus, Eq. (4.35) becomes

$$\frac{d^2 T}{dy^2} = -\frac{\mathrm{P_r}}{c_p \mu} \frac{J^2}{\sigma} - \frac{\mathrm{P_r}}{c_p} \left(\frac{du}{dy}\right)^2 \tag{4.38}$$

Substituting the nondimensional quantities

$$\eta = \frac{y}{L} \quad U = \frac{u}{u_m} \quad \mathcal{J} = \frac{J}{\sigma B_0 u_m} \quad R_E = \frac{E}{u_m B_0} \quad \theta = \frac{T - T_w}{u_m / c_p} \tag{4.39}$$

the equation is

$$\frac{d^2 \theta}{d\eta^2} = -\mathrm{P_r} \left[\left(\frac{dU}{d\eta}\right)^2 + H_a^2 \mathcal{J}^2\right] \tag{4.40}$$

and the boundary conditions are

$$\theta(\pm 1) = 0 \quad \text{and} \quad \frac{d\theta}{d\eta}(0) = 0 \tag{4.41}$$

The last condition is the requirement that the temperature profile is symmetric about the channel center line. Substituting Eq. (4.17) for the current and Eq. (4.33) for the velocity, the energy equation, becomes

$$\frac{d^2\theta}{d\eta^2} = -c_1 \mathrm{P_r}\left[c_2^2 + H_a^2 \cosh 2H_a\eta - 2H_a c_2 \cosh H_a\eta\right] \tag{4.42}$$

where

$$c_1 = \left(\frac{H_a}{H_a \cosh H_a - \sinh H_a}\right)^2$$

and (4.43)

$$c_2 = H_a(R_E + 1)\cosh H_a - R_E \sinh H_a$$

The solution satisfying the boundary conditions is

$$\theta = c_1 \mathrm{P_r}\left[\frac{c_2^2}{2}(1 - \eta^2) + \frac{1}{4}(\cosh 2H_a - \cosh 2H_a\eta) - \frac{2c_2}{H_a}(\cosh H_a - \cosh H_a\eta)\right] \tag{4.44}$$

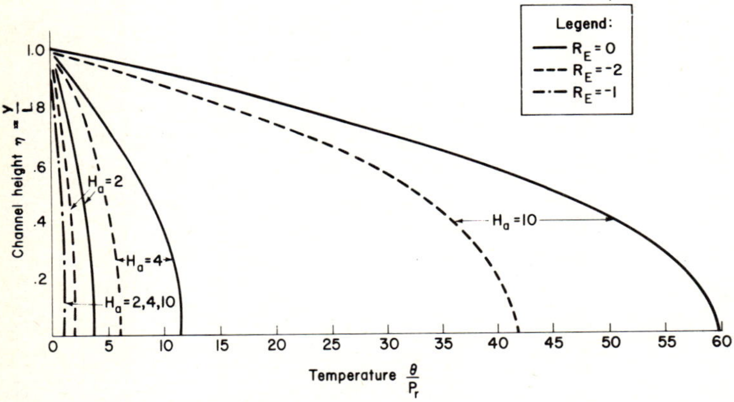

Fig. 4.8 Hartmann flow temperature distributions.

which is shown in Fig. 4.8 for several values of R_E and H_a. The viscous dissipation produces more heating than the Joule heating when $R_E = -1$ which is the lowest current or open circuit case. The magnetic field has only a very small influence as shown by the single curve for all the selected values of the Hartmann number. The largest current or short circuit case is represented by $R_E = 0$ and shows the considerable heating in the central or higher velocity portion of the channel which is strongly influenced by the magnetic field. Also, as the Joule heating is increased, the heat transferred to the walls, i.e., the wall temperature gradient, is significantly increased.

The other case that we will consider demonstrates the influence of the magnetic field on the channel wall temperatures when a uniform heating or cooling is applied to the plates.[5] The viscous and Joule heating are neglected in the example since they were featured in the first case. They may be included by obtaining a solution to Eq. (4.35) for a zero wall heat flux and then simply adding to the following solution. The energy equation and boundary conditions are

$$\rho u c_p \frac{\partial T}{\partial x} = \frac{c_p \mu}{\mathrm{P_r}} \frac{\partial^2 T}{\partial y^2} \tag{4.45}$$

and

$$\left.\frac{\partial T}{\partial y}\right|_{y=\pm L} = \pm \frac{q \mathrm{P_r}}{\mu c_p} = \pm \frac{q}{k}, \quad \left.\frac{\partial T}{\partial y}\right|_{y=0} = 0 \tag{4.46}$$

The last condition is again the symmetry requirement. Consider now a simple heat balance for the fluid with heat entering or leaving the wall between two stations along the channel. Thus, the change in temperature of the mean flow passing between the stations is caused by the heat flowing across the surface, i.e.,

$$\rho u_m L c_p \frac{\partial T}{\partial x} = q \text{ const} \tag{4.47}$$

Equation (4.47) can also be obtained by integrating Eq. (4.45) across half of the channel and applying the boundary condition Eq. (4.46). The fluid temperature will vary linearly with axial distance since integrating Eq. (4.47) gives

$$T - T_0 = \frac{q}{Lu_m \rho c_p} x + f(y) \tag{4.48}$$

where T_0 is some reference temperature at the initial station and $f(y)$ is a constant of integration which is a function of the transverse coordinate and must satisfy Eq. (4.45). Thus substituting Eq. (4.48) into Eq. (4.45) gives

$$\frac{d^2 f}{dy^2} = \frac{\mathrm{P_r}}{c_p \mu L} q\left(\frac{u}{u_m}\right) = \frac{\mathrm{P_r} q}{c_p \mu L} H_a \left(\frac{\cosh H_a - \cosh H_a \eta}{H_a \cosh H_a - \sinh H_a}\right) \tag{4.49}$$

and

$$f = \frac{\mathrm{P_r}\, qLH_a}{2c_p \mu}\left[\frac{\eta^2 \cosh H_a - (2/H_a^2)\cosh H_a \eta}{H_a \cosh H_a - \sinh H_a}\right] + \text{const} \tag{4.50}$$

which satisfies the boundary conditions Eq. (4.46). The integration constant must be determined from an overall energy balance from the origin to the x location of interest, i.e.,

$$\int_0^1 \rho u c_p L (T - T_0)\, d\eta = \int_0^x q\, dx = qx \tag{4.51}$$

Substituting Eq. (4.48) for the temperature difference, this requirement becomes

$$\int_0^1 \frac{u}{u_m} f(\eta)\, d\eta = 0 \tag{4.52}$$

The complete solution is

$$\frac{T - T_0}{qL\,\mathrm{P_r}/c_p \mu} = \frac{4\xi}{\mathrm{R_e P_r}} + \frac{1}{2}\left[\frac{\eta^2 \cosh H_a - (2/H_a^2)\cosh H_a \eta}{H_a \cosh H_a - \sinh H_a}\right]$$

$$- \frac{1}{2H_a^2}\left[\frac{(H_a^2/3 + 2)\cosh^2 H_a - (H_a/2 + 3/2H_a)\sinh H_a + 1}{(H_a \cosh H_a - \sinh H_a)^2}\right] \tag{4.53}$$

after the manipulations required by Eq. (4.52) and nondimensionalizing with

$$\xi = \frac{x}{L}, \quad R_e = \frac{4u_m L\rho}{\mu} \tag{4.54}$$

A convenient reference temperature of interest to designers is the bulk temperatures defined by

$$T_B = \frac{\int TudA}{\int udA} \tag{4.55}$$

which, in our problem, is

$$\frac{T_B - T_0}{qL\, P_r/c_p\mu} = \frac{4\xi}{R_e P_r} \tag{4.56}$$

Since the purpose of the problem was to determine the influence of the magnetic field on the wall temperature when the walls were heated or cooled, then evaluating Eq. (4.53) at $\eta = 1$ and using Eq. (4.56),

$$\frac{T_w - T_B}{qL\, P_r/c_p\mu} = \frac{1}{2}\,\frac{(1 - 2/H_a^2)\cosh H_a}{H_a \cosh H_a - \sinh H_a}$$

$$- \frac{1}{2H_a^2}\left[\frac{(H_a^2/3 + 2)\cosh^2 H_a - (H_a/2 + 3/2H_a)\sinh 2H_a + 1}{(H_a \cosh H_a - \sinh H_a)^2}\right] \tag{4.57}$$

This relation is shown in Fig. 4.9 for a range of Hartmann numbers and demonstrates that the difference between the wall and the bulk temperatures, for a selected wall heat flux, is reduced by an increased magnetic field strength. This result is evident since a strong magnetic field reduces the flow through the channel permitting a larger exchange of heat with the walls by thermal conduction; this decreases the temperature difference between the wall and the fluid bulk.

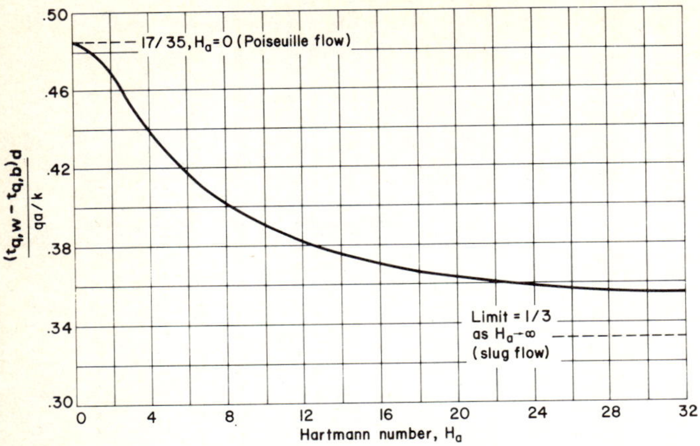

Fig. 4.9 Fully developed wall-to-bulk temperature difference for uniform wall heat flux and no internal heat sources.

4.3 UNSTEADY HARTMANN FLOW

Magnetofluid-dynamic flows starting from rest require time to obtain their steady velocity and temperature distributions. The influence of the magnetic field on this starting phase can be demonstrated for Hartmann flow[6] by assuming that the flow, initially at rest, is suddenly started by a force K in the direction of the channel axis. By assuming a small magnetic Reynolds number, the stretching of the magnetic field can be neglected. Equation (4.2), with a time-dependent term included, is

$$\frac{du^*}{dt^*} = \frac{\partial^2 u^*}{\partial \eta^2} - H_a^2 u^* + K \tag{4.58}$$

where

$$t^* = \frac{\nu}{L^2} t \quad \text{and} \quad K = R_e p_x^* + H_a^2 R_E$$

The boundary conditions for this problem are then

$$\begin{aligned} u^*(\eta, t^*) &= 0 \quad \text{at} \quad \eta = \pm 1 \quad \text{and} \quad t^* \geq 0 \\ u^*(\eta, t^*) &= 0 \quad \text{at} \quad -1 \leq \eta \leq 1 \quad \text{and} \quad t^* = 0 \end{aligned} \tag{4.59}$$

The Laplace transforms of this equation and boundary conditions are obtained by multiplying each term by e^{-st^*} and integrating the t^* dependence in each term between 0 and ∞ while applying the initial condition. The transformed equation is then an ordinary differential in the transformed variable $\overline{u}^*(s, \eta^*)$, that is,

$$\frac{d^2\overline{u}^*}{d\eta^2} - (s + H_a^2)\,\overline{u}^* = -\frac{K}{s} \tag{4.60}$$

and

$$\overline{u}^*(\pm 1) = 0$$

and have the solution

$$\overline{u}^* = \frac{-K}{s(s + H_a^2)}\left[\frac{\cosh(\sqrt{s + H_a^2}\,\eta)}{\cosh(\sqrt{s + H_a^2})} - 1\right] \tag{4.61}$$

A table of Laplace transforms provides the required inverse transforms which gives

$$u^*(\eta, t^*) = -(R_e\, p_x^* + H_a^2 R_E)\int_0^{t^*} e^{-H_a^2\tau}\, d\tau \int_1^{\eta} \theta\left(\frac{\lambda}{2}, \tau\right) d\lambda \tag{4.62}$$

where

$$\theta\left(\frac{\lambda}{2}, \tau\right) = 2\sum_{u=0}^{\infty}(-1)^n \exp\left[-\pi^2\tau\left(n + \frac{1}{2}\right)^2\right] \sin\frac{\pi(2n + 1)\lambda}{2}$$

When $t^* = 0$, then $u^* = 0$, as was required by the initial condition and as t^* becomes infinite, then

$$\lim_{t^* \to \infty} u^*(\eta, t^*) = -\left(R_E + \frac{R_e p_x{}^2}{H_a^2}\right)\left(1 - \frac{\cosh H_a \eta}{\cosh H_a}\right) \tag{4.63}$$

which is the steady solution [Eq. (4.13)] for stationary walls. The influence of the magnetic field is demonstrated by

$$\lim_{H_a \to \infty} (H_a^2 e^{-H_a^2\tau}) \to 0 \tag{4.64}$$

Thus, as the magnetic field strength becomes infinite, the velocity approaches zero and the magnetic field retards the development of the velocity field. The result is similar to the influence of viscosity since

$$\lim_{\nu \to \infty} \exp\left[-\pi^2\tau\left(n + \frac{1}{2}\right)^2\right] \to 0 \tag{4.65}$$

and then the velocity becomes negligible.

4.4 MAGNETOFLUID-DYNAMIC PIPE FLOW

The flow of a conducting fluid through round pipes may be subjected to a magnetic field if they connect components of MFD systems or when they are used for flow meters. MFD flow meters are constructed by placing two electrodes in opposite sides of the pipe wall on a radial line transverse to an imposed magnetic field. The induced potential difference is then a measure of the flow rate. In a fully developed MFD pipe flow, the ratio of induced potential difference to the flow rate is not influenced by the shape of the velocity profile if it is radially symmetric. However, if the transverse magnetic field is parallel rather than in a radial direction, then the velocity profile will be different at different angular positions. This asymmetry can produce considerable variations in the ratio of the induced potential difference to flow rate and depends on the magnetic field strength. In this section, a solution of the MFD equations[7] will be presented for the fully developed flow of a conducting fluid in a straight, nonconducting round pipe subjected to a transverse parallel magnetic field. The flow is in the positive z direction and the magnetic field is in the positive x direction.

Simple manipulation of the incompressible MFD equations of Chap. 2, assuming no displacement currents, will reduce them to

$$\rho(\mathbf{v}\cdot\nabla)\mathbf{v} - \mu_e(\mathbf{H}\cdot\nabla)\mathbf{H} = -\nabla\left(p + \frac{\mu_e\mathbf{H}^2}{2}\right) + \mu\nabla^2\mathbf{v} \tag{4.66}$$

$$\mu_e\sigma\nabla \times (\mathbf{v} \times \mathbf{H}) + \nabla^2\mathbf{H} = 0 \tag{4.67}$$

$$\nabla\cdot\mathbf{v} = 0 \tag{4.68}$$

and

$$\nabla \cdot \mathbf{H} = 0 \tag{4.69}$$

The boundary conditions are the usual requirement that the velocity and induced magnetic field are zero at the wall. With these conditions, the fully developed assumption, and Eqs. (4.68) and (4.69), only the pressure will be a function of the z direction. Thus, in cylindrical coordinates

$$\begin{aligned} &v_r = v_\theta = 0 \quad v_z = v_z(r, \theta) \\ &H_r = H_0 \cos\theta \quad H_\theta = -H_0 \sin\theta \quad H_z = H_z(r, \theta) \end{aligned} \tag{4.70}$$

where the angular distance is measured from the positive x axis. Substituting these in the equations in cylindrical coordinates, gives

$$\begin{aligned} p(r, \theta, z) &= -\frac{\mu_e}{2} H_z^2 + K_1 z + K_2 \\ \frac{\partial p}{\partial z} &= \text{const} = K_1 \end{aligned} \tag{4.71}$$

and

$$\mu\left(\frac{\partial^2 v_z}{\partial r^2} + \frac{1}{r}\frac{\partial v_z}{\partial r} + \frac{1}{r^2}\frac{\partial^2 v_z}{\partial \theta^2}\right) + \frac{\mu_e}{r} H_\theta \frac{\partial H_z}{\partial \theta} + \mu_e H_r \frac{\partial H_z}{\partial r} = K_1 \tag{4.72}$$

$$\frac{\partial}{\partial r}\left(r\frac{\partial H_z}{\partial r}\right) + \frac{1}{r}\frac{\partial^2 H_z}{\partial \theta^2} + \sigma\mu_e\left(H_r \frac{\partial r v_z}{\partial r} + \frac{\partial v_e H_\theta}{\partial \theta}\right) = 0 \tag{4.73}$$

$$v_z(a, \theta) = 0 \quad H_z(a, \theta) = 0 \tag{4.74}$$

Equations in (4.71), for the pressure variation, were obtained by a simple integration of Eq. (4.66) with z. The components of the current and electric field can be obtained from Ampère's law and Ohm's law when the above system is solved:

$$j_r = \frac{1}{r}\frac{\partial H_z}{\partial \theta} \quad j_\theta = -\frac{\partial H_z}{\partial r} \quad j_z = 0 \tag{4.75}$$

$$E_r = \frac{j_r}{\sigma} + \mu_e v_z H_\theta\ ,\ E_\theta = \frac{j_\theta}{\sigma} - \mu_e v_z H_r\ ,\ E_z = 0 \tag{4.76}$$

Equations (4.72) and (4.73) can be nondimensionalized by introducing

$$v = \frac{v_z}{v_0}\ ,\ H = \frac{H_z}{H_0}\ ,\ \rho = \frac{r}{a} \tag{4.77}$$

where v_0 is some characteristic velocity, H_0 is the imposed magnetic field and a is the pipe radius. The equations are now

$$\nabla^2 v - \frac{H_a^2}{R_\sigma}\left[\left(\frac{\sin\theta}{\rho}\right)\frac{\partial H}{\partial \theta} - \cos\theta\,\frac{\partial H}{\partial \rho}\right] = K \tag{4.78}$$

and

$$\nabla^2 H - R_\sigma\left[\left(\frac{\sin\theta}{\rho}\right)\frac{\partial v}{\partial \theta} - \cos\theta\,\frac{\partial v}{\partial \rho}\right] = 0 \tag{4.79}$$

where $K = K_1 a^2/\mu v_0$. These equations can be uncoupled with the transformation

$$\begin{aligned} f(\rho,\theta) &= e^{\alpha\rho\cos\theta}\left[v + \left(\frac{2\alpha}{R_\sigma}\right)H - \frac{K\rho\cos\theta}{2\alpha}\right] \\ g(\rho,\theta) &= e^{-\alpha\rho\cos\theta}\left[v - \left(\frac{2\alpha}{R_\sigma}\right)H + \frac{K\rho\cos\theta}{2\alpha}\right] \end{aligned} \tag{4.80}$$

which produces

$$\nabla^2 f - \alpha^2 f = 0 \tag{4.81}$$

$$\nabla^2 g - \alpha^2 g = 0 \tag{4.82}$$

and the boundary conditions

$$f(1,\theta) = -\frac{K}{2\alpha}\cos\theta\, e^{\alpha\cos\theta}$$
$$g(1,\theta) = \frac{K}{2\alpha}\cos\theta\, e^{-\alpha\cos\theta} \tag{4.83}$$

where

$$\nabla^2 \equiv \frac{\partial^2}{\partial\rho^2} + \frac{1}{\rho}\frac{\partial}{\partial\rho} + \frac{1}{\rho^2}\frac{\partial^2}{\partial\theta^2} \quad \text{and} \quad \alpha = \frac{H_a}{2} \tag{4.84}$$

The solution to these equations and boundary conditions are

$$v = -\frac{K}{2\alpha}\left[\cosh(\alpha\rho\cos\theta)\sum_{n=0}^{\infty}\epsilon_n\frac{I'_{2n}(\alpha)}{I_{2n}(\alpha)}I_{2n}(\alpha\rho)\cos(2n\theta)\right.$$
$$\left. - \sinh(\alpha\rho\cos\theta)\sum_{n=0}^{\infty}2\frac{I'_{2n+1}(\alpha)}{I_{2n+1}(\alpha)}I_{2n+1}(\alpha\rho)\cos(2n+1)\theta\right] \tag{4.85}$$

and

$$H = -\frac{R_\sigma K}{4\alpha^2}\left[\cosh(\alpha\rho\cos\theta)\sum_{n=0}^{\infty}\epsilon_n\frac{I'_{2n+1}(\alpha)}{I_{2n+1}(\alpha)}I_{2n+1}(\alpha\rho)\cos(2n+1)\theta\right.$$
$$\left. - \sinh(\alpha\rho\cos\theta)\sum_{n=0}^{\infty}2\frac{I'_{2n}(\alpha)}{I_{2n}(\alpha)}I_{2n}(\alpha\rho)\cos(2n\theta) - \rho\cos\theta\right] \tag{4.86}$$

where the I_n are modified Bessel functions of order n and

$$\epsilon_n = 1 \quad n = 0$$
$$\epsilon_n = 2 \quad n > 0 \tag{4.87}$$

The velocity and magnetic field profiles are shown in Figs. 4.10 and 4.11 for θ equal to zero and Fig. 4.12 for θ equal to $\pi/2$. These figures show that the anticipated asymmetry for different angular locations occurs in the region near the wall since

$$v(0,\theta) = -\frac{KI_1(\alpha)}{2\alpha I_0(\alpha)} \tag{4.88}$$

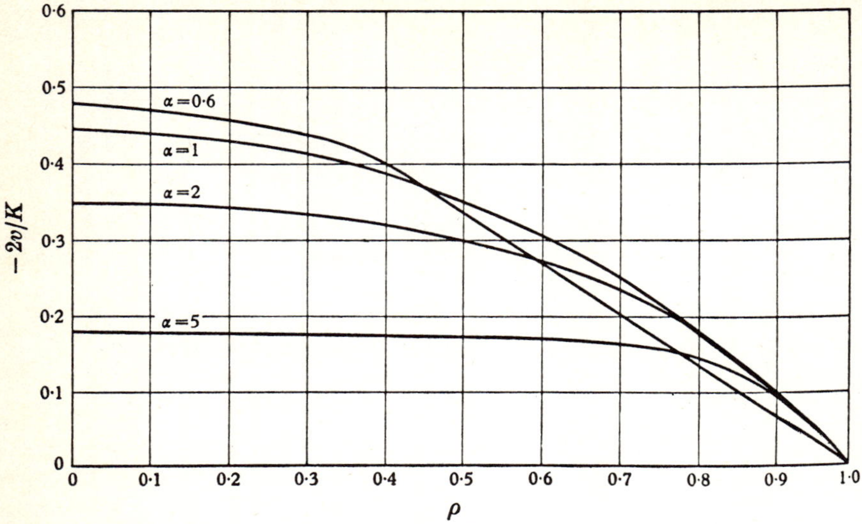

Fig. 4.10 Velocity distribution in the pipe, $\theta = 0$.

is independent of θ. Since

$$Q = \bar{v}A \tag{4.89}$$

then the mean velocity across the pipe is needed to determine the volume flow. Thus

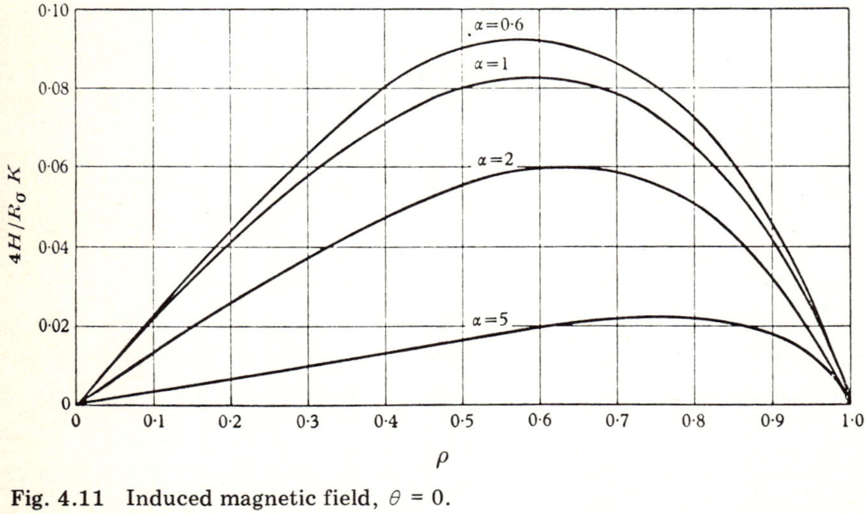

Fig. 4.11 Induced magnetic field, $\theta = 0$.

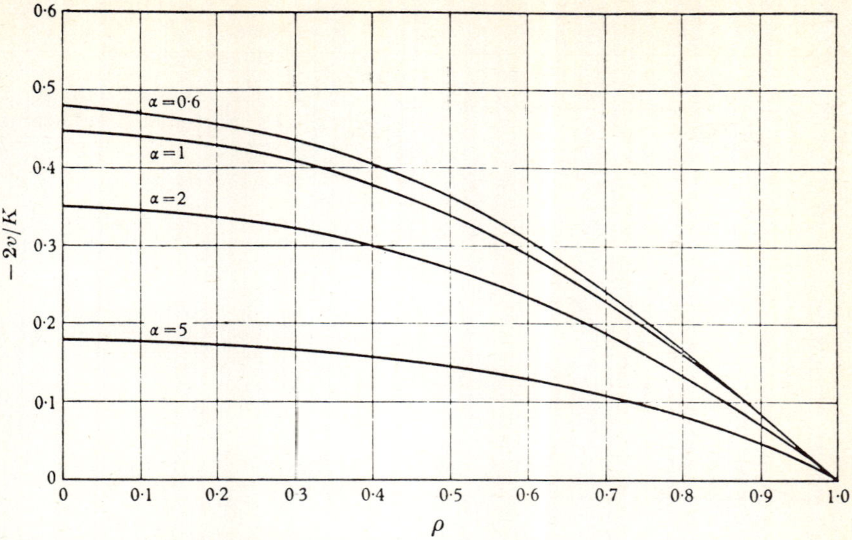

Fig. 4.12 Velocity distribution in the pipe, $\theta = \frac{1}{2}\pi$.

$$\overline{v}(\alpha) = \frac{1}{\pi}\int_{\rho=0}^{1}\int_{\theta=0}^{2\pi} \rho v(\rho,\theta)\, d\rho\, d\theta \tag{4.90}$$

after integrating, we get

$$\frac{\overline{v}_z}{ka^2} = -\frac{\overline{v}}{K} = \sum_{n=0}^{\infty} (-1)^n \epsilon_n \frac{I_n'(\alpha)}{2\alpha I_n(\alpha)}\left[\left(1 + \frac{n^2}{\alpha^2}\right) I_n^2(\alpha) - I_n'^2(\alpha)\right] \tag{4.91}$$

where

$$\frac{\partial p}{\partial z} = -k\mu \tag{4.92}$$

The mean velocity is shown in Fig. 4.13 for a range of Hartmann numbers. Now the induced voltage difference between θ equal to $\pi/2$ and $3\pi/2$ is required for designing an MFD flow meter, i.e.,

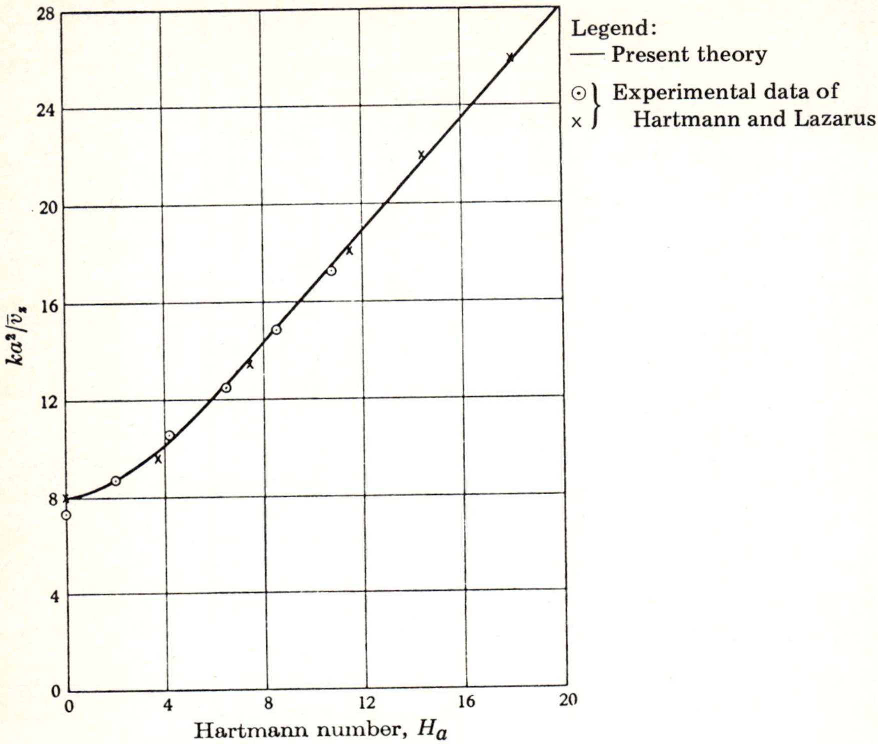

Fig. 4.13 Mean velocity versus Hartmann number.

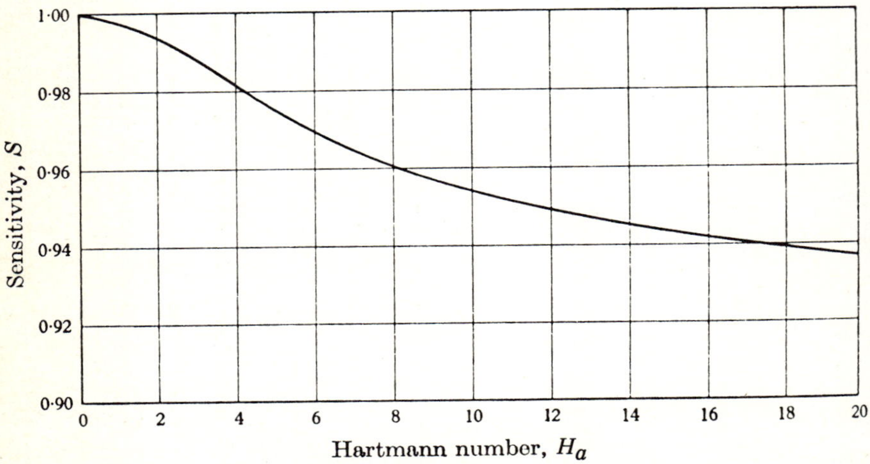

Fig. 4.14 Sensitivity versus Hartmann number.

$$V = 2\int_{r=0}^{a} E_r\left(r, \frac{\pi}{2}\right)dr = 2\mu_e H_0 v_0 a \int_{\rho=0}^{1} \left\{ \frac{1}{\rho R_\sigma}\left[\frac{\partial H(\rho,\theta)}{\partial\theta}\right]_{\theta=\pi/2} - v\left(\rho, \frac{\pi}{2}\right)\right\} d\rho \quad (4.93)$$

or

$$S = \frac{V}{2\mu_e H_0 v_0 a} = \frac{K}{2\alpha}\int_{\rho=0}^{1}\left\{\frac{1}{\alpha\rho}\sum_{n=0}^{\infty}(-1)^n(2n+1)\frac{I'_{2n+1}(\alpha)}{I_{2n+1}(\alpha)}I_{2n+1}(\alpha) + \frac{1}{2}\sum_{n=0}^{\infty}(-1)^n\epsilon_n\frac{I'_{2n}(\alpha)}{I_{2n}(\alpha)}I_{2n}(\alpha\rho) - \frac{1}{2\alpha}\right\}d\rho \quad (4.94)$$

which is called the *sensitivity* and is shown in Fig. 4.14 for the same range of Hartmann numbers. Thus, for a given magnetic field strength, the flow rate can be determined with Fig. 4.13 and the induced voltage across the electrodes from Fig. 4.14. The ratio of S to the flow rate will then be different for different Hartmann numbers.

4.5 MAGNETOFLUID-DYNAMIC BOUNDARY-LAYER ASSUMPTIONS

Boundary-layer phenomena occur when the influence of a physical quantity is restricted to small regions near confining boundaries. This phenomenon occurs when the nondimensional diffusion parameters—the Reynolds number, Peclet number, or magnetic Reynolds number—are large. The boundary layers are then the velocity and thermal or magnetic boundary layers; and each thickness is inversely proportional to the square root of the associated diffusion number. Prandtl fathered classical fluid-dynamic boundary theory by observing, from experimental flows, that for large Reynolds numbers, the viscosity and thermal conductivity appreciably influenced the flow only near a wall. When distant measurements in the flow direction are compared with a characteristic dimension in that direction, transverse measurements compared with the boundary-layer thickness, and velocities compared with the free stream velocity, the Navier-Stokes and energy equations can be considerably simplified by neglecting small quantities. The number of component equations is reduced to those in the flow direction and pressure changes across the boundary layer are negligible. The pressure is then only a function of the flow direction and can be determined from the inviscid flow solution. Also, the number of viscous terms is reduced to the dominant term and the heat conduction in the flow direction is negligible. MFD boundary-layer flows are separated into two types by considering the limiting cases of a very large or a negligibly small magnetic Reynolds number. When the magnetic field is oriented in an arbitrary direction relative to a confining surface and the magnetic Reynolds number is very small, the flow direction component of the magnetic interaction and the corresponding Joule heating is only a function of the transverse magnetic field component and the local velocity in the flow direction. Changes in the transverse magnetic field component and pressure across the boundary layer are negligible. The thickness of the magnetic boundary layer is very large and the induced magnetic field is negligible. However, when the magnetic Reynolds number is very large, the magnetic boundary-layer thickness is small and nearly the same size as the viscous and thermal boundary layers and then the MFD boundary-layer equations must be solved simultaneously. A unique attribute of MFD occurs when the magnetic pressure number in the flow direction is also greater than unity. The boundary layer produces an upstream, as well as a downstream wake, through its coupled action with the magnetic field. If the Reynolds number is also very large, the MFD flow can be described by three regions. The external flow will still be uninfluenced by the viscosity and thermal conductivity which are now important only in very thin layers. A third region that separates these inner and outer regions is produced by the mutual changes in the momentum and magnetic field.

Another unique feature of MFD can be demonstrated by considering a layer adjacent to a solid surface in which both the velocity and induced magnetic field change, called the Hartmann layer. In the limit of negligible viscosity and infinite conductivity the

Hartmann layer becomes a vortex sheet and a current sheet at the solid surface. The magnitude of the velocity slip and change in shear across this layer can be determined by considering the change of tangential stress across a thin layer of the electrically conducting fluid very near the solid surface, i.e.,

$$\Delta\left(\tau_x + \frac{B_x B_0}{\mu_0}\right) = 0 \tag{4.95}$$

Therefore

$$\mu \frac{du}{dy} + \frac{B_0}{\mu_0} B_x = \text{const} \tag{4.96}$$

since changes in the flow direction are considered to be small compared to transverse changes. Also from the tangential component of Ohm's law,

$$E_z = -\frac{1}{\mu_0 \sigma} \frac{dB_x}{dy} - uB_0 = \text{const} \tag{4.97}$$

for a constant tangential electric field. The solutions to these equations can be obtained by addition and subtraction and are

$$B_x \pm \mu_0(\mu\sigma)^{1/2} u = e^{\mp H_a y/l} + \text{const} \tag{4.98}$$

where l is a typical length and the sum approaches a constant at large distances from the surface. The exponential form cannot be accepted for the difference solution since it diverges at large distances. The solutions are then

$$B_x + \mu_0(\mu\sigma)^{1/2} u = e^{-H_a y/l} + \text{const} \tag{4.99}$$

$$B_x - \mu_0(\mu\sigma)^{1/2} u = \text{const} \tag{4.100}$$

and

$$u = \frac{e^{-H_a y/l}}{2\mu_0(\mu\sigma)^{1/2}} + \text{const} \tag{4.101}$$

$$B_x = \frac{e^{-H_a y/l}}{2} + \text{const} \tag{4.102}$$

Since the difference is constant across the Hartmann layer then

$$\Delta u = \frac{1}{\mu_0(\mu\sigma)^{1/2}} \Delta B_x \tag{4.103}$$

even when the viscosity vanishes and the conductivity becomes infinite. Consequently, the slip at the surface is not arbitrary but depends on the strength of the current sheet since the product $\mu\sigma$ can have any value as H_a becomes infinite. The corresponding change in shear stress is finite and balances the induced force.

$$\Delta\tau_x = -B_0(\mu\sigma)^{1/2}\Delta u \tag{4.104}$$

In the succeeding sections of this chapter, various MFD boundary-layer flows will be discussed for both limiting magnetic Reynolds number cases.

4.6 MAGNETOFLUID-DYNAMIC STAGNATION POINT FLOW

A blunt body that moves through the atmosphere at hypersonic speeds has a flow field between the body and detached shock wave that is nonviscous in the outer region and viscous in the wall boundary layer. The free stream flow is generally electrically nonconducting while the flow between the shock wave and body is conducting due to the ionization of the gas by the large temperature rise across the shock wave. Application of a transverse magnetic field fixed to the body should reduce the heat transfer at the stagnation point and increase the body's drag. Both these results are desirable for protecting a vehicle reentering the atmosphere.

The inviscid MFD flow over a spherical body with a spherical detached shock has been analyzed by Bush[8] and the results compared with the nonmagnetic solution. The magnetic field increases the shock detachment distance, reduces the pressure at any given station along the body, and decreases the tangential velocity gradient. This last effect contributes considerably to the decrease in the heat transfer at the stagnation point since the velocity distribution at the outer edge of the boundary layer directly influences the boundary-layer behavior. The magnetic field will also affect the boundary layer. This combined effect of the magnetic field on the stagnation point heat transfer and surface shear[9] will be calculated for a small magnetic Reynolds number with the MFD boundary layer equations which now appear in the form applicable for an axisymmetric body whose radius of curvature $r_0(x)$ is large, compared to the boundary-layer thickness. The equations are

$$\frac{\partial(\rho u r_0)}{\partial x} + \frac{\partial(\rho v r_0)}{\partial y} = 0 \tag{4.105}$$

$$u\frac{\partial u}{\partial x} + v\frac{\partial u}{\partial y} = \left(\frac{\rho_e}{\rho}\right)\left[u_e\frac{du_e}{dx} + \frac{u_e\sigma_e B_e^2}{\rho_e}\left(1 - \frac{\sigma u}{\sigma_e u_e}\right)\right] + \frac{1}{\rho}\frac{\partial}{\partial y}\left(\mu\frac{\partial u}{\partial y}\right) \tag{4.106}$$

$$u\frac{\partial h}{\partial x} + v\frac{\partial h}{\partial y} = \frac{1}{\rho}\frac{\partial}{\partial y}\left(\frac{k}{c_p}\frac{\partial h}{\partial y}\right) \tag{4.107}$$

where the effects of viscous and Joule heating have been neglected. The electric field E has been taken as zero since the current will form a short circuit loop around the body in a plane perpendicular to the body axis. The subscript e designates quantities evaluated at the outer edge of the boundary layer. These equations can be transformed to ordinary differential equations by the following assumptions and transformation relations for the coordinates, stream function, and total enthalpy. Let

$$X = \int_0^x \left(\frac{r_0}{L}\right)^2 \frac{\rho_e\mu_e}{\rho_0\mu_0}\,dx$$

$$Y = \int_0^y \left(\frac{r_0}{L}\right)\left(\frac{\rho}{\rho_0}\right)dy \tag{4.108}$$

$$u_e = C\left(\frac{X}{L}\right)^{1/3}$$

and assume

$$\frac{\rho\mu}{\rho_e\mu_e} = 1 \qquad \frac{\rho_e}{\rho} = \frac{h}{h_e} \qquad \frac{\sigma}{\sigma_0} = \left(\frac{h}{h_e}\right)^n \qquad \text{and} \qquad \Pr = 1 \tag{4.109}$$

then set

$$\psi = \sqrt{\frac{3\mu_0 u_e X}{2\rho_0}}\, f(\eta)$$

and

$$h = h_e g(\eta) \tag{4.110}$$

$$\eta = Y\sqrt{\frac{2}{3}\frac{\rho_0 u_e}{\mu_0 X}}$$

Defining the magnetic interaction parameter

$$Q_v = \frac{\sigma_e B_0{}^2}{\rho_e(du_e/dx)} \tag{4.111}$$

to be a constant, then a dipole magnetic field fixed to the body must have the form

$$B_0 = \pm\sqrt{\left(\frac{\rho_e}{\sigma_e}\right)\left(\frac{u_e}{3X}\right)\left(\frac{r_0}{L}\right)\left(\frac{\rho_e\mu_e}{\rho_0\mu_0}\right)Q_V} \tag{4.112}$$

The subscript zero defines reference values of the fluid properties and L is some fixed reference length. The velocity variation outside of the boundary layer in the region of a stagnation point is shown in Eq. (4.108) where C is a proportionality constant. The quantity h is the specific enthalpy, i.e., enthalpy per unit mass. Applying these assumptions and transformations to the boundary-layer equations and assuming that the total enthalpy is a constant at the wall gives

$$g'' + fg' = 0 \tag{4.113}$$

$$f''' + ff'' + \frac{1}{2}\left\{g[1 + Q_V(1 - g^n f')] - (f')^2\right\} = 0 \tag{4.114}$$

and

$$f(0) = f'(0) = 0 \quad f'(\infty) = 1$$
$$g(0) = g_w \quad g(\infty) = 1 \tag{4.115}$$

where the boundary conditions were determined by the no-slip condition at the wall and the condition at the outer edge of the boundary layer. By defining

$$(\dot{q}_w)_{sp} = \left(k \frac{\partial T}{\partial y}\right)_{w, sp} = (2\rho_e \mu_e)_{sp}^{1/2} h_e \left(\frac{du_e}{dx}\right)_{sp}^{1/2} g'_{w, sp} \tag{4.116}$$

the stagnation point heat-transfer rate can be compared with the nonmagnetic case that has the same conditions upstream of the shock wave and total enthalpy at the wall. The comparison relation is

$$\frac{\dot{q}_w}{(\dot{q}_w)_{NM}} = \left[\frac{(du_e/dx)_{sp}}{(du_e/dx)_{sp, NM}}\right]^{1/2} \left(\frac{g'_w}{g'_{w, NM}}\right) \tag{4.117}$$

and the first term on the right is the adjustment of the inviscid flow by the magnetic field. The second factor is the alterations of the total enthalpy profile in the boundary layer.

The differential equations were solved by Bush with a finite difference technique, which provided numerical values for the surface shear and Eq. (4.117). Bush[10] used his previous inviscid flow results (shown in Fig. 4.15) to determine the heat-transfer rate and surface shear (Fig. 4.16) for two electrical conductivity relations and for a cool wall ($g_w = 0.05$).

The results demonstrate that the inviscid solution predicts too large a decrease in the stagnation point heat transfer compared to either viscous solution. If the conductivity is constant, then a substantial reduction in heat transfer is still produced by the magnetic field. Whereas in the nonuniform conductivity case, the decrease in the outer velocity gradient is nearly offset by the increase in the change in the total enthalpy profile and the reduction is not as large. The magnetic field increases the surface shear in both cases, but is more effective when the conductivity is not uniform through the boundary layer.

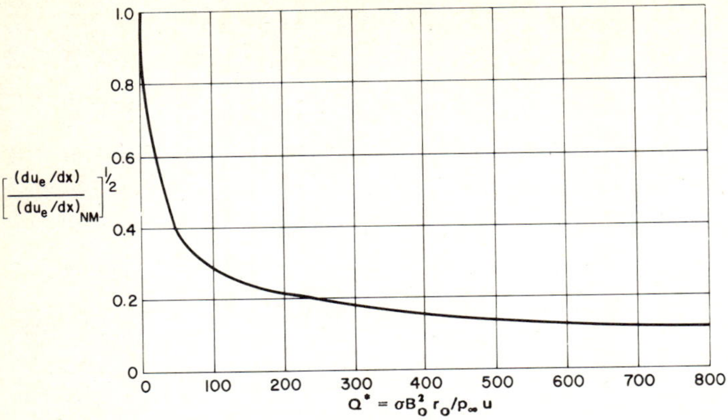

Fig. 4.15 Heat-transfer parameter versus magnetic parameter.

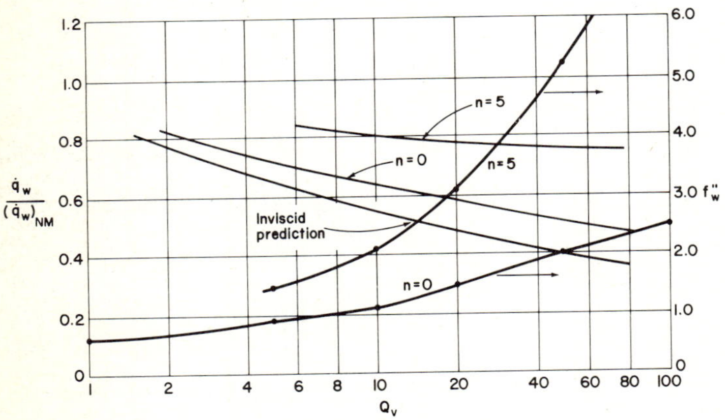

Fig. 4.16 Heat-transfer ratio and surface shear versus the magnetic parameter at the stagnation point of an axisymmetric blunt body.

4.7 MAGNETOFLUID-DYNAMIC FLAT PLATE FLOWS

The analytical results of MFD flow parallel to a flat plate[11] are used to determine the losses at flat or nearly flat surfaces such as thin airfoils and the side walls of large volume generators and accelerators. In most applications, the magnetic field is fixed to the surface although it may be fixed with the main flow which then produces different effects in the boundary layer. An example is a rotor vane moving in a fluid at rest in a passage with the magnetic field fixed to the passage. Both cases will be discussed in this section for a constant wall temperature and an insulated plate. The fluid is assumed to be incompressible and the magnetic field is perpendicular to the surface. The magnetic Reynolds number is taken to be small enough so that the induced magnetic field can be neglected. The pressure gradient in the flow direction is taken as zero and the boundary-layer flow is maintained by the main, nonviscous flow. The fluid properties are also assumed to be constant in a limited temperature range.

4.7.1 Magnetic Field Fixed to Plate. The appropriate MFD boundary-layer equations and boundary conditions are:

$$\frac{\partial u}{\partial x} + \frac{\partial v}{\partial y} = 0 \tag{4.118}$$

$$u\frac{\partial u}{\partial x} + v\frac{\partial u}{\partial y} = -\frac{\sigma_0 B_0{}^2}{\rho} u + \nu \frac{\partial^2 u}{\partial y^2} \tag{4.119}$$

$$\rho u c_p \frac{\partial T}{\partial x} + \rho v c_p \frac{\partial T}{\partial y} = k\frac{\partial^2 T}{\partial y^2} + \sigma_0 B_0{}^2 u^2 + \mu\left(\frac{\partial u}{\partial y}\right)^2 \tag{4.120}$$

$$J_z = \sigma_0 \mathbf{u} \times \mathbf{B}_0 \tag{4.121}$$

$$\begin{aligned} y = 0: u = 0 \quad T = T_w \text{ or } \left.\frac{\partial T}{\partial y}\right|_{y=0} = 0 \\ y \to \infty: \frac{\partial u}{\partial y} \to 0 \quad \text{and} \quad \frac{\partial u}{\partial x} \to -\frac{\sigma_0 B_0{}^2}{\rho} = -m_1 \end{aligned} \tag{4.122}$$

The boundary-layer flow must be restricted to the region of the plate since the continuity equation and boundary conditions require

$$\frac{\partial u}{\partial x} + \frac{\partial v}{\partial y} = -m_1 + \frac{\partial v}{\partial y} = 0 \tag{4.123}$$

or

$$v = my + \text{const} \tag{4.124}$$

Thus, as $y \to \infty$ then $v \to \infty$.

The equations can be transformed into ordinary differential equations by defining

and

$$\xi = x \qquad \eta = \sqrt{\frac{u_\infty}{\nu x}}\, y \qquad m = \frac{\sigma_0 B_0{}^2}{\rho u_\infty} \tag{4.125}$$

$$\psi = \sqrt{u_\infty \nu x}\,[f_0(\eta) + mxf_2(\eta) + (mx)^2 f_4(\eta) + \cdots]$$

where $\psi(\xi, \eta)$ is the stream function and u_∞ is the velocity of the undisturbed flow far upstream of the plate. Since

$$u = \frac{\partial\psi}{\partial y} = \frac{\partial\psi}{\partial\eta}\frac{\partial\eta}{\partial y}$$

and $$\tag{4.126}$$

$$v = -\frac{\partial\psi}{\partial x} = -\left(\frac{\partial\psi}{\partial\xi} - \frac{\partial\psi}{\partial\eta}\frac{\partial\eta}{\partial x}\right)$$

the momentum equations become, after collecting like powers of mx:

$$2f_0''' = -f_0''f_0 \tag{4.127}$$

$$f_2''' = f_0'f_2' - \tfrac{1}{2}f_0f_2'' - \tfrac{3}{2}f_2f_0'' + f_0' \tag{4.128}$$

$$f_4''' = 2f_0'f_4' + f_2'f_2' - \tfrac{1}{2}f_0f_4'' - \tfrac{3}{2}f_2f_2'' - \tfrac{5}{2}f_4f_0'' + f_2' \tag{4.129}$$

and the boundary conditions are

$$\begin{aligned} f_0 &= f_2 = f_4 = \cdots = 0 & &\text{at} & \eta &= 0 \\ f_0' &= f_2' = f_4' = \cdots = 0 & &\text{at} & \eta &= 0 \\ f_0' &= 1\,,\ f_2' = -1 & &\text{as} & \eta &\to \infty \\ f_4' &= f_6' = \cdots = 0 & &\text{as} & \eta &\to \infty \end{aligned} \tag{4.130}$$

Equation (4.127) is the Blasius equation for flat plate flow of a nonconducting fluid and the solution is available in tabulated form. Numerical results for f_2 and f_4 were obtained by Rossow, who provided the velocity profiles and relations for the skin friction coefficient and total drag on the magnetic field fixed to the plate, i.e.,

$$c_f = \frac{\mu(\partial u/\partial y)_{y=0}}{\rho(u_\infty{}^2/2)} = \frac{0.664 - 1.789\,mx + 0.706\,m^2x^2 - \cdots}{\sqrt{\mathrm{R}_{ex}}} \tag{4.131}$$

$$\frac{F}{A} = \sigma_0 B_0^2 \sqrt{u_\infty \nu x} \int_0^{\eta \to \infty} (f_0' + mxf_2' + \cdots) d\eta$$

$$= \sigma B_0^2 \sqrt{u_\infty \nu x} (f_0 + mxf_2 + \cdots)\Big|_{\eta \to \infty} \tag{4.132}$$

Thus, the force on the plate is finite only for a conducting layer of finite thickness.

Before transforming the energy equation to an ordinary differential equation, a simple calculation will reveal the influence of the magnetic field on the total energy for the boundary layer. Multiply the momentum equation by u and add the result to the energy equation and let the Prandtl number equal unity, then

$$u \frac{\partial E}{\partial x} + v \frac{\partial E}{\partial y} = \nu \frac{\partial^2 E}{\partial y^2} \tag{4.133}$$

where

$$E = c_p T + \frac{u^2}{2} \tag{4.134}$$

is the total energy. Equation (4.133) shows that the magnetic field does not change the total energy in the boundary layer. The kinetic energy lost from the flow due to the retardation by the magnetic field appears entirely as ohmic heating.

Returning to the energy equation, let

$$T(\eta, x) = T_0(\eta) + T_2(\eta)\, mx + T_4(\eta)\, m^2 x^2 + \cdots \tag{4.135}$$

and since

$$\frac{\partial T}{\partial x} = \frac{\partial T}{\partial \xi} + \frac{\partial T}{\partial \eta}\frac{\partial \eta}{\partial x} \qquad \frac{\partial T}{\partial y} = \frac{\partial T}{\partial \eta}\frac{\partial \eta}{\partial y} \tag{4.136}$$

substituting and collecting by powers of mx gives

$$\frac{T_0''}{\mathrm{P_r}} + \frac{T_0' f_0}{2} + \frac{u_\infty^2 (f_0'')^2}{c_p} = 0 \tag{4.137}$$

$$\frac{T_2''}{P_r} + \frac{T_2' f_0}{2} - f_0' T_2 + \frac{3}{2} f_2 T_0' + \frac{u_\infty^2}{c_p}(f_0''^2 + 2f_0'' f_2'') = 0 \tag{4.138}$$

and so forth, and the boundary conditions are

$$T_2 = 0\,, \quad T_0 = T_w \quad \text{or} \quad \frac{\partial T_0}{\partial \eta} = 0 \quad \text{at} \quad \eta = 0$$

and

$$T_0 = T_\infty \quad \text{as} \quad \eta \to \infty \tag{4.139}$$

$$T_2'' = T_2' = 0 \quad \text{as} \quad \eta \to \infty$$

Equation (4.137) was solved by E. Pohlhausen[29] for a nonconducting fluid, a Prandtl number equal to unity, and a constant wall temperature, i.e.,

$$T_0 = T_\infty + \left(T_w - T_\infty - \frac{u_\infty^2}{2c_p}\right)(1 - f_0') + \frac{u_\infty^2}{2c_p}(1 - f_0'^2) \tag{4.140}$$

The equation for T_2 was solved numerically by Rossow for the same conditions. The relation for the surface heat flux is then

$$h = \frac{k(\partial T/\partial \eta)|_{\eta=0}}{u_\infty^2/2c_p} = \frac{\rho u_\infty c_p}{2\sqrt{R_{ex}}}(0.664 - 0.704\,mx - \cdots) \tag{4.141}$$

When the fluid is nonconducting, m is zero and Eq. (4.141) becomes the Reynolds analogy which states that the heat-transfer coefficient is directly proportional to the skin friction coefficient and that changes in the skin friction coefficient imply changes in the heat-transfer coefficient for a constant temperature plate. This relationship is established with $m = 0$ and Eq. (4.131) for the skin friction coefficient.

Equation (4.137) was also solved by Pohlhausen for an insulated plate; i.e., $\partial T/\partial \eta$ is zero at the wall, and is called the *thermometer problem* since the wall will "measure" the recovery temperature:

$$T_0 = T_\infty + \frac{u_\infty^2}{2c_p}(1 - f_0'^2) \tag{4.142}$$

Rossow[11] used this result and numerically integrated Eq. (4.138) for the insulated plate. An immediate result is that the magnetic field will not affect the recovery temperature since the velocity is zero at the wall and then the magnetic interaction at the wall is zero.

The velocity profiles and temperature profiles for both thermal boundary conditions are shown in Fig. (4.17). The velocity profiles show a reversed flow at the surface for mx equal to one-half which is physically unacceptable, since the magnetic field can only retard and not produce flow in the reverse direction. Rossow[11] attributed this difficulty to the use of only two terms in Eq. (4.125), which restricts its applicability to values of mx less than 0.2. Hugelmann[12] extended the range of mx with a Karman-Pohlhausen integral method and obtained agreement with Rossow's results up to mx equal to 0.2 and showed that the skin friction coefficient asymptotically approaches zero for large mx values. Thus, the magnetic field does not produce flow separation or reverse flow. However, Rossow's results do demonstrate the correct influence of the magnetic field in the restricted range, i.e., the flow is retarded, the surface shear is decreased and the ohmic heating increases the fluid temperature away from the plate for both thermal conditions since the wall temperatures are maintained at fixed values. The recovery temperature for the insulated plate is not changed by the magnetic field as predicted by the theoretical results.

4.7.2 Magnetic Field Fixed to Fluid. The magnetic field now moves past the plate with the fluid and the fluid in the boundary layer that has been slowed by the viscosity will be dragged along. The induced current flow in the boundary layer is

$$J_z = \sigma B_0(u - u_\infty) \tag{4.143}$$

and the momentum equation is

$$u\frac{\partial u}{\partial x} + v\frac{\partial u}{\partial y} = -\frac{\sigma_0 B_0{}^2}{\rho}(u - u_\infty) + \nu\frac{\partial^2 u}{\partial y^2} \tag{4.144}$$

Using the transformed Equation (4.125) and collecting by powers of mx,

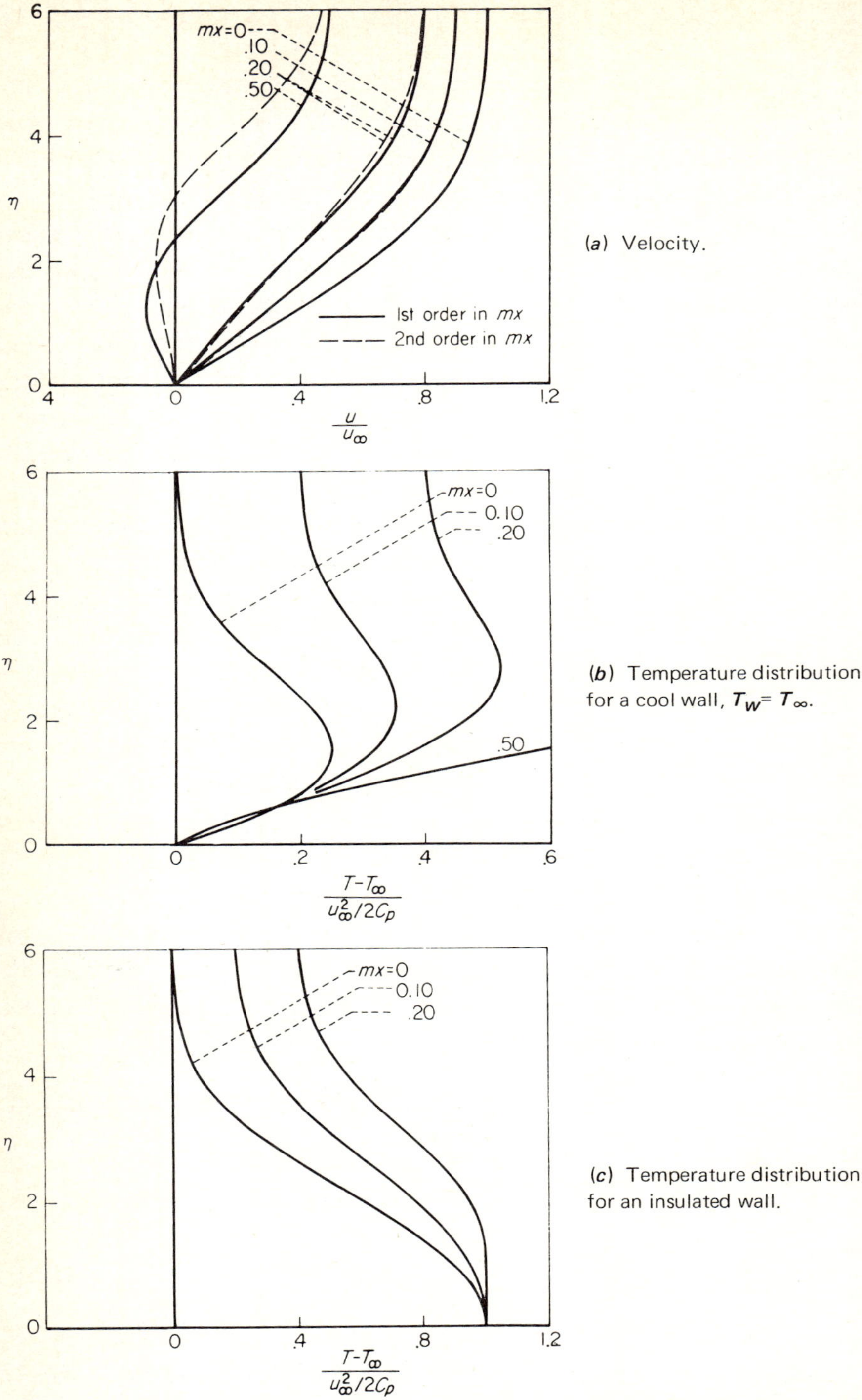

Fig. 4.17 Boundary-layer profiles for fluid flowing over a semi-infinite flat plate with a transverse magnetic field fixed relative to the plate.

$$2f_0''' = -f_0''f_0 \tag{4.145}$$

$$f_2''' = f_0'f_2' - \frac{1}{2}f_0'f_2'' - \frac{3}{2}f_2f_0'' + f_0' - 1 \tag{4.146}$$

$$f_4''' = 2f_0'f_4' + f_2'f_2' - \frac{1}{2}f_0f_4'' - \frac{3}{2}f_2f_2'' - \frac{5}{2}f_4f_0'' + f_2' \tag{4.147}$$

and the boundary conditions are

$$\begin{aligned} f_0 &= f_2 = f_4 = \cdots = 0 \quad \text{at} \quad \eta = 0 \\ f_0' &= f_2' = f_4' = \cdots = 0 \quad \text{at} \quad \eta = 0 \\ f_0' &= 1 \quad f_2' = f_4' = \cdots = 0 \quad \text{as} \quad \eta \to \infty \end{aligned} \tag{4.148}$$

Equation (4.145) is still the Blasius equation and Eqs. (4.146) and (4.147) were numerically integrated. The skin friction coefficient is

$$c_f = \frac{0.664 + 2.293\,mx - 2.768\,m^2x^2 + \cdots}{\sqrt{\mathrm{R}_{ex}}} \tag{4.149}$$

and the force exerted on the magnet is now a reaction to the dragging of the boundary-layer fluid.

$$\begin{aligned} \frac{F}{A} &= \int_0^{y\to\infty} \sigma_0 B_0^2(u_\infty - u)\,dy \\ &= \sigma_0 B_0^2\sqrt{u_\infty \nu x}\,(1.73 - 2.21\,mx + 5.22\,m^2x^2 - \cdots) \end{aligned} \tag{4.150}$$

The energy equation in its altered form is

$$\rho u c_p \frac{\partial T}{\partial x} + \rho v c_p \frac{\partial T}{\partial y} = k\frac{\partial^2 T}{\partial y^2} + \sigma_0 B_0^2(u - u_\infty)^2 + \mu\left(\frac{\partial u}{\partial y}\right)^2 \tag{4.151}$$

and the equation for the total energy is

$$\rho u \frac{\partial E}{\partial x} + \rho v \frac{\partial E}{\partial y} + u_\infty(u - u_\infty)\sigma_0 B_0^2 = \mu \frac{\partial^2 E}{\partial y^2} \tag{4.152}$$

The magnetic field now adds energy to the boundary layer by the work done in dragging the retarded fluid. Again transforming Eq. (4.151) and collecting by powers of mx,

$$\frac{T_0''}{P_r} + \frac{T_0' f_0}{2} + \frac{u_\infty^2 (f_0'')^2}{c_p} = 0 \tag{4.153}$$

$$\frac{T_2''}{P_r} + \frac{f_0}{2} T_2' - f_0' T_2 + \frac{3}{2} f_2 T_0' + \frac{u_\infty^2}{c_p}[(1 - f_0')^2 + 2 f_0'' f_2''] = 0 \tag{4.154}$$

and the boundary conditions are

$$\begin{aligned} &T_2 = 0\,, \quad T_0 = T_w \quad \text{or} \quad \frac{\partial T}{\partial \eta} = 0 \quad \text{at} \quad \eta = 0 \\ &T_2 = 0 \quad \text{as} \quad \eta \to \infty \\ &T_2'' = T_2' = 0 \quad \text{as} \quad \eta \to \infty \end{aligned} \tag{4.155}$$

Equation (4.153) is the same as Eq. (4.137), and Eq. (4.154) was integrated numerically for $T = T_w$ and a Prandtl number equal to unity. The wall heat-transfer coefficient is

$$h = \frac{\rho u_\infty c_p}{2\sqrt{R_{ex}}}(0.664 + 4.044\,mx + \cdots) \tag{4.156}$$

The insulated wall results are also obtained by using Pohlhausen's results [Eq. (4.142)], and numerically integrating Eq. (4.154).

The results for the magnetic field fixed to the fluid are shown in Fig. 4.18 which are again restricted to values of mx less than 0.2. The dragging of the boundary-layer fluid is demonstrated by the increase in velocity, wall shear, and fluid temperature. The recovery temperature is also increased by the ohmic heating, since now the magnetic interaction at the wall is not equal to zero.

4.8 UNSTEADY MAGNETOFLUID-DYNAMIC FLAT PLATE FLOW

In classical fluid dynamics, Rayleigh flow is the simplest unsteady flat plate flow and demonstrates the impulsive development of a flow field from rest. The MFD Rayleigh problem will then show the influence of the magnetic field on this flow development and the two cases for the magnetic field fixed to the plate or fixed to the fluid will be explained.

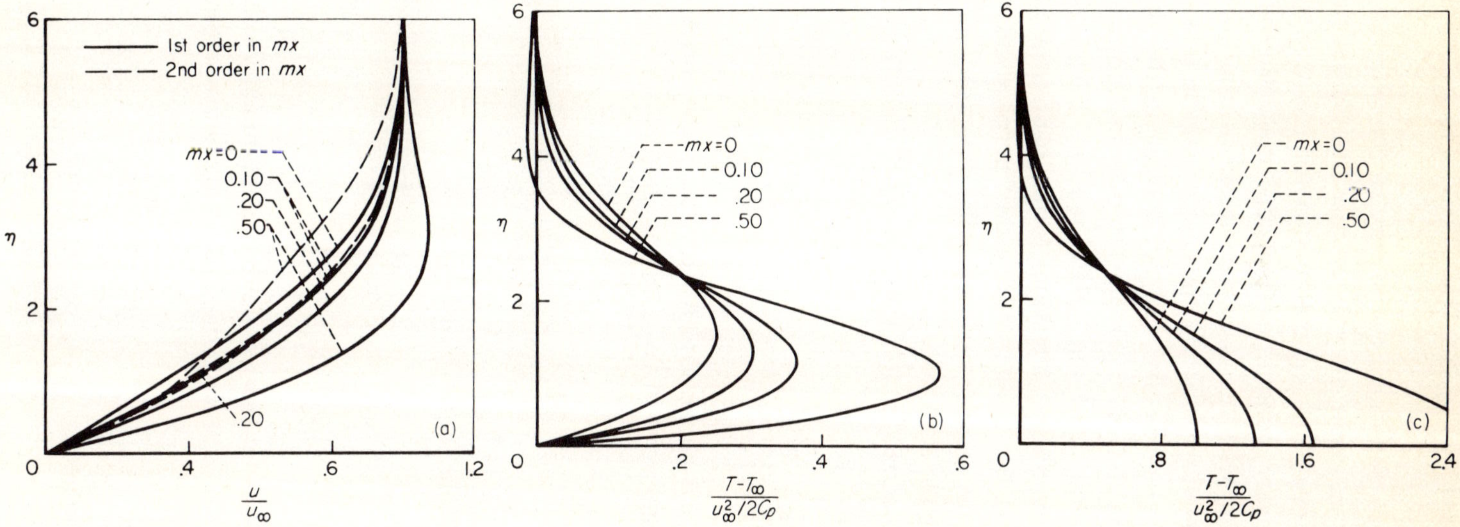

(*a*) Velocity.

(*b*) Temperature distribution for a cool wall, $T_W = T_\infty$.

(*c*) Temperature distribution for an insulated wall.

Fig. 4.18 Boundary-layer profiles for fluid flowing over a semi-infinite flat plate with a transverse magnetic field fixed relative to the fluid.

The problem is to determine the history of the velocity profile and skin friction for any time after an infinite flat plate is suddenly set in motion in its own plane and maintained at a constant velocity. The fluid is electrically conducting and the magnetic field will produce an interaction with the fluid that crosses it. Since the plate is infinite in all directions and the magnetic field is taken to be perpendicular to it, the velocity will not change in the direction parallel to the motion and will be zero in the direction normal to the plane of the magnetic field and plate motion. These physical assumptions and the continuity equations for an incompressible fluid provide the result that the velocity component perpendicular to the plate is also zero. The velocity in the parallel direction will change only in the perpendicular direction.

The momentum equation for the magnetic field fixed to the plate is

$$\frac{\partial u}{\partial t} + \frac{\sigma B_0{}^2}{\rho}(u - u_\infty) = \nu \frac{\partial^2 u}{\partial y^2} \tag{4.157}$$

since the magnetic field moves through the fluid. The corresponding equation for the magnetic field fixed to the fluid at a large distance from the plate is

$$\frac{\partial u}{\partial t} + \frac{\sigma B_0{}^2}{\rho} u = \nu \frac{\partial^2 u}{\partial y^2} \tag{4.158}$$

The boundary conditions for both cases are

$$\begin{aligned} t &= 0 \quad u = 0 \quad \text{for } y \geq 0 \\ y &= 0 \quad u = u_\infty \quad \text{for } t > 0 \end{aligned} \tag{4.159}$$

The analytical solutions for these equations and boundary conditions were obtained by Rossow[11] with the Laplace transform method and are

$$(B_0 \text{ fixed to plate}) \quad u = u_\infty \left[1 - e^{-m_1 t} \operatorname{erf}\left(\frac{y}{2\sqrt{\nu t}}\right)\right] \tag{4.160}$$

$$(B_0 \text{ fixed to fluid}) \quad u = \frac{u_\infty}{2}\left[\exp\left(-\sqrt{\frac{m_1}{\nu}}\,y\right)\operatorname{erfc}\left(\frac{y}{2\sqrt{\nu t}} - \sqrt{m_1 t}\right) + \exp\left(\sqrt{\frac{m_1}{\nu}}\,y\right)\operatorname{erfc}\left(\frac{y}{2\sqrt{\nu t}} + \sqrt{m_1 t}\right)\right] \quad (4.161)$$

The corresponding skin friction coefficients are

$$(B_0 \text{ fixed to plate}) \quad c_f = \frac{\mu\,(\partial u/\partial y)_{y=0}}{\rho u_\infty^{\,2}/2} = \frac{2\nu}{u_\infty\sqrt{\pi\nu t}}\,e^{-m_1 t} \quad (4.162)$$

$$(B_0 \text{ fixed to fluid}) \quad c_f = \frac{2\nu}{u_\infty}\left(\sqrt{\frac{m_1}{\nu}}\,\operatorname{erf}\sqrt{m_1 t} + \frac{e^{-m_1 t}}{\sqrt{\pi\nu t}}\right) \quad (4.163)$$

All these expressions reduce to their classical form for a nonconducting fluid. The results are shown in Figs. 4.19 and 4.20. The magnetic field accelerates the fluid to larger velocities than velocities produced by the viscosity when it is fixed to the plate. The surface shear is then correspondingly reduced. The entire flow field approaches u_∞ and the surface shear approaches zero for infinite times. However, accelerating the fluid from rest to u_∞ produces an

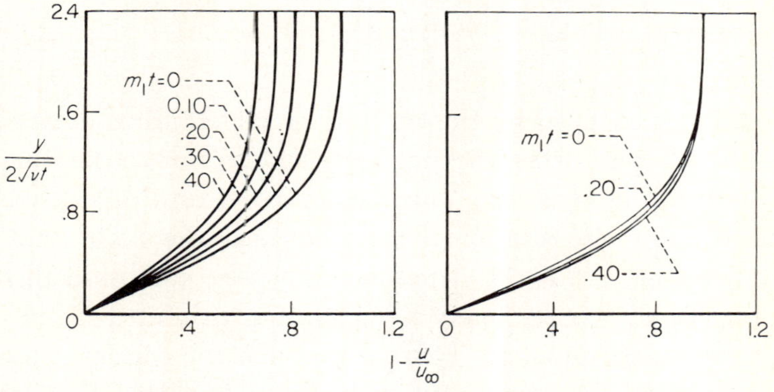

(*a*) Magnetic field fixed relative to plate.

(*b*) Magnetic field fixed relative to fluid.

Fig. 4.19 Boundary-layer velocity profiles on a flat plate set into motion impulsively.

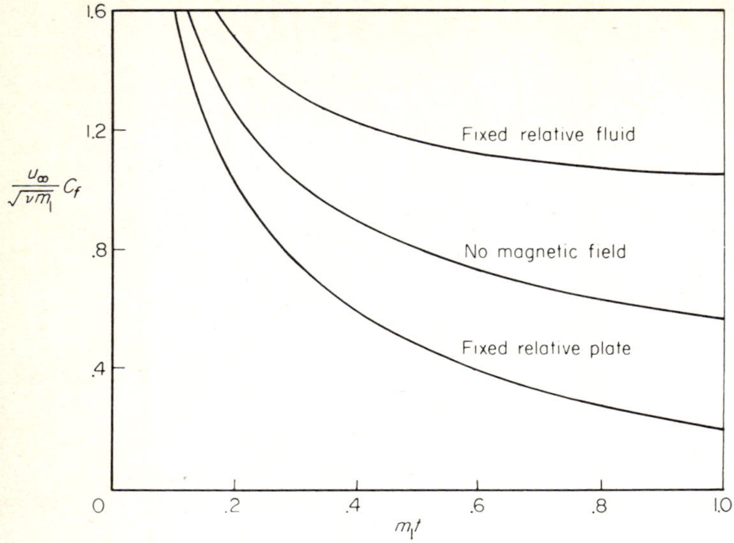

Fig. 4.20 Skin friction parameter as a function of a magnetic parameter.

infinite reaction force on the magnet since the fluid and magnetic field extend an infinite distance from the plate, i.e.,

$$\frac{F}{A} = \int_0^{y \to \infty} \sigma B_0^2 (u_\infty - u)\, dy = \sigma B_0^2 e^{-m_1 t} \int_0^{y \to \infty} \operatorname{erf}\left(\frac{y}{2\sqrt{\nu t}}\right) dy \tag{4.164}$$

The fluid velocity produced by the viscosity is not disturbed at large distances from the plate when the magnetic field is fixed to the fluid since the interaction there is zero. The magnetic field now retards the acceleration by the viscosity in the region near the plate, and, consequently, increases the skin friction coefficient compared to a nonconducting fluid. A completely new result is exhibited by this case since the velocity does not approach u_∞ and the surface shear does not approach zero for infinite times unless the fluid is nonconducting. The reaction force on the magnet will be finite in contrast to the previous case since the effect of the magnetic field is restricted to the region of the plate.

4.9 MAGNETOFLUID-DYNAMIC WAKES

In previous sections, we have discussed the influence of magnetic fields for a variety of flows and obtained results that were qualitatively similar. In contrast, entirely new results not experienced in ordinary fluid flow will be produced when the induced magnetic field becomes important. When the magnetic Reynolds number is sufficiently large, but finite, the thickness of the magnetic and ordinary boundary layers are the same order of magnitude. Disturbances can then propagate upstream by the combined response of the magnetic field and fluid to produce wakes upstream and downstream. The conditions for their occurrence can be demonstrated by examining the steady two-dimensional flow of a uniform stream over a body under a uniform magnetic field[13] $\mathbf{H}_0 = iH_x + jH_g + k0$. We have to solve simultaneously the equation of the magnetic field with the flow equations. The magnetic Reynolds number and magnetic Mach number will have a significant influence on the boundary-layer flow. We assume that the compressibility effect is negligible and the deviations of the velocity and magnetic field from the uniform states are small. Hence, we have the following velocity and magnetic field components (Fig. 4.21)

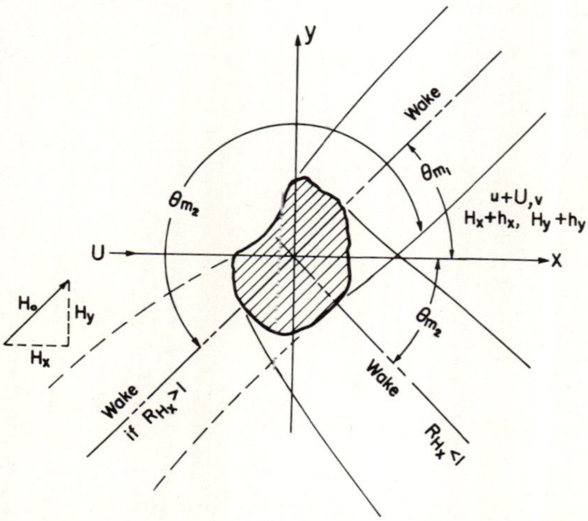

Fig. 4.21 Wakes around a body in a conducting fluid in a magnetic field.

$$u_x = U + u \quad u_y = V \quad H_x = H_{x0} + h_x \quad H_y = H_{y0} + h_y \tag{4.165}$$

where

$$u, v \ll U, h_x, h_y \ll H_{x0}, H_{y0}$$

Starting with the MFD equations in the form:

$$\nabla^* \mathbf{v}^* = 0 \tag{4.166}$$

$$\nabla^* \times (\mathbf{v}^* \times \mathbf{H}^*) - \frac{1}{R_\sigma}\{\nabla^* \times [\nu_H^*(\nabla^* \times \mathbf{H}^*)]\} = 0 \tag{4.167}$$

$$\rho^*\mathbf{v}^* \cdot \nabla^*\mathbf{v}^* - R_H(\mathbf{H}^* \cdot \nabla^*)\mathbf{H}^* = -\nabla^*\left(p + R_H\frac{H^{*2}}{2}\right) + \frac{1}{R_e}(\nabla^* \cdot \pi^*) \tag{4.168}$$

$$\rho^*T^*\mathbf{v}^* \cdot \nabla^*S^* = \frac{1}{P_e}\nabla^* \cdot (k^*\nabla^*T^*) + R_H(\nabla^* \times \mathbf{H}^*) \times \left[\frac{\nu_H^*}{R_\sigma}(\nabla^* \times \mathbf{H}^*) - \mathbf{v}^* \times \mathbf{H}^*\right] \tag{4.169}$$

and by neglecting the higher order terms of u, v, h_x, and h_y and with a number of manipulations (see Problems, Chap. 4), the linearized differential equation for u^* is:

$$\nabla^{*2}\left[\left(\frac{\partial}{\partial x^*} - \frac{1}{R_e}\nabla^{*2}\right)\left(\frac{\partial}{\partial x^*} - \frac{1}{R_\sigma}\nabla^{*2}\right) - \left(R_{H_x}\frac{\partial}{\partial x^*} - R_{H_y}\frac{\partial}{\partial y^*}\right)^2\right]u^* = 0 \tag{4.170}$$

where

$$\nabla^{*2} = \frac{\partial^2}{\partial x^{*2}} + \frac{\partial^2}{\partial y^{*2}} \qquad R_{H_x} = \frac{\mu_e^* H_{x_0}^{*\,2}}{\rho^* U^2} \qquad R_{H_y} = \frac{\mu_e^* H_{y_0}^{*\,2}}{\rho^* U^2}$$

The operator of the square bracket may be considered as a product of two Oseen-type operators. For instance, if we assume that $R_e = R_\sigma$, Eq. (4.170) becomes

$$\nabla^{*2}\left[(1 + R_{H_x})\frac{\partial}{\partial x^*} + R_{H_y}\frac{\partial}{\partial y^*} - \frac{1}{R_e}\nabla^{*2}\right]\left[(1 - R_{H_x})\frac{\partial}{\partial x^*} - R_{H_y}\frac{\partial}{\partial y^*} - \frac{1}{R_e}\nabla^{*2}\right]u^* = 0 \quad (4.171)$$

The two operators in the square brackets are the Oseen types of operators which may be written in the general form:

$$\left(A_1\frac{\partial}{\partial x^*} + B_1\frac{\partial}{\partial y^*} - \frac{1}{R_e}\nabla^{*2}\right)u^* = 0 \quad (4.172)$$

The solution of Eq. (4.172) is

$$u = \text{const}\cdot\exp\left[\frac{(A_1x^* + B_1y^*)R_e}{2}\right]K_0\left(\frac{A_1^2 + B_1^2}{2}R_e r^*\right) \quad (4.173)$$

where $r^{*2} = x^{*2} + y^{*2}$ and K_0 is the Bessel function of 0th order. For a given radial distance from the origin, the maximum value of u^* occurs at

$$\frac{y^*}{x^*} = \frac{B_1}{A_1} \quad (4.174)$$

Thus we have a wake along the direction

$$\theta = \theta_m = \tan^{-1}\left(\frac{B_1}{A_1}\right) \quad (4.175)$$

Since there are two Oseen operators in Eq. (4.171), we have two wakes which are respectively along the lines (Fig. 4.21):

$$\theta_{m1} = \tan^{-1}\left(\frac{R_{H_y}}{1 + R_{H_x}}\right) \qquad \theta_{m2} = \tan^{-1}\left(\frac{-R_{H_y}}{1 - R_{H_x}}\right) \quad (4.176)$$

If $R_{H_x} > 1$ we may have upstream inclined wake.

From the solution, Eq. (4.173), we see that the rate of decrease of u^* with r^* increases with the Reynolds number, and we have boundary-layer phenomena for the large value of R_e. If $R_{H_x} > 1$, we can show with the boundary-layer equations that there is an upstream influence of the boundary-layer flow. The case of $R_{H_x} > 1$ corresponds to $M_m < 1$ which is sub-Alfvén flow, i.e.,

$$U < V_H = \left(\frac{\mu_e^*}{\rho^*}\right)^{1/2} H_0^* \tag{4.177}$$

and disturbance can propagate upstream. Conversely, $R_{H_x} < 1$ corresponds to $M_m > 1$ and super-Alfvén flow and disturbances won't propagate upstream. The startling new result and upstream wakes have been observed experimentally by Ahlstrom[14] for a semi-infinite Rankine body (axisymmetric body defined by a source in a uniform stream), and is produced by disturbances propagating upstream in the fluid and magnetic field by Alfvén waves, described briefly in Chap. 1 and discussed thoroughly in Chap. 5.

4.10 MAGNETOFLUID-DYNAMIC FREE-CONVECTION FLOW

When forced flow through a parallel plate heat exchanger ceases, a free-convection flow pattern is established since the plates or the fluid remain at an elevated temperature. If the fluid is a conductor of electricity, this flow can be influenced by an imposed magnetic field. Examination of simple representation flow models will reveal the influence of magnetic fields on the velocity profile, temperature profile, wall shear, and the local heat-transfer rate. The strength of the required magnetic field can also be determined.

This section contains an analysis of laminar free-convection flow over a vertical flat plate with an imposed transverse, uniformly distributed, constant magnetic field with linearly increasing surface temperature.[15] The analytical results are evaluated, for sample, liquid metals and the required field strengths determined.

The basic equations, Eqs. (4.178)–(4.180), describe steady, two-dimensional, laminar, free-convection, boundary-layer–type flow of an incompressible, conducting fluid through a constant uniformly distributed, transverse magnetic field of strength B_0. The direction of

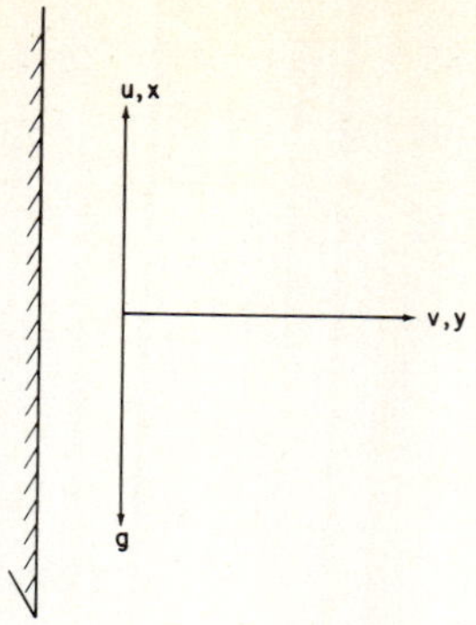

Fig. 4.22 Free-convection flow model.

the various velocities, forces, and coordinates is shown in Fig. 4.22:

$$\frac{\partial u}{\partial x} + \frac{\partial v}{\partial y} = 0 \tag{4.178}$$

$$u\frac{\partial u}{\partial x} + v\frac{\partial u}{\partial y} = g\beta(T - T_\infty) - \frac{\sigma_0 B_0{}^2}{\rho}u + \nu\frac{\partial^2 u}{\partial y^2} \tag{4.179}$$

$$u\frac{\partial T}{\partial x} + v\frac{\partial T}{\partial y} = \alpha\frac{\partial^2 T}{\partial y^2}, \quad \alpha = \frac{k}{\rho c_p} \tag{4.180}$$

These equations contain the additional assumptions that the density is a variable only in the buoyancy term and the work against gravity is neglected along with the viscous and Joule heating.

Equations (4.179) and (4.180) are reduced to ordinary differential equations by the similarity transformation technique which requires the introduction of the stream function ψ through the relations given in Eq. (4.181). These yield the expressions to be transformed, Eqs. (4.182) and (4.183):

$$u = \frac{\partial \psi}{\partial y}, \quad v = -\frac{\partial \psi}{\partial x} \tag{4.181}$$

$$\frac{\partial \psi}{\partial y}\frac{\partial^2 \psi}{\partial y \partial x} - \frac{\partial \psi}{\partial x}\frac{\partial^2 \psi}{\partial y^2} = g\beta(T - T_\infty) - \frac{\sigma_0 B_0{}^2}{\rho}\frac{\partial \psi}{\partial y} + \nu\frac{\partial^3 \psi}{\partial y^3} \tag{4.182}$$

$$\frac{\partial \psi}{\partial y}\frac{\partial T}{\partial x} - \frac{\partial \psi}{\partial x}\frac{\partial T}{\partial y} = \alpha\frac{\partial^2 T}{\partial y^2} \tag{4.183}$$

With a new set of independent and dependent variables, defined by Eq. (4.184), the partial differential equations transform into ordinary differential equations, Eqs. (4.185) and (4.186), only for a linearly increasing surface temperature of the form, Eq. (4.187):

$$\psi = (g\beta N)^{1/4}\nu^{1/2}xF(\eta) \qquad \eta = (g\beta N)^{1/4}\nu^{-1/2}y$$
$$\theta(\eta) = \frac{T - T_\infty}{T_w - T_\infty} \qquad A^2 = \frac{\sigma_0 B_0{}^2}{\rho(g\beta N)^{1/2}} \tag{4.184}$$

$$F''' + FF'' - (A^2 + F')F' + \theta = 0 \tag{4.185}$$

$$\theta'' + \mathrm{P_r}(F\theta' - F'\theta) = 0 \tag{4.186}$$

$$T_w - T_\infty = Nx \tag{4.187}$$

The appropriate flat plate, free-convection boundary conditions, Eq. (4.188), are also transformed into the applicable form, Eq. (4.189):

$$\begin{aligned} &y = 0 \quad u = v = 0 \quad T = T_w \\ &y \to \infty \quad u = 0 \quad T = T_\infty = \text{const} \end{aligned} \tag{4.188}$$

$$\begin{aligned} &\eta = 0 \quad F(0) = F'(0) = 0 \quad \theta(0) = 1 \\ &\eta \to \infty \quad F'(\infty) = 0, \; \theta(\infty) = 0 \end{aligned} \tag{4.189}$$

The resulting differential equations contain two arbitrary parameters, the Prandtl number $\mathrm{P_r}$ and the influence parameter A, which are composed of the various fluid properties, the magnetic field strength, and the buoyant force. The influence parameter A, the ratio of the Hartmann number to the fourth root of the Grashof number, is a measure of the relative influence of the magnetic and buoyant forces on the temperature and flow fields.

Solution of the resulting semi-infinite domain, nonlinear equations is accomplished with a three-part series method.

The employed power series, Eq. (4.190), contains a term K that satisfies the boundary conditions and differential equations at infinity, a second term that satisfies the boundary conditions at zero and is the solution to the initial homogeneous differential equation, and additional terms that are utilized to obtain increased numerical accuracy. This accuracy is limited by the number of terms that will not initiate divergence of the numerical results:

$$F = K + pF_1 + p^2F_2 + p^3F_3 + \cdots \tag{4.190}$$

$$\theta = p\theta_1 + p^2\theta_2 + p^3\theta_3 + \cdots \tag{4.191}$$

$$\begin{aligned} &\eta = 0 \quad F_n'(0) = F(0) = 0 \quad \theta_1(0) = \frac{1}{p} \quad \theta_j(0) = 0 \quad j \geq 2 \\ &\eta \to \infty \quad F_n'(\infty) = F_n(\infty) = 0 \quad F(\infty) = K \quad \theta_n(\infty) = 0 \end{aligned} \tag{4.192}$$

Equation (4.191), the temperature representation, along with Eq. (4.190) and the associated boundary conditions, Eq. (4.192), contain an undetermined parameter p which aids in the collection of terms for each set of the resulting linear differential equations. In some problems, it will have a physical meaning which results in a power series in that parameter. However, in the present case p equals unity.

Substitution of the series representation into the differential equations and collection of terms by like powers of p result in a family of linear differential equations, and the first three sets are:

$$\begin{aligned} &F_1''' + KF_1'' - A^2F_1' + \theta_1 = 0 \\ &\theta_1'' + K\,\mathrm{P_r}\theta_1' = 0 \end{aligned} \tag{4.193}$$

$$\begin{aligned} &F_2''' + KF_2'' - A^2F_2' + \theta_2 = (F_1')^2 - F_1F_1'' \\ &\theta_2'' + K\,\mathrm{P_r}\,\theta_2' = \mathrm{P_r}(F_1'\theta_1 - F_1\theta_1') \end{aligned} \tag{4.194}$$

$$\begin{aligned} &F_3''' + KF_3'' - A^2F_3' + \theta_3 = 2F_2'F_1' - F_2F_1'' - F_1F_2'' \\ &\theta_3'' + K\,\mathrm{P_r}\theta_3' = \mathrm{P_r}(F_1'\theta_2 + F_2'\theta_1 - F_2\theta_1' - F_1\theta_2') \end{aligned} \tag{4.195}$$

The solutions to the first two sets, Eqs. (4.196)-(4.199), when substituted into Eqs. (4.190) and (4.191), provide the required representations for F and θ. The constant K is determined by satisfying the boundary condition $F(0)$ and is a function of Pr and A:

$$F_1 = c_1[m(\mathrm{P_r}K)^{-1}e^{-\mathrm{P_r}K\eta} - e^{-m\eta}] \tag{4.196}$$

$$\theta_1 = e^{-\mathrm{P_r}K\eta} \tag{4.197}$$

$$F_2 = c_2e^{-m\eta} - c_3e^{-\mathrm{P_r}K\eta} - c_4e^{-(m+\mathrm{P_r}K)\eta} \tag{4.198}$$

$$\theta_2 = c_5[e^{-(m+\mathrm{P_r}K)\eta} - e^{-\mathrm{P_r}K\eta}] \tag{4.199}$$

The various constants are:

$$m = \frac{K + \sqrt{K^2 + 4A^2}}{2} \qquad c_1 = \frac{1}{m(\mathrm{P_r}^2K^2 - \mathrm{P_r}K^2 - A^2)},$$

$$c_2 = c_1c_5 - \frac{1}{m}\left[\frac{c_5 - mc_1^2(m^2/\mathrm{P_r}K - 2m + \mathrm{P_r}K)}{(m + \mathrm{P_r}K)^2 - (m + \mathrm{P_r}K)K - A^2}\right] \qquad c_3 = \frac{mc_1c_5}{\mathrm{P_r}K},$$

$$c_4 = \frac{m}{m + \mathrm{P_r}K}(c_2 - c_1c_5) \qquad c_5 = \frac{\mathrm{P_r}c_1(m - \mathrm{P_r}K)}{m(m + \mathrm{P_r}K)} \tag{4.200}$$

The series for θ, its first derivative $\theta'_{(0)}$ (the wall temperature gradient), F' (the velocity profile), and $F''(0)$ (the wall velocity gradient) were evaluated for a range of the influence parameter and two values of the Prandtl number and are presented in Figs. 4.23–4.26. The chosen Prandtl numbers (0.01 and 0.005) are generally representative of liquid metals and provide a measure of the Prandtl number effect on the results. Increasing the parameter A produces a flattening of the velocity and temperature profiles which also corresponds to a reduction in the local wall shear and heat transfer and increased temperatures in the fluid. This qualitatively agrees with expectations, since the magnetic field exerts a retarding

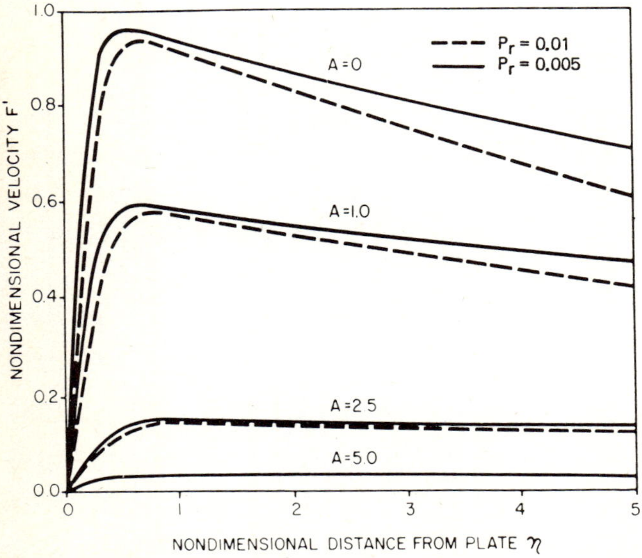

Fig. 4.23 Flat plate velocity profiles.

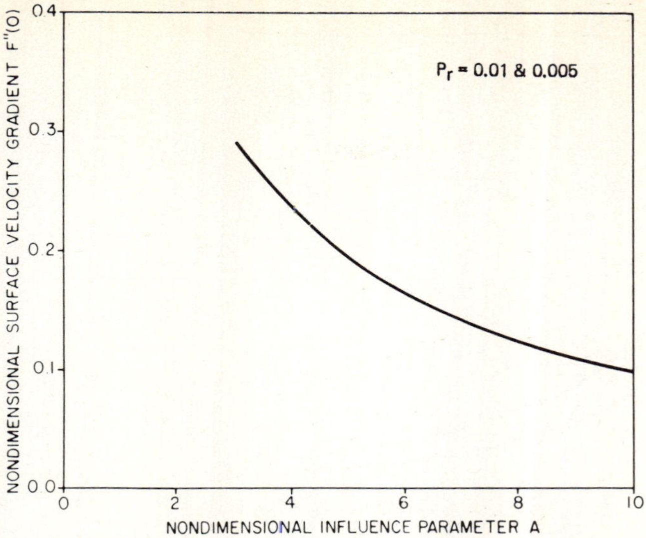

Fig. 4.24 Flat plate surface velocity gradient.

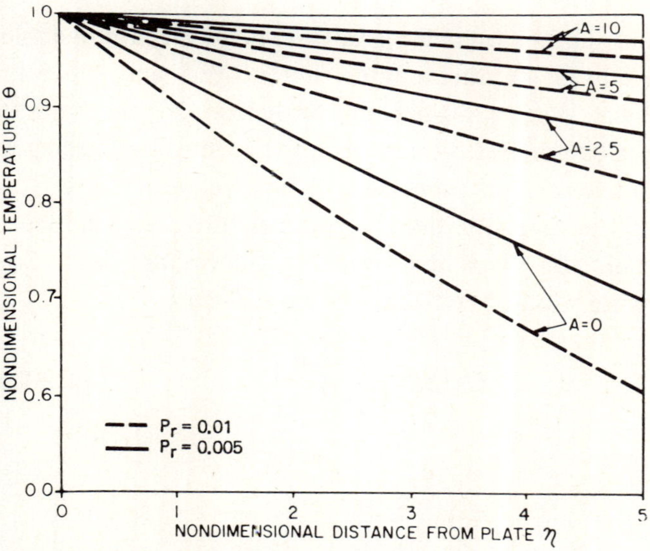

Fig. 4.25 Flat plate temperature profiles.

force on the free-convection flow. This force is limited to only retardation; thus, separation will never occur and the heat transfer is never terminated, as is verified by the asymptotic trend exhibited in Figs. 4.24 and 4.26 for large A values. Variation of the Prandtl number by a factor of 2 produces only minor differences in the

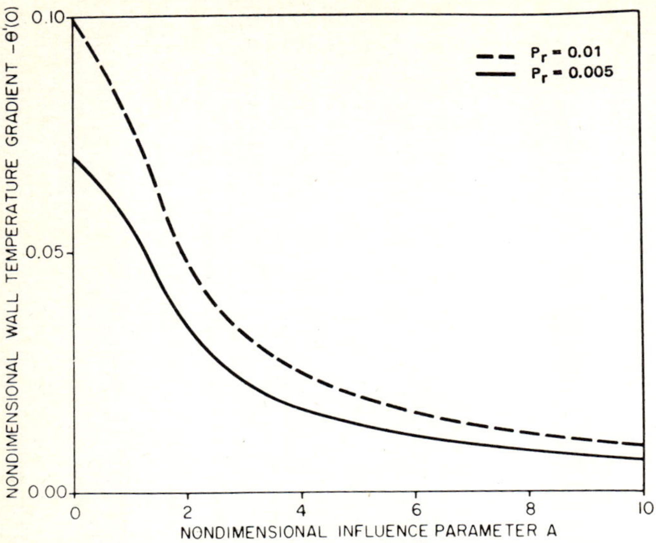

Fig. 4.26 Flat plate surface temperature gradient.

velocity profile and surface velocity gradient; however, appreciable differences are produced in the temperature profile, and the wall temperature gradient; e.g., a 50-percent reduction in the Prandtl number results in a 30-percent reduction in the surface heat transfer and an appreciable temperature increase at large η values.

An estimate of the magnetic-field strength required to produce a sizable influence on the flow field can be determined with Eq. (4.201) and the assumption of constant fluid properties

$$B_0 = A(g\beta N)^{1/4}\left(\frac{\rho}{\sigma_0}\right)^{1/2} \tag{4.201}$$

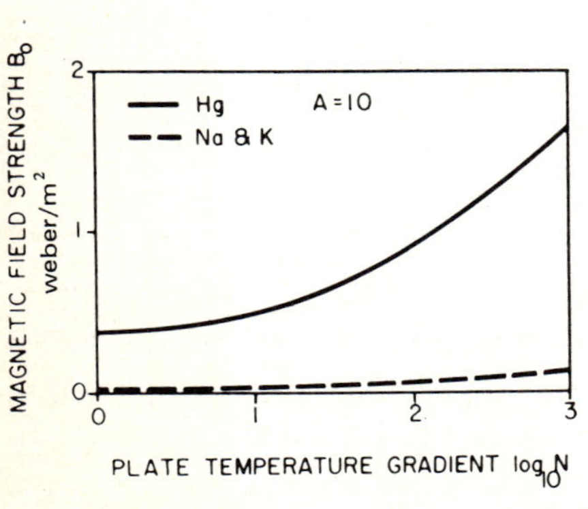

Fig. 4.27 Estimated required magnetic field strength.

β is the volumetric expansion coefficient.

Figure 4.27 demonstrates that considerable influence can be exerted on a liquid metal, free-convection flow field by magnetic fields of practical strength, i.e., under 2 weber/m^2.

For engineering applications, the local heat-transfer rate q and wall shear τ_w can be determined for a single surface with Eqs. (4.202)–(4.203):

$$q = -k\nu^{-1/2}(g\beta N)^{1/4}(T_w - T_\infty)\theta'(0) \tag{4.202}$$

$$\tau_w = -\rho\nu^{1/2}(g\beta N)^{3/4}xF''(0) \tag{4.203}$$

The overall heat-transfer rate Q and drag D can be determined with Eqs. (4.204) and (4.205), which were obtained by integrating Eqs. (4.202) and (4.203) over a plate surface of width b and length L

$$T_w - T_\infty = Nx \qquad Q_x = \frac{Q_L}{2} \tag{4.204}$$

$$D = \frac{\tau_w\, bL^2}{2x} \tag{4.205}$$

Solutions have been obtained for a variety of MFD free-convection problems and wall conditions. They include the flow over impervious and porous, semi-infinite flat plates, flow between infinite and finite, open-ended and closed, parallel plates with and without heat sources, and flow through an open-ended pipe. These flows were subjected to transverse magnetic fields and wall temperatures that were constant or varied as a fractional power of the distance in the flow direction. The general results of these investigations are that the imposed magnetic field diminished the velocity field, wall shear, flow rate, and wall heat transfer; also the onset of free convection was retarded while the fluid temperatures and the time required for the flow to reach steady state are increased. In addition, sizable influences on the flow and thermal fields can be produced with moderate magnetic field strengths only for liquid metal flows while the effects of induced magnetic fields and Joule heating are very small.

4.11 RELAXED ASSUMPTIONS

The solutions presented in this chapter contain simplifying assumptions that limit the extent of their application. The parallel plate problems were solved with the assumption that the plates were infinite, the magnetic fields were transverse to the wall and flow direction, the plates were electrically insulated, the fluid's electrical conductivity was constant, and free convection was neglected in the heat-transfer studies. The purpose of this section is to present the results of solutions that were obtained with these assumptions removed to provide an estimate of their importance for particular applications.

4.11.1 Channel Aspect Ratio. Shercliff[16] determined the influence of channel side walls on the Hartmann velocity profile for various channel aspect ratios, i.e., width/height. The results are presented in Fig. 4.28 for the nondimensional pressure gradient producing the flow. The figure compares Hartmann's results with Shercliff's calculation for a square channel and the experiments of Hartmann and Lazarus[17] with a square channel and a round pipe. The effect of side walls is to increase the required pressure gradient to maintain the flow for a given Hartmann number. The square channel represents the maximum influence and thus, the maximum error that will be

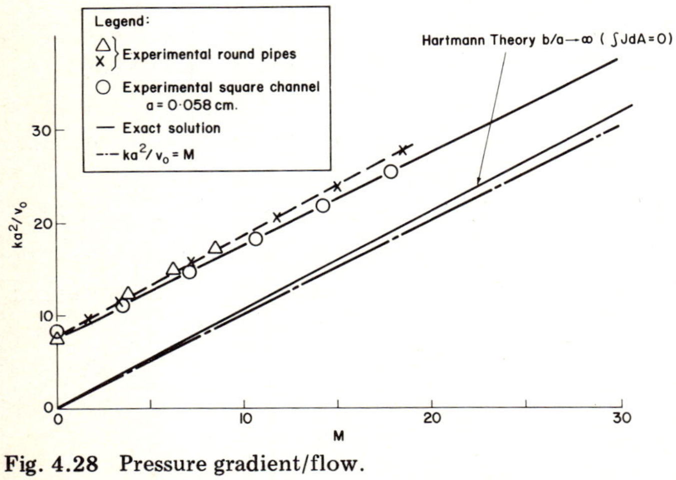

Fig. 4.28 Pressure gradient/flow.

TABLE 4.1 Entrance Length

H_a	$\ell H_a/aR_e$			
4	0.077	0.078	0.067	0.080
6	—	—	—	0.050
10	0.073	0.030	—	0.031
50	0.014	—	—	—
Ref.	18	18	19	20
Starting profile	Parabolic	Uniform	Uniform	Uniform
Method	Finite difference	Finite difference	Approx.	Exact

introduced by applying Hartmann's solution to a channel with side walls.

4.11.2 Entrance Region. The flow entering the magnetic field portion of the passage requires a distance to adjust to the fully developed flow that was examined earlier in this chapter. This entrance length; i.e., 99 percent of the developed distance, has been calculated by several authors for the flow between infinite plates for both parabolic and uniform entrance velocity profiles and subjected to a constant magnetic field. Table 4.1 contains their results for a range of Hartmann numbers. However, in an actual device, the flow will enter a fringing magnetic field and Table 4.2 contains Khozhainov's experimental measurements of the entrance length for a square channel.[21] His entrance lengths were measured from the front edge of the magnetic pole piece. Thus, when applying the fully developed results, an additional passage length is required in which the pressure

TABLE 4.2 Experimental Entrance Lengths

H_a	$\ell H_a/aR_e$	
5	0.050	0.026
10	0.042	0.029
14	0.040	0.029
19	0.036	—
R_e	2100	4300

drop continually changes until the flow is fully developed and then drops linearly, according to Hartmann's theory.

4.11.3 Magnetic Field Orientation. Eraslan[22] has provided a solution for insulated channel flows with different aspect ratios and with the magnetic field at different angles to the short side of the channel in the plane perpendicular to the flow direction. The velocity profiles and current density loops rotate with the magnetic field and the centerline value of the velocity increases to a peak when the magnetic field is parallel to the long side. The surface shear on the long sides also increases to its peak value when the magnetic field is parallel to the long side. Considerable differences in the surface shear occur for large Hartmann numbers and for different aspect ratios; e.g., for $H_a = 10$, 25 percent for a square duct, and 200 percent for

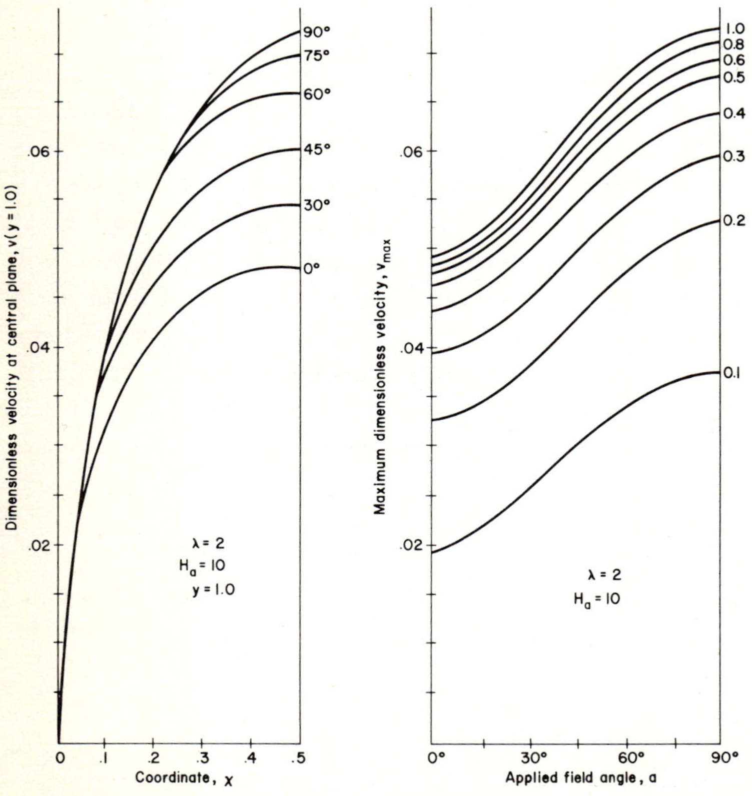

Fig. 4.29 Velocity variations in the duct.

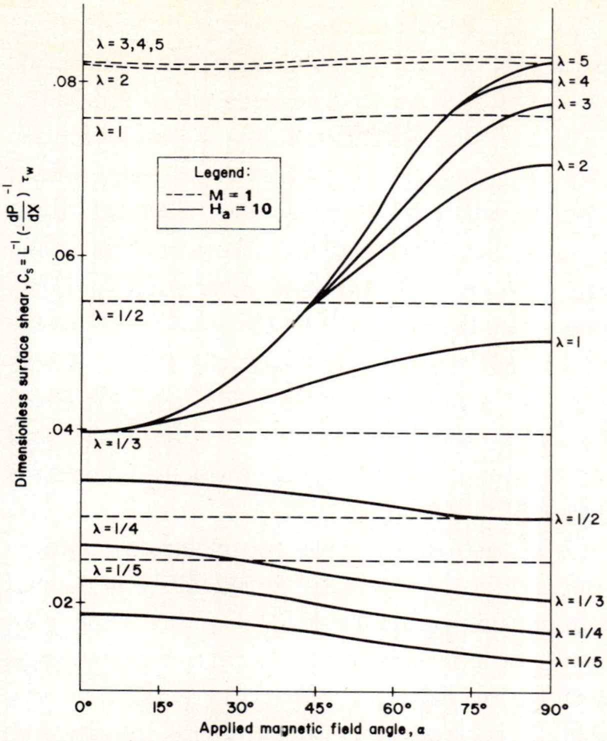

Fig. 4.30 Wall shear variation versus applied magnetic field orientation.

an aspect ratio of 5. Only small, additional increases occur for aspect ratios greater than 5 and thus, parallel plate flow is approximated by $\lambda \geq 5$. Eraslan's results for the surface shear and the velocity profiles for a channel with an aspect ratio of 2 are shown in Figs. 4.29 and 4.30.

4.11.4 Conducting Walls. The Hartmann flow problem with walls of different conductances was solved by Chang and Yen[23] for zero total current in the passage. The velocity profiles and current distributions remain symmetrical for any combination of the two wall conductances and the effect of increasing their sum on the velocity profiles is similar to increasing the Hartmann number. Since the total current flowing in the passage is required to be zero, increased wall conductances permit an increased current to return through the walls, resulting in a greater current flow in the fluid in the $v \times B$ direction and a larger $J \times B$ interaction. The voltage gradient E is then controlled by the resistances of the conducting walls. The surface shear and flow rates are decreased compared to the insulated wall case for an increased sum of the wall conductances. These results are shown in Figs. 4.31 and 4.32. Yen[24] later solved the corresponding constant wall heat flux problem with the result that increased wall conductances increases the Joule heating in the fluid and the temperature field is only influenced by the sum of the two wall conductances. Snyder[25] generalized Yen's results for arbitrary external circuit loading and found that the effect of wall conductance is opposite for generators and accelerators. Increased wall conductance increases the difference between the wall temperature and fluid bulk temperature in generators and decreases the difference in accelerators.

4.11.5 Variable Fluid Electrical Conductivity. Since the electrical conductivity is temperature dependent, it may vary considerably in devices that produce large temperature gradients. MFD Couette flow with the conductivity represented as a simple polynomial profile across the passage was examined in Ref. 26 with the result that any reduction in the conductivity from the uniform case produces a smaller MFD interaction. Calculations based on a constant conductivity, can greatly underestimate the fixed surface shear,

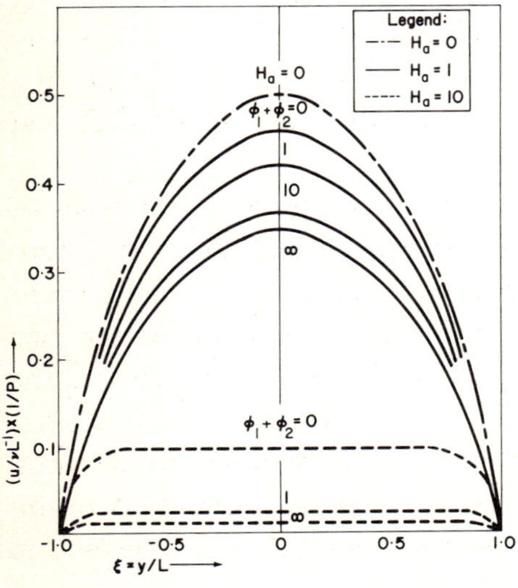

Fig. 4.31 Velocity profiles as influenced by wall conditions and Hartmann number.

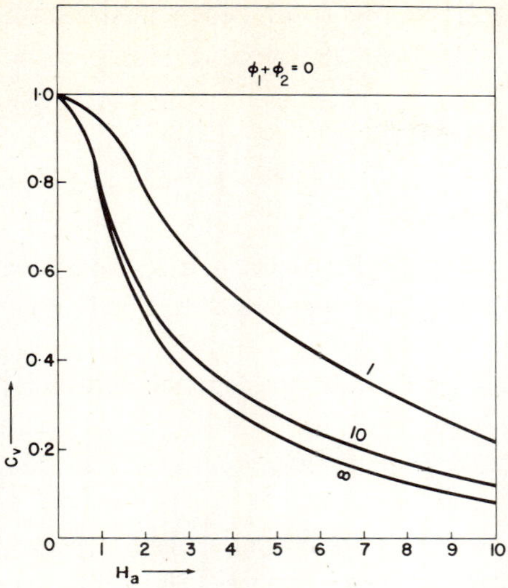

Fig. 4.32 Coefficient of viscous drag C_V versus Hartmann number H_a for different wall conditions.

underestimate the flow rate, and overestimate the moving surface shear. For example, when $\sigma/\sigma_0 = 1/4 + 3/4\,\eta^2$, the error in the fixed surface shear introduced by assuming a constant conductivity is 32 percent for $H_a = 4$. A highly cooled surface where $\sigma/\sigma_0 = \eta^2$ would have a fixed surface shear 245 percent larger than that computed with a uniform conductivity. The corresponding increases in flow rate are 6 percent and 35 percent, respectively. Similar results were obtained by Kieffer[27] for Hartmann flow. Thus, experimental results and theoretical predictions must include a detailed specification of the conductivity when sizable temperature gradients exist, even at moderate Hartmann numbers. The electrical conductivity is also a function of pressure and magnetic field strength, and at low pressures and high magnetic field strength, it acquires a tensor character. These effects are discussed in the last chapter.

4.11.6 Free Convection. The forced convection heat-transfer studies in this chapter neglected the free convection due to gravity. However, Singer[28] analyzed combined forced and free-convection MFD flow in a vertical channel. Since a magnetic field strongly inhibits free convection, then if

$$H_a > 5 + \frac{R_a}{40} \tag{4.206}$$

the contribution of free convection to the total heat-transfer rate between the fluid and wall is less than 10 percent and can be neglected.

REFERENCES

1. S. I. Pai, A Critical Survey of Magnetofluid Dynamics, part II, *Aerospace Res. Lab. Rept.* ARL 63-175, Office of Aerospace Research, USAF, May, 1964.
2. C. C. Chang and T. S. Lundgren, Duct Flow in Magnetohydrodynamics, *ZAMP*, vol. 12, no. 2, pp. 100–114, 1961.
3. S. I. Pai, Plane Couette Flow in Radiation Magnetogasdynamics, *Proc. 6th Intern. Conf. Ionization Phenomena Gas*, Paris, 1963.
4. A. Sherman, Exact Solutions for Magnetohydrodynamics Channel Flows, *Gen. Elec. Tech. Inform. Series* R62SD991, Space Sciences Lab., 1962.
5. M. Perlmutter and R. Siegel, Heat Transfer to an Electrically Conducting Fluid Flowing in a Channel with a Transverse Magnetic Field, *NASA* TN D-875, August, 1961.
6. J. Carstoiu, Parallel Unsteady Flow of a Conducting Fluid, *Sylvania Elect. Prod. Eng. Note* 78, Applied Research Lab., September, 1959.
7. R. Gold, Magnetohydrodynamic Pipe Flow, part I, *J. Fluid Mech.*, vol. 13, p. 505, 1962.
8. W. B. Bush, Magnetohydrodynamic-Hypersonic Flow Past a Blunt Body, *J. Aerospace Sci.*, vol. 25, no. 11, pp. 685–690, November, 1958.
9. W. B. Bush, The Stagnation-Point Boundary Layer in the Presence of an Applied Magnetic Field, *J. Aerospace Sci.*, vol. 28, no. 8, pp. 610–611, August, 1961.
10. W. B. Bush, A Note on Magnetohydrodynamic-Hypersonic Flow Past a Blunt Body, *J. Aerospace Sci.*, vol. 26, no. 8, pp. 536–537, August, 1959.
11. V. J. Rossow, On Flow of Electrically Conducting Fluids Over a Flat Plate in the Presence of a Transverse Magnetic Field, *NACA* Rept. 1358, 1958.
12. R. D. Hugelman, An MHD Boundary-Layer Compatibility Condition, *AIAA J.*, vol. 3, no. 7, pp. 1367–1368, July, 1965.
13. H. P. Greenspan and G. F. Carrier, The Magnetohydrodynamic Flow Past a Flat Plate, *J. Fluid Mech.*, vol. 6, p. 77, 1959.
14. H. G. Ahlstrom, Experiments on the Upstream Wake in Magnetofluid Dynamics, part II, *J. Fluid Mech.*, vol. 15, pp. 205–222, 1963.
15. K. R. Cramer, Several Magnetohydrodynamic Free Convection Solutions, ASME, Series C, *J. Heat Trans.*, pp. 35–40, February, 1963.

16. J. A. Shercliff, Steady Motion of Conducting Fluids in Pipes Under Transverse Magnetic Fields, *Proc. Cambridge Phil. Soc.* 49, pp. 126–144, 1953.

17. J. Hartmann and F. Lazarus, Hg-Dynamics, part II—Experimental Investigations on the Flow of Mercury in a Homogeneous Magnetic Field, *Kgl. Danske Videnskab. Selskab, Mat.-Fys. Medd.*, vol. 15, no. 7, 1937.

18. C. L. Hwang, K. C. Li, and L. T. Fan, Magnetohydrodynamic Channel Entrance Flow With Parabolic Velocity at the Entry, *Phys. Fluids*, vol. 9, no. 6, June, 1966.

19. M. Roidt and R. D. Cess, An Approximate Analysis of Laminar Magnetohydrodynamic Flow in the Entrance Region of a Flat Duct, *J. Appl. Mech.* Trans. ASME 84, pp. 171–176, 1962.

20. R. Manohar, An Exact Analysis of Laminar Magnetohydrodynamic Flow in the Entrance Region of a Flat Duct, *ZAMM*, vol. 46, no. 2, pp. 111–117, 1966.

21. A. I. Khozhainov, Experimental Study of the Initial Region of a Square Magnetohydrodynamic Channel Taking Account of the Inhomogeneity of the Magnetic Field in the Longitudinal Direction, *Sov. Phys.-Tech. Phys.*, vol. 10, no. 9, March, 1966.

22. A. H. Eraslan, Duct-Flow of Conducting Fluids Under Arbitrarily Oriented Applied Magnetic Fields, *AIAA J.*, vol. 4, no. 4, April, 1966.

23. C. C. Chang and J. T. Yen, Magneto-Hydrodynamic Channel Flow as Influenced by Wall Conductance, *ZAMP*, vol. 13, pp. 266–272, 1962.

24. J. T. Yen, Effect of Wall Electrical Conductance on Magnetohydrodynamic Heat Transfer in a Channel, ASME, Series C, *J. Heat Trans.*, vol. 85, pp. 371–377, 1963.

25. W. T. Snyder, The Influence of Wall Conductance on Magnetohydrodynamic Channel-Flow Heat Transfer, ASME, Series C, *J. Heat Trans.*, vol. 86, pp. 552–558, 1964.

26. K. R. Cramer, Variable Conductivity MFD Couette Flow, ARL 65-13, Aerospace Research Labs, Office of Aerospace Research, USAF, January, 1965.

27. P. C. Kieffer, Variable Conductivity Fully Developed MHD Channel Flow, *Aerospace Res. Lab. Rept.*, ARL 67-46, Office of Aerospace Research, USAF, February, 1967.

28. R. M. Singer, A Study of Convective Magnetohydrodynamic Channel Flow, *Argonne Natl. Lab. Rept.* ANL-6967, March, 1965.

29. E. Pohlhausen, Der Wärmeaustausch zwischen festen Körpern und Flüssigkeiten mit kleiner Reibung und kleiner Wärmeleitung, *ZAMM*, vol. 1, pp. 121, 1921.

Chapter 5

MAGNETOFLUID-DYNAMIC WAVE PHENOMENA

5.1 INTRODUCTION

The study of wave propagation in an ionized gas or liquid has both academic interest and practical applications. The wave motion may differentiate magnetofluid dynamics from ordinary gas dynamics. The practical applications of wave propagation are numerous. Some of the important applications are

1. radio wave propagation in ionosphere
2. space communication systems
3. magnetofluid-dynamic power generation
4. traveling-wave tubes
5. many geophysical problems such as geomagnetic storms, auroras, and other ionospheric disturbances
6. many astrophysical problems such as solar flares, generation of cosmic rf radiation, etc.
7. diagnostic and confinement schemes in nuclear fusion devices and
8. other problems of plasma dynamics associated with wave phenomena such as plasma jets.

Because of the many applications, wave motion in a plasma has been extensively investigated. In this chapter, we are going to discuss the essential points of wave motion in electrically conducting fluids with special emphasis on the new phenomena that differ from those wave motions in neutral gases.

The properties of the wave in an electrically conducting fluid depend on the amplitude of the wave. The simplest type of wave is the wave of infinitesimal amplitude. Ordinary sound waves and radio waves belong to such groups. Both of these waves may be considered as a special case of magnetogas-dynamic waves. In fact, the magnetogas-dynamic wave is the resultant wave of the sound wave and electromagnetic waves by the interaction through an externally applied magnetic field. Such an interaction will give us many new phenomena that are not found in ordinary gas dynamics nor in ordinary electrodynamics. One of the main features of these waves of infinitesimal amplitude is that the superposition principle is applicable to these waves. Mathematically speaking, we may linearize the equations that govern these waves. Since these equations are linear, the sum of two solutions of the fundamental linearized equations is a solution of these equations. Hence we may study any typical solution of these wave equations which will give the general features of the wave propagation.

For waves of finite amplitude, the shape of the wave will distort as the wave propagates, while the wave of infinitesimal amplitude will maintain its shape when it propagates. We shall discuss some of the distortions of finite wave in Sec. 5.7. When the distortion is large, the ordinary wave will develop into a shock wave in which a large change of physical variables occurs in a very thin region. For a first approximation, we may consider that the shock wave is a surface of discontinuity. The speed of propagation of a shock wave is faster than the corresponding wave of small amplitude. We shall discuss the properties of shock waves in Sec. 5.8.

Inside the shock wave, we have to consider the diffusion phenomena due to viscosity, heat conduction, and other transport phenomena. Outside the shock structure, these transport phenomena usually have little effect on the wave motion. If we want to find the exact influence of these transport phenomena on an ordinary wave, we may add the corresponding linearized terms due to these transport terms into the fundamental equations. In general, the transport coefficients will introduce a damping to the wave and change slightly the speed of propagation. Sometimes, these transport properties will introduce their own wave which will not occur when they are zero. We shall mention a few in Secs. 5.2–5.5. One of the new waves of magnetofluid dynamics is known as Alfvén's wave. The interaction of Alfvén's wave with the ordinary sound wave gives many new phenomena which will be discussed in Sec. 5.6.

The wave phenomena depend on the various forces in the plasma. If we want to study the flow phenomena in a plasma in detail, we may find some new wave phenomena which cannot be obtained by the macroscopic treatment of a single fluid theory. We shall mention some of these waves in Sec. 5.9.

5.2 WAVES OF SMALL AMPLITUDE IN MAGNETOGAS[1,2,3] DYNAMICS

We assume that originally the electrically conducting fluid is at rest with a pressure p_0, temperature T_0, and density ρ_0 and that it is subjected to an externally applied uniform magnetic field $H_0 = \mathbf{i}\, H_x + \mathbf{j}\, H_y + \mathbf{k}.0$ where $\mathbf{i}$, $\mathbf{j}$, and $\mathbf{k}$ are respectively the x-, y-, and z-wise unit vectors; H_x and H_y are constant. There is no electric current, nor excess electric charge, nor applied electric field in the flow field. The conducting fluid is perturbed by a small disturbance so that in the resultant disturbed motion, we have

$$
\begin{aligned}
u &= u(x,t);\; v = v(x,t);\; w = w(x,t);\; p = p_0 + p'(x,t);\\
&\quad T = T_0 + T'(x,t);\\
\rho &= \rho_0 + \rho'(x,t),\; \mathbf{E} = \mathbf{E}(x,t),\; \mathbf{J} = \mathbf{J}(X,t),\; \rho_e = \rho_e(x,t);\\
&\quad \mathbf{H} = \mathbf{H}_0 + \mathbf{h}(x,t)
\end{aligned}
\tag{5.1}
$$

where u, v, and w are respectively the perturbed x, y, and z components of velocity; prime refers to perturbed quantities; $\mathbf{E}$, $\mathbf{J}$, ρ_e, $\mathbf{h}$ are the perturbed electromagnetic quantities with their usual meanings; the vector quantities have three components each (and we shall use the subscripts x, y, and z to denote these components), e.g., h_x is the x component of the perturbed magnetic field, etc. For simplicity, we assume that all the perturbed quantities are functions of one spatial coordinate x and time t only. Thus we discuss the wave propagation in the direction of x axis. Substituting Eq. (5.1) into the fundamental equations of magnetofluid dynamics of Chap. 2 and neglecting the higher order terms, we have 16 linear partial differential equations for the 16 perturbed magnetofluid-dynamic variables (Prob. 5.1). When we examine these linearized equations, we find that the variables may be divided into three independent groups:

Group 1. The x-wise component of the magnetic field h_x is independent of all the other variables. From the linearized equation of h_x and the condition that the divergence of h is zero, we conclude that h_x is a constant. Without loss of generality, we may put $h_x = 0$. In other words, we may absorb the value of h_x in the definition of H_x. We do not need to consider h_x any more.

Group 2. The second group consists of the perturbed quantities: w, h_z, J_x, J_y, E_x, E_y, and ρ_e. This group is called *transverse waves*, which deal with the components of velocity and magnetic field perpendicular to the applied magnetic field.

Group 3. The third group consists of the perturbed quantities: u, v, p', T', ρ', h_y, J_z, and E_z. The group is called *longitudinal waves* in which the ordinary sound wave is a special case.

One of the most important type of waves is known as the *monochromatic wave* which is a wave of a given frequency ω. All the waves of small amplitude may be considered as a series of such monochromatic waves. If we know the properties of such monochromatic waves, we may deduce the properties of a general wave. For a monochromatic wave of frequency ω, traveling in the direction of x, all the perturbed quantities are proportional to

$$\exp[i(\omega t - \lambda x)] = \exp[-i(\lambda_R x - Vt)] \exp(\lambda_i x) \tag{5.2}$$

where ω is a given real angular frequency, $\lambda = \lambda_R + i\lambda_i$ is the complex wave number, $i = \sqrt{-1}$ and

$$V = \frac{\omega}{\lambda_R} = \text{speed of wave propagation} \tag{5.3}$$

If we substitute the expressions of Eq. (5.2) for all the perturbed quantities, into the governing linearized equations, we obtain the dispersion relation $\lambda(\omega)$ for both the transverse and the longitudinal waves. From these dispersion relations, we know the properties of these waves such as their speed of propagation, whether they are damped or not, and other properties.

Before we discuss the magnetogas-dynamic waves, we consider first the electromagnetic waves and the gas-dynamic wave. The interaction of these basic waves gives the magnetogas-dynamic waves. If there is no externally applied magnetic field, $\mathbf{H}_0 = 0$, the electromagnetic variables are decoupled from the gas-dynamic variables and we have the corresponding electromagnetic waves from the equations of electromagnetic variables and the gas-dynamic waves from the equation of gas-dynamic variables.

5.3 ELECTROMAGNETIC WAVES

If there is no externally applied magnetic field $\mathbf{H}_0 = 0$, the electromagnetic variables are not affected by the gas-dynamic variables. For the transverse waves of Group (2), the electromagnetic variables are governed by the following equations:

$$J_x = \sigma E_x \qquad J_x + \frac{\partial \epsilon E_x}{\partial t} = 0 \qquad \frac{\partial \rho_e}{\partial t} + \frac{\partial J_x}{\partial x} = 0 \tag{5.4}$$

and

$$J_y = \sigma E_y \qquad \frac{\partial \mu_e h_z}{\partial t} = -\frac{\partial E_y}{\partial x} \qquad J_y + \frac{\partial \epsilon E_y}{\partial t} = -\frac{\partial h_z}{\partial x} \tag{5.5}$$

We have two types of electromagnetic waves given by Eqs. (5.4) and (5.5).

The first type of electromagnetic wave consists of variables J_x, E_x, and ρ_e. For finite electric conductivity, it is a damped wave. The x component of electric field may be written as

$$E_x = E_{x0} \exp(-i\lambda x) \cdot \exp\left(-\frac{\sigma t}{\epsilon}\right) \tag{5.6}$$

Since the wave can occur at one frequency $\omega = i\sigma/\epsilon$ and is damped, it is of little practical interest.

The most interesting electromagnetic wave is the transverse wave given by Eq. (5.5). From Eq. (5.5), we obtain an equation for E_y as follows:

$$\frac{1}{c^2}\frac{\partial^2 E_y}{\partial t^2} + \frac{1}{\nu_H}\frac{\partial E_y}{\partial t} - \frac{\partial^2 E_y}{\partial z^2} = 0 \tag{5.7}$$

This is the well-known telegraphic equation for electromagnetic waves. Since there is no external applied magnetic field, we cannot distinguish the y direction from z direction. Hence Eq. (5.7) holds true for E_z and h_y, too.

If we substitute the expression of Eq. (5.2) for E_y into Eq. (5.7), we have the dispersion relation:

$$\lambda^2 - \frac{\omega^2}{c^2} + \frac{i\omega}{\nu_H} = 0 \tag{5.8}$$

For a given real value of frequency ω, λ is complex. Hence we have a damped wave. In general, the speed of propagation and the rate of damping depend on the frequency, speed of light, and the electrical conductivity. When the electrical conductivity is small, we have approximately

$$\lambda \simeq \frac{\omega}{c} - i\frac{\sigma}{2c\epsilon} \tag{5.9}$$

The electromagnetic wave will propagate at the speed of light c and the damping is proportional to the electrical conductivity.

It is of interest to see what kind of wave we may have under the magnetogas-dynamic approximations in which the displacement current is neglected. The magnetogas-dynamic approximation corresponds to the assumption of $c \to \infty$. The dispersion relation [Eq. (5.9)] becomes

$$\lambda^2 = -i\frac{\omega}{\nu_H} \tag{5.10}$$

We simply have a damped magnetic wave. In the analysis of wave motion in a plasma, we should not neglect the displacement current, because we should consider high frequency phenomena in which the time scale is no longer proportional to L/U.

Let us look into more detail of the wave motion, particularly when there is no dissipation mechanism. In the present case, we assume $\sigma = 0$. Equation (5.7) reduces then to the wave equation:

$$\frac{1}{c^2}\frac{\partial^2 E_y}{\partial t^2} - \frac{\partial^2 E_y}{\partial x^2} = 0 \tag{5.11}$$

The general solution of Eq. (5.11) may be written as

$$E_y = f(x - ct) + g(x + ct) \tag{5.12}$$

where f and g are arbitrary functions determined by the initial conditions. The function $f(x - ct)$ represents a traveling plane wave propagated in the positive x direction with a speed of c while the function $g(x + ct)$ represents a traveling plane wave propagated in the negative x direction with a speed of light c.

5.4 GAS-DYNAMIC WAVES

If there is no externally applied magnetic field $H_0 = 0$, the gas-dynamic variables are not affected by the electromagnetic variables. The linearized equations for the gas-dynamic variables give the various modes of gas-dynamic waves. For the transverse gas-dynamic wave, we simply have

$$\rho_0 \frac{\partial w}{\partial t} = \mu \frac{\partial^2 w}{\partial x^2} \tag{5.13}$$

and a similar equation for v. The dispersion relation of Eq. (5.13) is

$$\lambda^2 = -i\frac{\omega}{\nu} \tag{5.14}$$

where $\nu = \mu/\rho_0$ is the coefficient of kinematic viscosity of the fluid. It is a well-known damped wave in a viscous fluid. This wave is independent of the compressibility effect of the medium.

For the longitudinal waves, the dispersion relation is

$$\kappa\left(\frac{1}{\rho_0} + \frac{i4\omega\nu}{3p_0}\right)\lambda^4 - \left[\frac{\omega^2\kappa}{p_0} + \frac{4\nu\omega^2}{3T_0(\gamma - 1)} - i\omega c_p\right]\lambda^2 - \frac{i\omega^3}{T_0(\gamma - 1)} = 0 \tag{5.15}$$

There are two solutions for λ^2 from Eq. (5.15) that represent two modes of longitudinal waves. These are the well-known sound waves in a viscous and heat-conducting medium. One is the ordinary sound wave while the other is sometimes known as the *heat wave*. The heat wave depends on the heat conductivity κ. If $\kappa = 0$, there is no heat wave. For an inviscid and non-heat-conducting fluid $\kappa = \nu = 0$, Eq. (5.15) becomes

$$\lambda^2 = \frac{\omega^2}{a_0^2} \tag{5.16}$$

where $a_0^2 = \gamma RT_0$ and a_0 is the ordinary sound speed in the medium at rest with a temperature T_0. The coefficient of viscosity and the coefficient of heat conductivity do affect the speed of propagation of sound and introduce some damping of sound waves. In general we should solve Eq. (5.15) for the two roots of λ^2 and then determine the speed of wave propagation and damping from these two roots (Prob. 5.2).

5.5 MAGNETOGAS-DYNAMIC WAVES

If there is an externally applied magnetic field, the gas-dynamic and the electromagnetic waves discussed in previous sections would interact. Now we first consider the case where the magnetogas-dynamic approximations hold, so that the displacement current and the excess electric charge are negligible. For the transverse waves, we consider only the variables w and h_z because in the present case $J_x = 0$ and J_y, E_x, and E_y can be expressed in terms of w and h_z. The dispersion relation for the transverse waves is then:

$$\nu\nu_H\lambda^4 + [V_x^2 + i(\nu + \nu_H)\omega]\lambda^2 - \omega^2 = 0 \tag{5.17}$$

where

$$V_x = H_x \sqrt{\frac{\mu_e}{\rho_0}} = \text{the } x \text{ component} \tag{5.18}$$

of the velocity of Alfvén's wave.

Since Eq. (5.17) has two roots for λ^2, there are two different modes of magnetogas-dynamic transverse waves. If $H_x = 0$, these two modes simply reduce to the viscous wave of Eq. (5.14) and magnetic wave of Eq. (5.10). Hence we may consider the viscous wave and the magnetic wave as two basic transverse waves. When there is external applied magnetic field $H_x \neq 0$, these two basic waves will interact and two new transverse waves will result.

There is a special new wave occurring in magnetogas-dynamics. If we consider an inviscid and infinitely electrical conducting medium, Eq. (5.17) gives

$$\lambda^2 = \frac{\omega^2}{V_x^2} \tag{5.19}$$

This is known as Alfvén's wave which has a speed of propagation of V_x. It was Alfvén who first showed that if there is a homogeneous magnetic field H_x in an incompressible and inviscid fluid of density ρ_0 and of infinite electric conductivity $\sigma \to \infty$, the disturbances in this fluid will propagate as a wave in the direction of H_x with a speed of V_x. We now know that Alfvén's wave and the corresponding velocity of Alfvén's wave play an important role in magnetogas dynamics and we shall discuss this in the following section.

In magnetogas-dynamic approximation, the transverse wave is independent of the y component of the magnetic field. But if we do not neglect the displacement current nor the excess electric charge, the transverse waves do depend on the y component as well as on the x component of the magnetic field. The corresponding dispersion relation is then

$$\left(i\omega - \nu_H \frac{\omega^2}{c^2}\right)\left[(i\omega + \nu\lambda^2)\left(i\omega + \nu_H\lambda^2 - \nu_H \frac{\omega^2}{c^2}\right) + V_x^2\left(\lambda^2 - \frac{\omega^2}{c^2}\right)\right]$$
$$- \frac{\omega^2}{c^2} V_y^2 \left(i\omega + \nu_H\lambda^2 - \nu_H \frac{\omega^2}{c^2}\right) = 0 \tag{5.20}$$

where $V_y = H_y \sqrt{\mu_e/\rho_0}$ = the y component of velocity of Alfvén's wave. If we set $c \to \infty$, Eq. (5.20) reduces to Eq. (5.17). For an inviscid and infinitely electric conducting fluid, Eq. (5.20) gives

$$\lambda^2 = \frac{\omega^2}{V_x^2}\left(1 + \frac{V_x^2 + V_y^2}{c^2}\right) \tag{5.21}$$

If the velocity of the Alfvén wave $\sqrt{V_x^2 + V_y^2}$ is much smaller than the speed of light. Eq. (5.21) reduces to Eq. (5.19), the value of ordinary magnetogas dynamics. If the magnetic field H_x is very large, the maximum value of the speed of propagation is c instead of H_x.

The dispersion relation of magnetogas-dynamic longitudinal waves is

$$\left[\kappa\left(\frac{1}{\rho_0} + \frac{i4\omega\nu}{3p_0}\right)\lambda^4 - \left(\frac{\omega^2\kappa}{p_0} + \frac{4\nu\omega^2}{3T_0(\gamma - 1)} - i\omega c_p\right)\lambda^2 - \frac{i\omega^3}{T_0(\gamma - 1)}\right]$$
$$[(\nu_H\lambda^2 + i\omega)(\nu\lambda^2 + i\omega) + V_x^2\lambda^2]$$
$$- \lambda^2 V_y^2 (i\omega + \nu\lambda^2)\left[\frac{\omega^2}{T_0(\gamma - 1)} - \frac{i\omega\kappa\lambda^2}{p_0}\right] = 0 \tag{5.22}$$

The first bracket of Eq. (5.22) represents the sound waves in a viscous and heat-conducting medium [cf. Eq. (5.15)], and the second bracket of Eq. (5.22) represents the magnetogas-dynamic transverse wave. If there is no transverse magnetic field $H_y = V_y = 0$, there is no coupling between the sound waves and the magnetic waves. If $V_y \neq 0$, there are couplings between the four basic waves: ordinary sound wave, heat wave, viscous wave, and magnetic wave. Equation (5.22) gives us four modes of these magnetogas-dynamic waves.

The effect of H_y on the sound waves may be seen clearly by considering an ideal plasma in which $\kappa = \nu = \nu_H = 0$, and Eq. (5.22) becomes

$$\left(a_0^2 - \frac{\omega^2}{\lambda^2}\right)\left(V_x^2 - \frac{\omega^2}{\lambda^2}\right) - \frac{\omega^2}{\lambda^2} V_y^2 = 0 \tag{5.23}$$

If there is no x component of magnetic field $H_x = 0$, Eq. (5.23) gives an effective sound speed of magnetogas dynamics a_e which is governed by the relation:

$$a_e^2 = a_0^2 + V_y^2 \tag{5.24}$$

If we have both H_x and H_y, Eq. (5.23) gives two modes of these sound waves, or longitudinal magnetogas-dynamic waves in an electrically conducting medium. Their speeds of propagation are different. We may call one a *fast wave* whose speed of propagation is V_{fast} and the other a *slow wave* whose speed of propagation is V_{slow}. Equation (5.23) gives the inequality:

$$V_{\text{slow}} \leq a_0 \leq V_{\text{fast}} \tag{5.25a}$$

and

$$V_{\text{slow}} \leq V_x \leq V_{\text{fast}} \tag{5.25b}$$

For an ideal plasma, we have three characteristic velocities of wave propagations, i.e., V_x, V_{fast}, and V_{slow}. The wave pattern in magnetogas dynamics is then considerably different from that in ordinary gas dynamics. For instance, in ordinary gas dynamics, we have only one set of downstream inclined waves when the steady flow is supersonic. In magnetogas dynamics, we may have zero, one, or two sets of wave patterns over a body in a steady flow. The standing wave may incline downstream or upstream. These wave patterns can be easily studied by the help of Friedrichs' diagrams which will be discussed in the next section.

Equation (5.22) is a quartic equation for λ^2. Hence we have four different modes of magnetogas-dynamic longitudinal waves in general and they represent the interaction of the four basic waves: sound wave, heat wave, viscous wave, and magnetic wave.

5.6 FRIEDRICHS DIAGRAMS—SUB AND[4,5] SUPER-ALFVÉN WAVES

The wave motion in magnetogas dynamics of an ideal plasma may be understood with the help of Friedrichs diagrams. Before we discuss

Friedrichs diagrams, it is useful to recall a similar situation in ordinary gas dynamics. In ordinary gas dynamics of an ideal fluid, i.e., inviscid and non-heat-conducting gas, we have only one speed of wave propagation, the sound speed a. The sound speed is the same in all directions. The disturbances produced by a source are on the surface of a sphere with center at the source and with a radius at the place where a is the sound speed. When the source is moving (Fig. 5.1), we may choose the coordinate system fixed on the source. If the source moves with a subsonic speed ($U < a$) [Fig. 5.1(a)], the disturbance will soon cover the whole space. If the source moves with a supersonic speed ($U > a$) [Fig. 5.1(b)], the disturbance will be limited within the Mach cone and the Mach cone is the envelope of the front of the disturbance, i.e., the envelope of the spheres of disturbances [Fig. 5.1(c)].

Similar situations as those in Fig. 5.1 occur in magnetogas dynamics but in a much more complicated manner. For simplicity, we consider the two-dimensional case and we have to consider only the fast and slow waves of magnetogas dynamics. The Friedrichs diagram for the shape of the disturbance that propagates from a

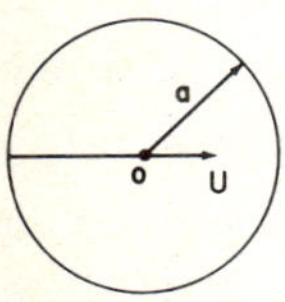

(a) Subsonic flow
$U < a$

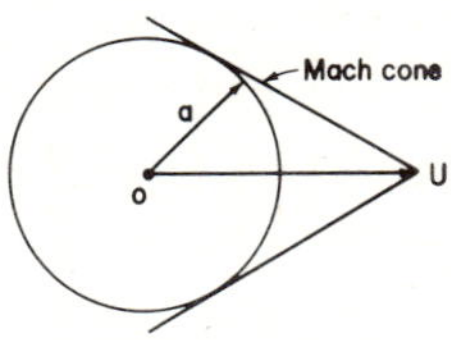

(b) Supersonic flow
$U > a$

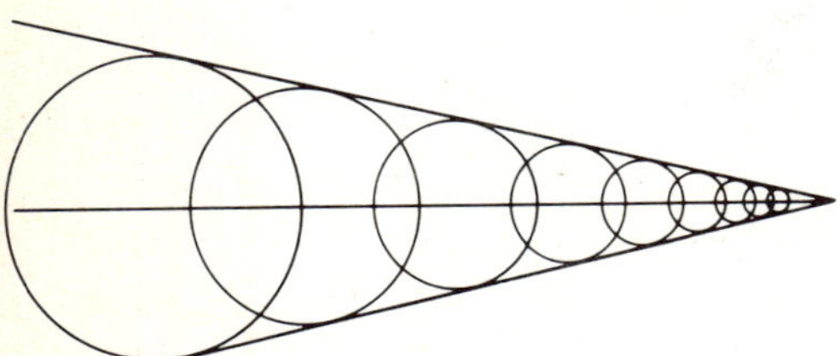

(c) Characteristics of supersonic flow envelope of the circles gives Mach line.

Fig. 5.1 Shape of wave front from a point disturbance in ordinary gas dynamics.

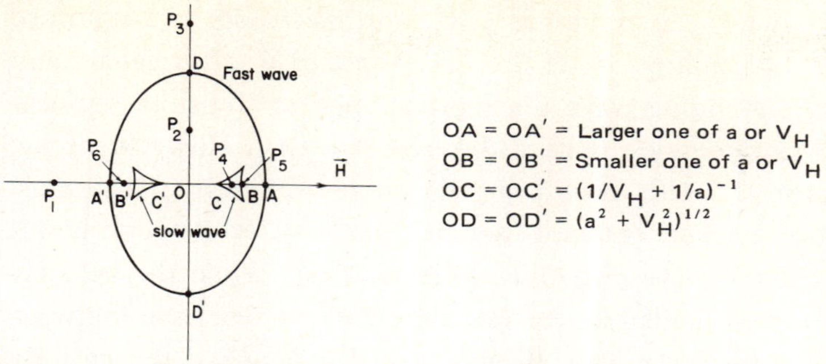

Fig. 5.2 Friedrichs diagram showing shape of wave front from a point disturbance.

point disturbance at the origin is shown in Fig. 5.2. The shape is not a circle and the shape of fast waves differs greatly from those of slow waves. In Fig. 5.2, the abscissa is in the direction of the magnetic field and the flow direction may be in arbitrary direction. Basically we have two characteristic speeds: one is the ordinary sound speed $a = \sqrt{\gamma RT}$ and the other is the Alfvén wave speed $V_H = H(\mu_e/\rho)^{1/2}$ where H is the magnitude of the magnetic field. The speed of wave propagation in magnetogas dynamics V may be expressed in terms of a and V_H as follows:

$$V = \left[\frac{(a^2 + V_H{}^2) \pm \sqrt{(a^2 + V_H{}^2)^2 - 4a^2V_H{}^2 \cos^2\theta}}{2}\right]^{1/2} \tag{5.26}$$

where θ is the angle between the direction of the magnetic field H and the velocity of the point source $OP = U$. The plus sign is for the fast wave while the minus sign is for the slow wave. The wave front for the fast wave is different from that for the slow wave. The velocity of the point source may be in any direction such as OP_1, OP_2, and so forth. The wave pattern depends on the location of the point P, i.e., it depends on both the direction and the amplitude of the velocity U. Figure 5.3 shows some typical cases for the wave patterns. When the velocity U is OP_1, i.e., the velocity U is in the same direction as H but its magnitude is larger than both a we may call it *super-Alfvén flow*, and we have a similar as that in ordinary gas dynamics, i.e., we have a set of backward inclined waves [Fig. 5.3(a)]. If the magnitude of U is between a and V_H, there is no standing wave that corresponds to the subsonic case of ordinary gas dynamics. This is the case $U = OP_2$. However, if the velocity vector $U = OP_4$ is within the curved triangle of the slow wave, we have a set of forward inclined wave patterns which is not possible in ordinary gas dynamics [Fig. 5.3(b)]. When the velocity vector U is perpendicular to the magnetic field H, we have the wave patterns shown in Fig. 5.3(c). When $U = OP_3 > \sqrt{a^2 + V_H{}^2}$, we have two sets of wave patterns, and when $OP_6 < \sqrt{a^2 + V_H{}^2}$, we have only one set of wave patterns. The case of two sets of waves is also not possible in ordinary gas dynamics. Finally if $U = OP_5$ where P_5 is inside the curved triangle but OP_5 is not parallel to H, we have again two sets of wave patterns: One set is inclined forward while the other is inclined backward [Fig. 5.3(d)]. This is also a case which cannot occur in ordinary gas dynamics. The forward inclined wave may be obtained by drawing the envelope of the disturbance shape of the slow waves. (Fig. 5.4).

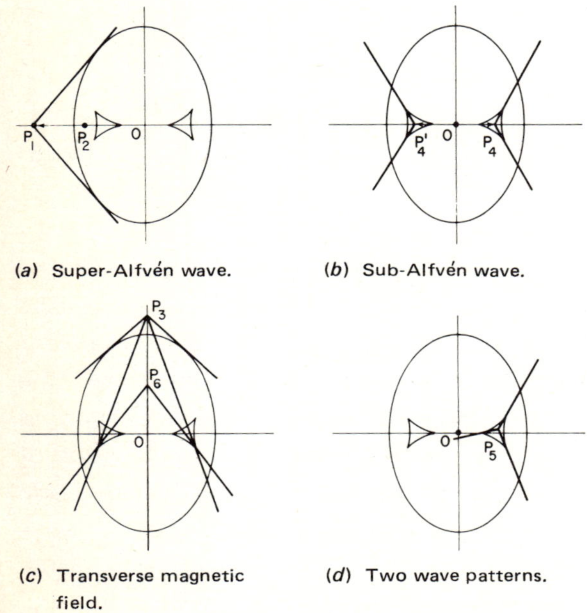

Fig. 5.3 Wave patterns in magnetogas dynamics.

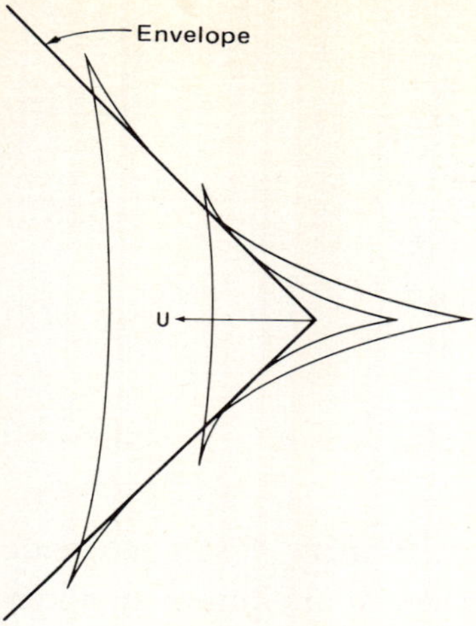

Fig. 5.4 Envelope of disturbance of a slow wave—forward inclined wave.

These wave patterns are mathematical facts but they have not been verified experimentally. Experimental observation of these phenomena will be most interesting and enlightening. Not only are there fast and slow waves of infinitesimal amplitude, there are waves of finite amplitude and shock waves of both types.

5.7 WAVES OF FINITE AMPLITUDE

For waves of finite amplitude, the speeds of wave propagation are the same as for waves of infinitesimal amplitude except that these speeds of propagation should be computed according to the local values of the state variables and magnetic field strength. Hence for an ideal plasma, we have fast wave, slow wave, and transverse wave just as in the case of waves of infinitesimal amplitude except that the speeds of propagation of these waves vary with the variables of magnetogas dynamics. Because of the variation of the speed of propagation of these magnetogas-dynamic waves of finite amplitude, the forms of these waves would be distorted as the waves propagate. In the case of waves of infinitesimal amplitude, the wave preserves its form as it propagates. To show this distortion, let us consider one-dimensional unsteady flow of an ideal plasma under the transverse magnetic field. The fundamental equations of this [illegible] are:

$$p = \rho RT \tag{5.27a}$$

$$\frac{\partial \rho}{\partial t} + u \frac{\partial \rho}{\partial x} = -\rho \frac{\partial u}{\partial x} \tag{5.27b}$$

$$\rho \frac{\partial u}{\partial t} + \rho u \frac{\partial u}{\partial x} + \frac{\partial p}{\partial x} + \mu_e H \frac{\partial H}{\partial x} = 0 \tag{5.27c}$$

$$\frac{\partial H}{\partial t} + u \frac{\partial H}{\partial x} = -H \frac{\partial u}{\partial x} \tag{5.27d}$$

$$\rho \frac{\partial h_0}{\partial t} + \rho u \frac{\partial h_0}{\partial x} - \frac{\partial p}{\partial t} + \mu_e u H \frac{\partial H}{\partial x} = 0 \tag{5.27e}$$

where u is the x-wise velocity component, H is the y-wise magnetic field, and h_0 is the stagnation enthalpy. The rest are known symbols. All the unknowns are variables for one spatial coordinate x and time t only.

Since Eqs. (5.27b) and (5.27d) are identical in form, we may write

$$H = A\rho = \frac{H_0}{\rho_0} \rho \tag{5.28}$$

where A is a constant determined by the initial conditions. Subscript 0 refers to the values at some reference point. Equation (5.28) shows that in the present problem, the functional forms of density ρ and magnetic field H are the same.

From Eqs. (5.27a)–(5.27e), we can show that along any line of flow, the entropy is a constant (see Prob. 5.5).

In order to show the distortion of waves of finite amplitude, we consider the special case where the flow is isentropic and the density is a function of velocity u only, that is,

$$\rho = f(u) \tag{5.29}$$

where the function f will be determined as follows:

Substituting Eq. (5.29) into Eq. (5.27b), we have

$$\frac{\partial u}{\partial t} + u \frac{\partial u}{\partial x} = -\frac{f}{f'} \frac{\partial u}{\partial x} \tag{5.30}$$

where $f' = df/du = d\rho/du$. Similarly, from Eq. (5.27c) we have

$$\frac{\partial u}{\partial t} + u \frac{\partial u}{\partial x} = -\frac{a_e^2}{f} f' \frac{\partial u}{\partial x} \tag{5.31}$$

where a_e is the effective sound speed of the present problem and is defined by

$$a_e^2 = \gamma RT + \frac{\mu_\rho H^2}{\rho} \tag{5.32}$$

From our previous study, we know that in the present case, we have the fast wave only and the fast wave has the speed of propagation of a_e. Furthermore, in the present case, a_e is a function of density only.

If Eq. (5.29) is a solution to our problem, Eqs. (5.30) and (5.31) must be the same. Hence, we have

$$\frac{f'}{f} = \frac{1}{\rho}\frac{d\rho}{du} = \pm\frac{1}{a_e} \tag{5.33}$$

Substituting Eq. (5.33) into Eq. (5.30), we have

$$\frac{\partial u}{\partial t} + (u \pm a_e)\frac{\partial u}{\partial x} = 0 \tag{5.34}$$

The general solution of Eq. (5.34) is

$$u = F[x - (u \pm a_e)t] \tag{5.35}$$

where F is an arbitrary function. For waves of infinitesimal amplitude, a_e is determined for uniform flow and is a constant. For waves of finite amplitude, a_e varies with the flow field. Let us consider the solution $F[x - (u + a_e)t]$. This solution means that the variation of u with respect to a point moving with velocity $a_e + u$ is zero, i.e., the disturbance is propagated with an instantaneous velocity $u + a_e$. If the velocity u is much smaller than a_e, we have $F(x - a_e t)$. The speed of propagation of a wave of infinitesimal amplitude is thus a_e. For waves of finite amplitude, the vel
propagation is different at different points. In Fig. 5.5,

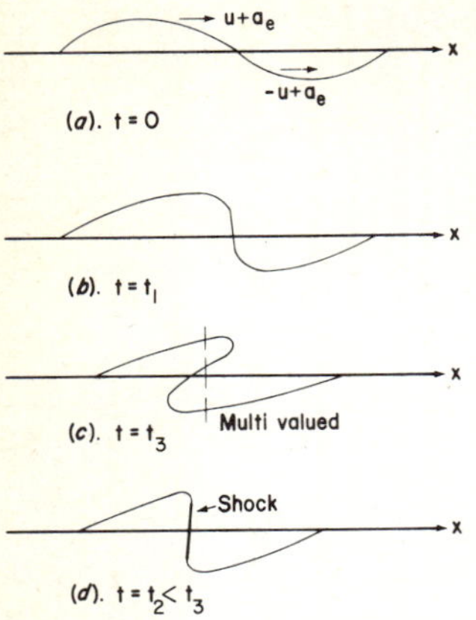

Fig. 5.5 Propagation of magnetogas-dynamic wave of finite amplitude.

assume there is a sinusoidal disturbance of finite velocity [Fig. 5.5(a)]. As the wave propagates to $t = t_1$, the crest moves at a higher speed while the trough moves at a lower speed and the shape of this wave distorts [as sketched in Fig. 5.5(b)]. Theoretically, if we continue this process, at a much later time $t = t_3$, the wave will be distorted in the form of Fig. 5.5(c). For a longitudinal wave, the form of Fig. 5.5(c) is impossible because it has multivalues of u at a given location x. Hence before the wave develops into the form of Fig. 5.5(c), a shock wave will be formed at some earlier time $t = t_2$, Fig. 5.5(d). Hence it shows that a compressible wave will develop into a shock wave. Similarly, we can easily show that an expansion wave will smooth itself out.

In solving for the wave motion, the method of characteristics may be used. For the one-dimensional unsteady flow, there are four sets of characteristics, i.e., double roots of flow lines

$$\frac{\partial x}{\partial t} = u \tag{5.36}$$

and characteristic lines

$$\frac{\partial x}{\partial t} = u \pm a_e$$

Along the flow lines, $dx/dt = u$, we have

$$\frac{H}{\rho} = \text{const} \tag{5.37}$$

and

$$S = \text{entropy} = \text{const} \tag{5.38}$$

Along the characteristic lines $dx/dt = u \pm a_e$, we have, respectively,

$$\mu_e H\left[\frac{\partial H}{\partial t} + (u \pm a_e)\frac{\partial H}{\partial x}\right] \pm \rho a_e\left[\frac{\partial u}{\partial t} + (u \pm a_e)\frac{\partial u}{\partial x}\right] + \left[\frac{\partial p}{\partial t} + (u \pm a_e)\frac{\partial p}{\partial x}\right] = 0 \quad (5.39)$$

The well-known numerical method of characteristics may be used for any given initial values.

For the more general case of two- or three-dimensional flow, we may also find the corresponding characteristics and use the method of characteristics to solve such problems of wave motion because the resultant equations are of the hyperbolic type (Prob. 5.6).

5.8 SHOCK WAVES

Since the shock wave is developed from the compression wave of magnetogas dynamics, one would expect that there will be shock waves corresponding to the fast wave, slow wave, and transverse wave. In studying the shock waves, two things are of interest: one is the Rankine-Hugoniot relations across a shock wave and the other is the shock structure. Both of these characteristics are affected by the magnetic field and there are many new features of these magnetogas-dynamic shocks. Let us first consider the normal magnetogas-dynamic shock under the influence of a transverse magnetic field which is the shock developed in the one-dimensional unsteady flow discussed in the last section. The Rankine-Hugoniot relation of this shock gives the ratio of the velocity behind the shock u_2 to that in front of the shock u_1 and is

$$\frac{u_2}{u_1} = \frac{1}{2}\left(\frac{\gamma - 1}{\gamma + 1} + \frac{2}{(\gamma + 1)M_1^2} + \frac{2\gamma h_1^2}{\gamma + 1}\right) + \frac{1}{2}\left\{\left[\left(\frac{\gamma - 1}{\gamma + 1}\right)^2 + \frac{2}{(\gamma + 1)M_1^2} + \frac{2\gamma h_1^2}{\gamma + 1}\right]^2 + \frac{8(2 - \gamma)h_1^2}{\gamma + 1}\right\}^{1/2} \quad (5.40)$$

where M_1 is the Mach number (u_1/a_1) in front of the sh[illegible] $h_1 = H_1/(2\rho_1 u_1^2/\mu_e)^{1/2}$ the nondimensional magnetic field the shock. When there is no magnetic field, $h_1 = 0$, Eq. (5.40) reduces to the relation for a normal shock in ordinary gas dynamics. For a given value of M_1, the ratio (u_2/u_1) increases or the strength of the shock decreases with the increase of h_1. This occurs since the effective Mach number $M_{e1} = u_1/a_{e1}$ decreases as h_1 increases. When $M_{e1} = 1$, $u_2/u_1 = 1$. For a given Mach number, if $h_1^2 > h_c^2 = \frac{1}{2}\,[1 - (1/M_1^2)]$, $M_{e1} < 1$ and no shock will exist.

For the flow in the transition region for u_1 to u_2, we have to consider the diffusion phenomena due to viscosity, heat conductivity, and magnetic diffusivity. We have the so-called *dispersion shock* in magnetogas dynamics. There is a critical value of the magnetic field $h_1 = h_b$ such that $0 > h_b > h_c$ which occurs at $u_2/u_1 = \xi_c$ such that $1 > \xi_c > (u_2/u_1)_0$ and subscript 0 refers to the value at $h_1 = 0$. When the value of h_1 is smaller than h_b, we have a partially dispersed shock and the velocity in the transitional region decreases gradually from $u_2/u_1 = 1$ to the value near ξ_c and then decreases rapidly to the value near that given by the Rankine-Hugoniot relation and finally decreases to the value given by Rankine-Hugoniot relation. We may call it a *partially dispersed shock*. If h_1 lies between h_b and h_c, the rate of decrease in velocity u in the transitional region is small in the whole flow field and we may calculate the variation of the velocity u in the transitional region by neglecting the viscosity and heat conduction. This is known as a *fully dispersed shock*.

As the velocity decreases, both the temperature and the magnetic field increase in the transitional region.

For a magnetogas-dynamic oblique shock, we have to consider both the direction of the magnetic field and that of the flow velocity with respect to the shock front. In ordinary gas dynamics, by proper choice of the coordinate system it is always possible to reduce the oblique shock to a corresponding normal shock case. However, it is, in general, not possible to do so in magnetogas dynamics. Since the shock wave is developed from an ordinary wave, and there are fast, slow, and transverse waves in magnetogas dynamics, we would expect that several different types of shock waves exist in magnetogas dynamics.

We shall assume that both the velocity vector $\mathbf{u}$ and the magnetic field $\mathbf{H}$ have arbitrary directions with respect to the shock front whose normal vector is $\mathbf{n}$. The jump of any physical quantity Q across the shock front may be denoted by

$$[Q] = Q_1 - Q_0 \tag{5.41}$$

where subscript 1 refers to the value on the side to which the normal $\mathbf{n}$ points and subscript 0 refers to the value on the other side of the shock. The conservation laws across the surface of discontinuity are:

$$[H_n] = 0 \tag{5.42a}$$

$$[(u_n - U)\mathbf{B} - B_n\mathbf{u}] = 0 \tag{5.42b}$$

$$\left[(u_n - U)\rho\mathbf{u} + \left(p + \frac{\mathbf{B}\cdot\mathbf{H}}{2}\right)\mathbf{n} - B_n\mathbf{H}\right] = 0 \tag{5.42c}$$

$$[(u_n - U)\rho] = 0 \tag{5.42d}$$

$$\left[(u_n - U)\left(\frac{\rho u^2}{2} + \rho U_m + \frac{\mathbf{B}\cdot\mathbf{H}}{2}\right) + u_n\left(p + \frac{\mathbf{B}\cdot\mathbf{H}}{2}\right) - H_n(\mathbf{B}\cdot\mathbf{u})\right] = 0 \tag{5.42e}$$

where subscript n refers to the component of a vector in the direction of $\mathbf{n}$ and U is the normal velocity of the surface of discontinuity. In order that there exists such a surface of discontinuity, there must be a surface current along this surface of discontinuity with a value of

$$\mathbf{J}^* = \mathbf{n} \times [\mathbf{H}] \tag{5.43}$$

From Eq. (5.42b), we see that if we choose the coordinate system such that the flow velocity $\mathbf{u}$ is parallel to the magnetic field $\mathbf{H}$ on one side of the shock wave front and $(u_{n1} - U)\mathbf{B}_1 - B_n\mathbf{u}_1 = 0$, these two vectors $\mathbf{u}$ and $\mathbf{H}$ are automatically parallel on the other side of the shock front and $(u_{n0} - U)\mathbf{B}_0 - B_n\mathbf{u}_0 = 0$.

Let us introduce the following notations:

$$\tau = \rho^{-1}$$

$$\widetilde{Q} = \tfrac{1}{2}(Q_0 + Q_1) \tag{5.45}$$

$$m = \rho(u_n - U) \tag{5.46}$$

where the normal **n** is so chosen that across a shock wave

$$u_n - U > 0 \tag{5.47}$$

From Eq. (5.42a), we have

$$H_{n0} = H_{n1} = H_n = \text{constant} \tag{5.48}$$

From Eqs. (5.42b)–(5.42d), we have

$$\tilde{\tau}^2 m(\tilde{\tau} m^2 - B_n H_n)\{\tilde{\tau} m^4 + (\tilde{\tau}[\tau]^{-1}[p] - \widetilde{\mathbf{B}}\cdot\widetilde{\mathbf{H}})m^2 - [\tau]^{-1}[p]B_n H_n\} = 0 \tag{5.49}$$

Equation (5.49) shows that there are four different types of surface of discontinuity in magnetogas dynamics according to the four roots of m in this equation:

m = 0. This is a contact discontinuity. If $H_n \neq 0$, the only jump for this case is $[\rho] = 0$, because $[S] \neq 0$. If $H_n = 0$, we can have $[p + \mathbf{B}\cdot\mathbf{H}/2] = 0$.

Transverse shock or Alfvén's shock A. This corresponds to the root

$$m = m_{\text{tr}} = (\tilde{\tau}^{-1} B_n H_n)^{1/2} \tag{5.50}$$

In this case, both the density and entropy are continuous across the surface of continuity but the tangential projection of the magnetic field undergoes a rotation through an arbitrary angle on crossing the surface of discontinuity.

The most interesting magnetogas-dynamic shocks are the fast and slow shocks which correspond to the two roots given by the curl bracket of Eq. (5.49). The oblique shock relations have been calculated by various authors.[6,7] We shall give here some descriptions of the general features of such magnetogas-dynamic shocks.

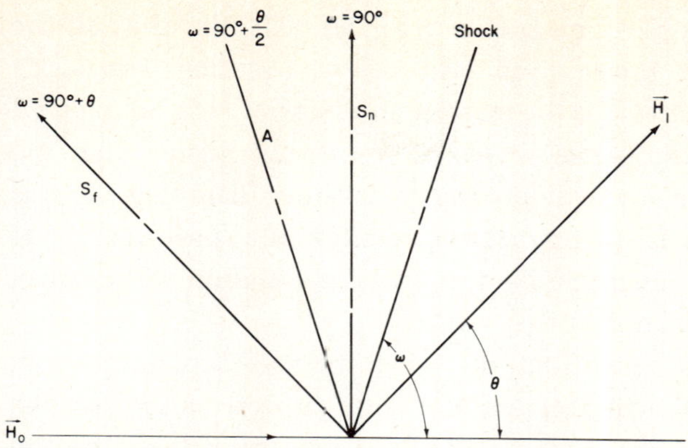

Fig. 5.6 Magnetogas-dynamic shock when u‖H.

Let us follow Ref. 6 by considering the case where the velocity and the magnetic field are parallel on both sides of the shock front. Let θ be the turning angle of the magnetic field and ω be the shock angle. Figure 5.6 shows the orientations of fast and slow shocks.

1. When $90^\circ > \omega$, we have the *fast shock.*

2. When $90^\circ < \omega < 180^\circ$, we have the *slow shock.* The inclination of a slow shock is upstream which is not possible in ordinary gas dynamics.

3. When the shock angle is 90°, the tangential component of the magnetic field is zero in front of the shock but different from zero behind the shock. It is known as *switch-on shock Sn*.

4. When $\omega = 90^\circ + \theta$, the tangential magnetic field is different from zero in front of the shock but vanishes behind the shock. It is known as *switch-off shock Sf* which is a special slow shock.

5. $\omega = 90 + \theta/2$, it is the *Alfvén's shock A*.

5.9 OTHER WAVE PHENOMENA IN MAGNETOGAS DYNAMICS

In the study of the wave motion (Sec. 5.8), we consider the plasma as a single fluid and neglect the effect of charge separation. In an actual flow field, charge separation may occur. One of the wave phenomena due to charge separation is the plasma oscillation which will be discussed in Chap. 7, Sec. 7.5. For instance, in a plasma without an applied magnetic field, we will find that all the waves of frequency lower than the electron-plasma frequency $\omega_e = e(n_0/\epsilon m_e)^{1/2}$ where m_e is the mass of an electron, will be damped out and only those transverse waves of frequency higher than the electron-plasma frequency will propagate without damping. This is the blackout of radio frequency during reentry of a space vehicle. If we apply an external magnetic field, the whole situation will change. (This will be discussed later in this text.)

Thermal radiation can affect the wave motion, too. In general, the radiation pressure will increase the sound speed of the plasma. The jump in temperature across a shock is greatly reduced by the thermal radiation effects.

REFERENCES

1. S. I. Pai, "Magnetogasdynamics and Plasma Dynamics," Springer-Verlag, New York, 1962.
2. S. I. Pai, A Critical Survey of Magnetofluid Dynamics, parts I-IV, *Aerospace Res. Lab. Repts.* ARL 63-175 and ARL 66-0060, Wright-Patterson Air Force Base, Ohio, 1963-1966.
3. S. I. Pai, "Radiation Gasdynamics," Springer-Verlag, New York, 1966.
4. W. R. Sears and E. L. Resler, Jr., "Sub- and Super-Alfvénic Flows past Bodies." Advances in Aeronautical Science, vol. 4, pp. 657-673, Pergamon Press, Oxford, 1961.
5. E. L. Resler, Jr. and J. E. McCune, Compressibility Effects in Magnetoaerodynamic flows, *J. Aero. Sci.*, vol. 27, no. 7, pp. 493-503, 1960.
6. J. Bazer and W. B. Ericson, Oblique Shock Waves in a Steady Two Dimensional Hydromagnetic Flow, *Proc. Symposium on Electromagnetics and Fluid Dynamics of Gaseous Plasma*, vol. XI, Interscience Publishers, Inc., New York, 1961.
7. H. Cabannes, Attached Stationary Shock in Ionized Gases, *Rev. Mod. Phys.*, Oct. 1960, pp. 973-976.

Chapter 6

APPLICATIONS OF MAGNETOFLUID DYNAMICS

6.1 INTRODUCTION

This chapter has the purpose of presenting selected, imaginative applications of magnetofluid-dynamic theory to the explanation of natural phenomena and to the design of MFD devices. However, the desire to demonstrate the versatility and wide-ranging applicability of magnetofluid dynamics restricts the discussion of the many topics to their elementary features. A thorough discussion of a specific topic would require more than the desirable quantity of material for this chapter. Indeed, most of the selected items warrant separate books.

This chapter contains three different subject areas describing astrophysical, geophysical, and engineering applications of MFD. The larger number of engineering applications reflects only an arbitrary choice and is not a measure of the relative activity or importance of the MFD research in the other two topic areas.

The simple MFD flow meter and pump are two applications that were selected for exercises in the problem section and the following is a partial list of the many other MFD phenomena and devices.

Radio wave propagation in ionosphere
Space communication system
Traveling wave tubes
Geomagnetic storms
Auroras and ionospheric disturbances
Solar flares
Generation of cosmic rf radiation

Diagnostic methods
Plasma jets
Reentry ICBM problems
Electromagnetic shock tubes
MFD wind tunnel drives
Electric arc heaters
Space power generators

6.2 ASTROPHYSICAL APPLICATIONS

An exploding star is called a *novae* or a *supernovae*, depending on its magnitude. If the sudden energy release is at the star's center, a hydrodynamic shock wave will propagate symmetrically through the star toward its outer edge. The detailed description of this ideal explosion has been provided by Sedov[1] who applied classical blast wave theory to obtain a solution of the gas-dynamic equations when magnetic fields were absent. However, magnetic fields have been observed in many stars and appear to be described by a dipole configuration. On such a gigantic scale, the magnetic field will be effectively frozen in the medium and thus, it should be distorted by the explosion. Yu and Lykoudis[2] have analyzed a simple model of this problem with the added simplification that the motion of the gas is not affected by the magnetic field on the basis of an estimate that the maximum magnetic force is less than 10 percent of the inertia force even for an enormous magnetic field strength of 3×10^{11} gauss. Since Sedov showed that the shock strength decreases as it progresses to the outer region of the star, then at some point, the magnetic force will equal the inertia force. Thus, Yu and Lykoudis' analysis is limited to the inner region. They also assumed that the magnetic field before the explosion was a spherical dipole with the form:

$$H_{r\infty} = \frac{2m}{r^3} \cos\theta \tag{6.1}$$

$$H_{\theta\infty} = \frac{m}{r^3} \sin\theta \tag{6.2}$$

and

$$H_{\phi\infty} = 0 \tag{6.3}$$

and the velocity field was spherically symmetric, that is,

$$v_{\theta} = v_{\phi} = 0 \quad \text{and} \quad \frac{\partial}{\partial\theta} = 0 \tag{6.4}$$

The equations in spherical coordinates with these simplifications, are then

$$\frac{\partial\rho}{\partial t} + \frac{1}{r^2}\frac{\partial(r^2\rho v_r)}{\partial r} = 0 \tag{6.5}$$

$$\frac{\partial v_r}{\partial t} + v_r\frac{\partial v_r}{\partial r} = -\frac{1}{\rho}\frac{\partial p}{\partial r} - g\frac{M}{r^2} \tag{6.6}$$

$$\frac{\partial(p\rho^{-\gamma})}{\partial t} + v_r\frac{\partial(p\rho^{-\gamma})}{\partial r} = 0 \tag{6.7}$$

$$\mathfrak{M} = \int_0^r\int_0^\pi 2\pi r^2 \sin\theta\rho\, d\theta\, dr \tag{6.8}$$

$$\frac{\partial H_r}{\partial t} + \frac{v_r}{r^2}\frac{\partial(r^2H_r)}{\partial r} = 0 \tag{6.9}$$

$$\frac{\partial H_\theta}{\partial t} + \frac{1}{r}\frac{\partial(rv_rH_\theta)}{\partial r} = 0 \tag{6.10}$$

$$\frac{\partial H_\phi}{\partial t} + \frac{1}{r}\frac{\partial(rv_rH_\phi)}{\partial r} = 0 \tag{6.11}$$

Equation (6.7) is the adiabatic energy equation and Eq. (6.8) is the total mass of the star inside of the radius r. The last three equations were obtained by combining Ampère's law, Ohm's law, and the fact that $\nabla \cdot B$ is always zero. The associated jump conditions were determined by integrating the flow equations across the shock wave and applying the requirements for the tangential component of the electric field and the normal component of the magnetic field. They are:

$$v_{rs} = \frac{2c}{\gamma + 1}\, 1 - \frac{1}{M_a^2} \tag{6.12}$$

$$\rho_s = \frac{\gamma + 1}{\gamma - 1}\left[\frac{\rho_\infty}{1 + [2/(\gamma - 1)](1/M_a^2)}\right] \tag{6.13}$$

$$p_s = \frac{2}{\gamma + 1}\rho_\infty c^2 \left(1 - \frac{\gamma - 1}{2\gamma}\frac{1}{M_a^2}\right) \tag{6.14}$$

$$\mathfrak{M}_s = \mathfrak{M}_\infty \tag{6.15}$$

$$H_{\theta s} = \frac{\gamma + 1}{\gamma - 1}\left[\frac{H_{\theta\infty}}{1 + [2/(\gamma + 1)](1/M_a^2)}\right] \tag{6.16}$$

$$H_{rs} = H_{r\infty} \tag{6.17}$$

where $c = dr_s/dt$ is the shock velocity, $M_a = c/\sqrt{\gamma p_\infty/\rho_\infty}$ is the shock Mach number, and the subscript s refers to quantities after the shock.

The partial differential equations were reduced to ordinary differential equations by the following similarity transformation. With

$$H_r = \frac{m}{r^3} h_r(\lambda) \tag{6.18}$$

$$H_\theta = \frac{m}{r^3} h_\theta(\lambda) \tag{6.19}$$

$$v_r = \frac{r}{t} V(\lambda) \tag{6.20}$$

and

$$\lambda = \frac{r}{r_s} = \frac{r}{(200\,\pi A g)^{0.4}\, t^{0.8}} \tag{6.21}$$

the magnetic field equations become

$$(V - 0.8)\lambda \frac{dh_r}{d\lambda} - Vh_r = 0 \tag{6.22}$$

$$(V - 0.8)\lambda \frac{dh_\theta}{d\lambda} - \left(V - \frac{dV}{d\lambda}\right) h_\theta = 0 \tag{6.23}$$

and were solved by using Sedov's similarity solution of Eqs. (6.5)-(6.8) for the velocity obtained with the initial conditions

$$\begin{aligned} \rho &= \frac{A}{r^{2.5}} \quad V = 0 \quad M = 8\pi A r^{1/2} \\ p &= \frac{8\pi A^2 g}{3r^3} \quad \text{and} \quad RT = \frac{8\pi A g}{3r^{1/2}} \end{aligned} \tag{6.24}$$

and for $\gamma = 4/3$ and $M_a = 6$. In dimensional terms, Sedov's results are

$$v_r = \frac{2}{3}\frac{r}{t} \tag{6.25}$$

$$\rho = \frac{3}{100\pi g t^2} \tag{6.26}$$

$$\mathfrak{M} = \frac{r^3}{25 g t^2} \tag{6.27}$$

$$p_s = \frac{41}{15{,}000\pi}\frac{r^2}{g t^4} \tag{6.28}$$

$$r_s = \left(\frac{50\pi A g}{9}\right)^{2/5} \left(\frac{t}{M_a}\right)^{4/5} \tag{6.29}$$

$$c = \frac{4}{5}\left(\frac{50\pi A g}{9 r_s M_a^2}\right)^{1/4} \tag{6.30}$$

where r_s is the position of the shock wave. The solutions for the magnetic field components are then

$$H_r = \frac{2m \cos\theta}{r^3 \lambda^5} \tag{6.31}$$

and

$$H_\theta = \frac{1519}{216} \frac{m \sin\theta}{r^3 \lambda^5} \tag{6.32}$$

A sketch of the magnetic field before and after the explosion is shown in Fig. 6.1 for a chosen shock location and demonstrates that the magnetic lines are pushed outward by the gas motion and the field strength is increased behind the shock wave. These results provide some physical insight into the distortion of the magnetic field in an exploding star. However, Yu and Lykoudis stated that such a simple initial magnetic field would be unlikely in the interior of a star.

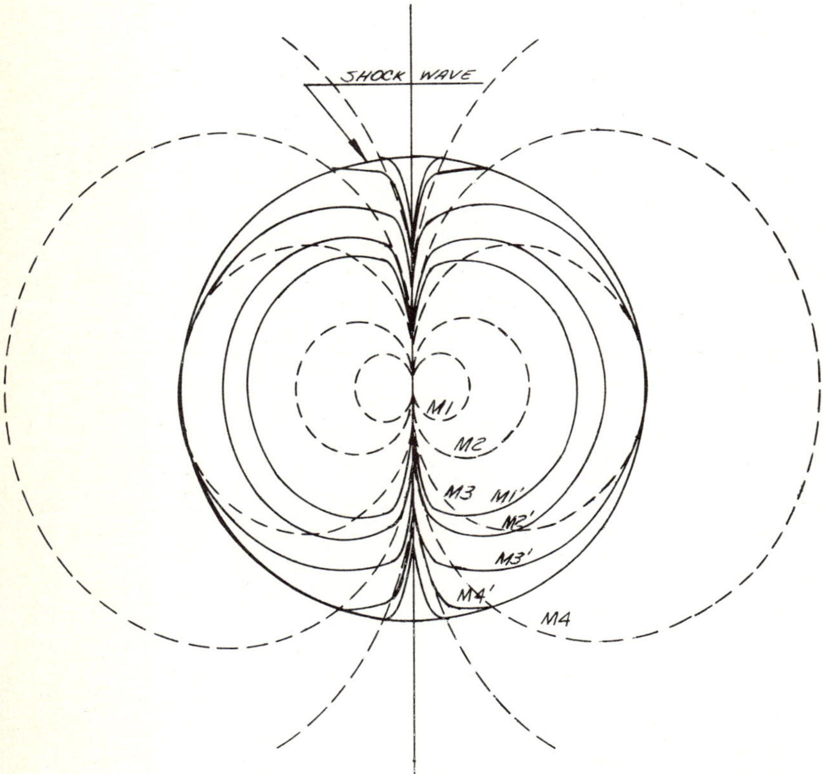

M1, M2, M3, M4: Magnetic lines before explosion
M1′, M2′, M3′, M4′: Magnetic lines after explosion

Fig. 6.1 Pre- and post-explosion magnetic field of an exploding star.

6.3 GEOPHYSICAL APPLICATIONS

Earth satellites have provided considerable information about the region of space near the earth. Kemmel and Petschek[3] have presented a brief summary of the details of the interaction of the earth's magnetic field and the solar wind. The earth and its magnetic field effectively form a blunt body in the highly supersonic solar wind flow and produce a detached shock wave in which the dissipation is not clearly understood. However, the total gas pressure behind the shock wave is sufficient to confine the earth's magnetic field to an inner region called the *magnetosphere*. Figure 6.2, provided by Kemmel and Petschek, presents a schematic description of all the available information on this interaction. Estimates of the gas pressure and magnetic pressure in the magnetosphere and the region behind the shock wave permit a simple description of both regions. The magnetosphere is considered a vacuum, since there $p \ll H^2/2$ and the magnetic pressure is negligible in the shock layer, since there $p \gg H^2/2$. Spreiter, Summers, and Alksne[4] have analyzed the shock layer on the basis of the flow of a dissipationless perfect gas with the MFD equations. They had reservations about applying continuum equations to the flow of a collisionless gas, but they were attempting to provide a simple analysis of the average bulk properties of the solar wind as it passes through the bow shock and around the forward portion of the magnetosphere. Their form of the MFD equations are

$$\nabla \cdot (\rho \mathbf{v}) = 0 \tag{6.33}$$

$$\rho(\mathbf{v}\cdot\nabla)\mathbf{v} + \nabla p = -\mathbf{H}\times\nabla\times\mathbf{H} = -\frac{\nabla \mathbf{H}^2}{2} + (\mathbf{H}\cdot\nabla)\mathbf{H} \tag{6.34}$$

$$\nabla \times (\mathbf{H}\times\mathbf{v}) = 0 \tag{6.35}$$

$$\nabla \cdot \mathbf{H} = 0 \tag{6.36}$$

$$(\mathbf{v}\cdot\nabla) S = 0 \tag{6.37}$$

$$p = e^{S/c_v}\rho^\gamma \tag{6.38}$$

with the auxiliary equations

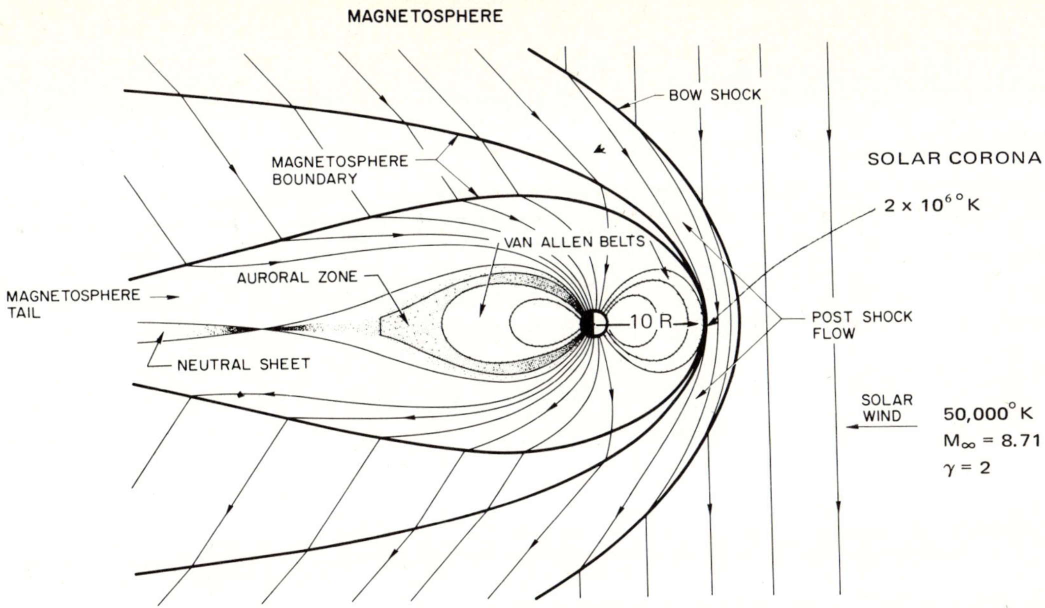

Fig. 6.2 Magnetosphere. This figure schematically summarizes our present knowledge of the earth–solar wind interaction. Shown is an idealized steady-state magnetosphere. The magnetosphere boundary is split up into two discontinuities, corresponding to the standing waves which are part of the field annihilation region (where postshock solar wind and earth magnetic lines are connected) (1964). Note the different locations of the auroral and Van Allen zones, considered to be distinct in this paper.

$$\gamma = \frac{N+2}{N} \qquad p = \frac{\rho R T}{m_m} = nkT$$

$$a = \left(\frac{\partial p}{\partial \rho}\right)^{1/2} = \left(\frac{\gamma p}{\rho}\right)^{1/2} = \left(\frac{\gamma R T}{m_m}\right)^{1/2} \tag{6.39}$$

$$e = c_v T \qquad h = c_p T = e + \frac{p}{\rho}$$

where N is the number of degrees of freedom of the gas particles; $R = (c_p - c_v)/m_m$; m_m is the mean molecular weight (equal to 1/2 for a fully ionized hydrogen plasma); n is the number of particles per cc = $2n_p$ where n_p is the number of protons per cc and k is the Boltzman constant equal to 1.38×10^{-16} ergs/°K.

The associated jump conditions across the shock wave and the discontinuity separating the shock layer and the magnetosphere are

$$[\rho v_n] = 0 \tag{6.40}$$

$$\left[\rho v_n \mathbf{v} + \left(p + \frac{H^2}{2}\right)\mathbf{n} - H_n \mathbf{H}\right] = 0 \tag{6.41}$$

$$[H_n v_t - H_t v_n] = 0 \tag{6.42}$$

$$[H_n] = 0 \tag{6.43}$$

$$\left[\rho v_n\left(\frac{v_H{}^2}{2} + h\right) + v_n H^2 - H_n \mathbf{v}\cdot\mathbf{H}\right] = 0 \tag{6.44}$$

Where, as usual, the brackets refer to the change in the quantity across the discontinuity, and the subscripts n and t refer to the normal and tangential components. These conditions may be simplified further for the discontinuity between the shock layer and the magnetosphere, since the earth's magnetic field does not appear to be connected to the interplanetary fields, that is,

$$v_n = H_n = 0 \quad [v_t] \neq 0 \quad [H_t] \neq 0$$
$$[\rho] \neq 0 \quad \text{and} \quad \left[p + \frac{H^2}{2}\right] = 0 \tag{6.45}$$

The gas pressure around the edge of the magnetosphere was then approximated by the simple newtonian flow relation

$$p = K\rho_\infty v_\infty^2 \cos^2 \psi \tag{6.46}$$

where the subscript ∞ refers to the undisturbed incident stream, ψ is the angle between a particular location and the undisturbed velocity vector, and K is a constant equal to 0.84 for $\gamma = 2$ and 0.88 for $\gamma = 5/3$. Also, estimates of the quantity of mass flow through the bow wave require it to be a fast shock wave.

Since the magnetic pressure is very small in the shock layer, the magnetic terms can be neglected in the flow equations and the problem is simplified to the calculation of the aerodynamic flow. The deformation of the magnetic field lines in the shock layer is then determined from the equations:

$$\begin{aligned} \nabla \times (\mathbf{H} \times \mathbf{v}) &= 0 \\ \nabla \cdot \mathbf{H} &= 0 \\ [H_n] &= 0 \end{aligned} \tag{6.47}$$

i.e., the magnetic field lines are frozen in the fluid.

The shape of the magnetosphere boundary was approximated by an axisymmetric body which was determined by rotating the magnetosphere boundary in the equatorial plane about the axis parallel to the undisturbed solar wind. The flow field streamlines in the shock layer were calculated, shown in Fig. 6.3, for a free-stream Mach number of 8.71 and a specific heat ratio of 2, which were considered reasonable choices based on representative values of the density, velocity and magnetic field in the solar wind. The length scale shown in the figure is one unit for the distance from the center of the earth to the magnetosphere nose which is the order of 10 earth radii. The associated constant velocity, mass flux, density, and temperature contours are shown in Fig. 6.4. The stagnation point temperature ratio is shown to be approximately 40. Thus, for a free-stream temperature of 50,000°K, the temperature will be

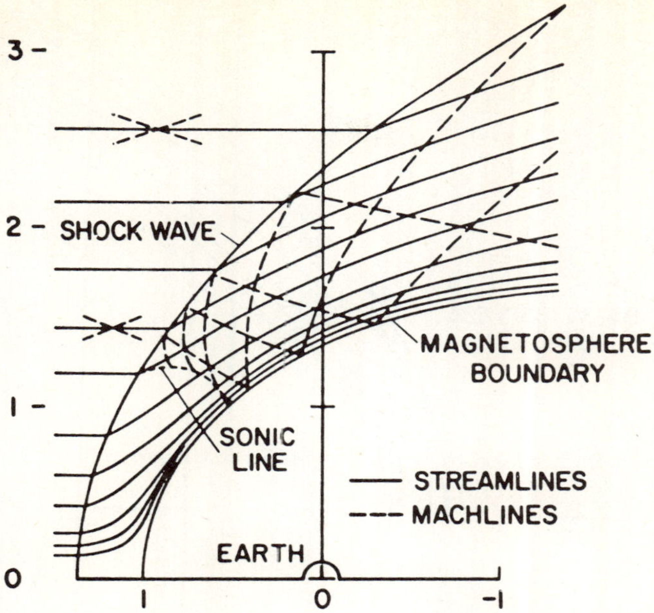

Fig. 6.3 Streamlines and wave patterns for supersonic flow past the magnetosphere; $M_\infty = 8.71$, $\gamma = 2$.

increased to about 2 million°K, which is the temperature of the gas before it leaves the solar corona. The shock detachment distance was calculated by Spreiter and Jones[5] as a function of the density ratio across the bow shock and is shown in Fig. 6.5. For the selected conditions, it is then approximately 40 percent of the distance from the center of the earth to the magnetosphere boundary, i.e., about 4 earth radii.

The distortion of the magnetic field in the shock layer was calculated from the equations in Eq. (6.47) after the velocity field was determined; Fig. 6.6 shows that the lines bend sharply as they cross the shock wave, except in the normal direction, and are constrained to the shock layer which produces a crowding in the stagnation region. All these results give a description of the average bulk properties of the flow which qualitatively agree with satellite measurements. The shock standoff distance has been used successfully in interpreting data from a number of satellites.

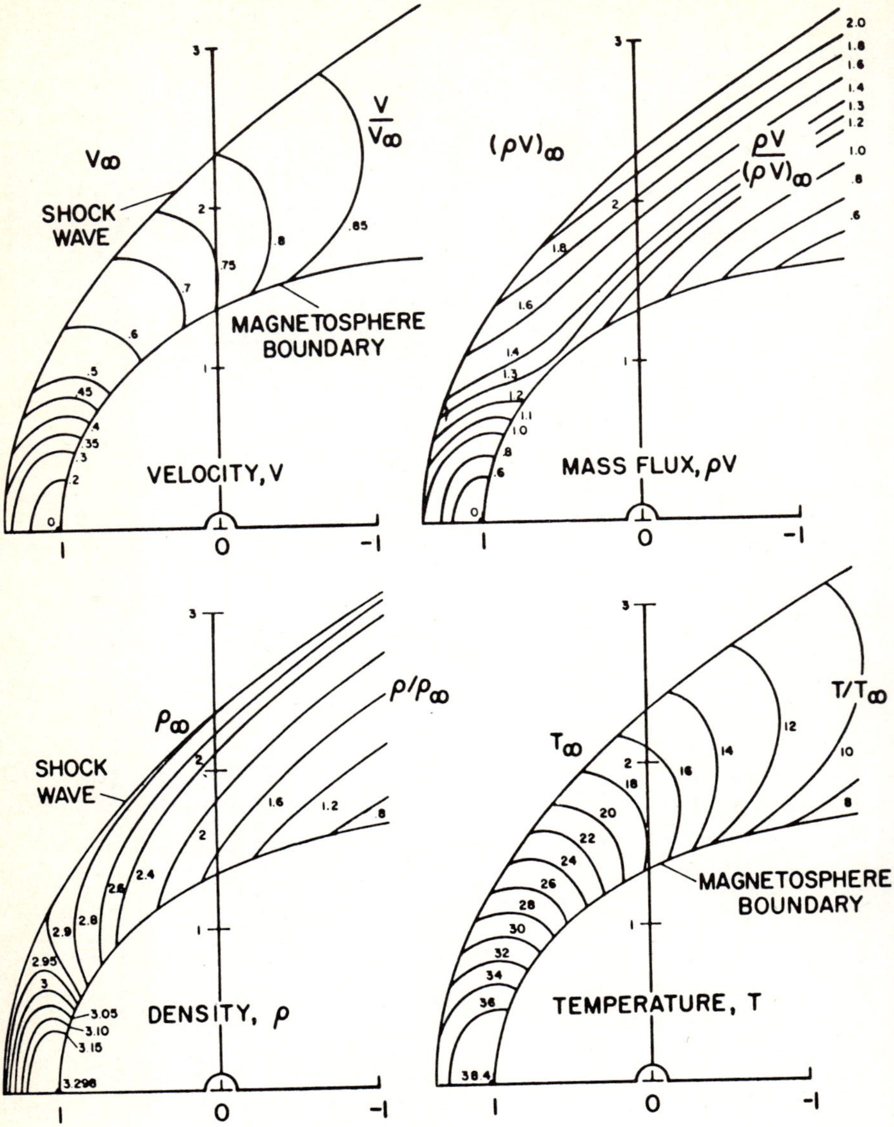

Fig. 6.4 Contour maps of constant velocity, mass flux, density, and temperature for supersonic flow past the magnetosphere; $M_\infty = 8.71$, $\gamma = 2$.

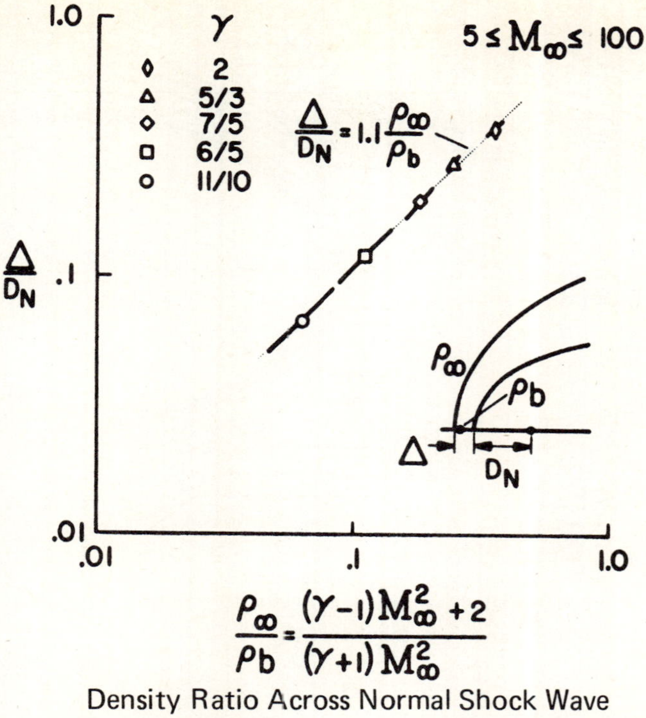

Fig. 6.5 Variation of standoff distance with density ratio across the nose of the bow shock wave.

6.4 ENGINEERING APPLICATIONS

In this section and those that follow, we will discuss a number of devices that have been designed to exploit MFD phenomena and one example in which the electrical conductivity of the medium produces a detrimental effect.

6.4.1 Propulsors. In the previous chapters we demonstrated that applying an electric field normal to both the magnetic field and flow directions could produce an increased flow rate. This MFD pumping can be used to produce thrust for the propulsion of a vehicle or to accelerate a secondary fluid in an ejector and several interesting applications follow.

Magnetofluid-dynamic submarines. Large submarines are being considered to transport cargo since at high speeds and at depths greater than four hull diameters, they are superior to surface ships

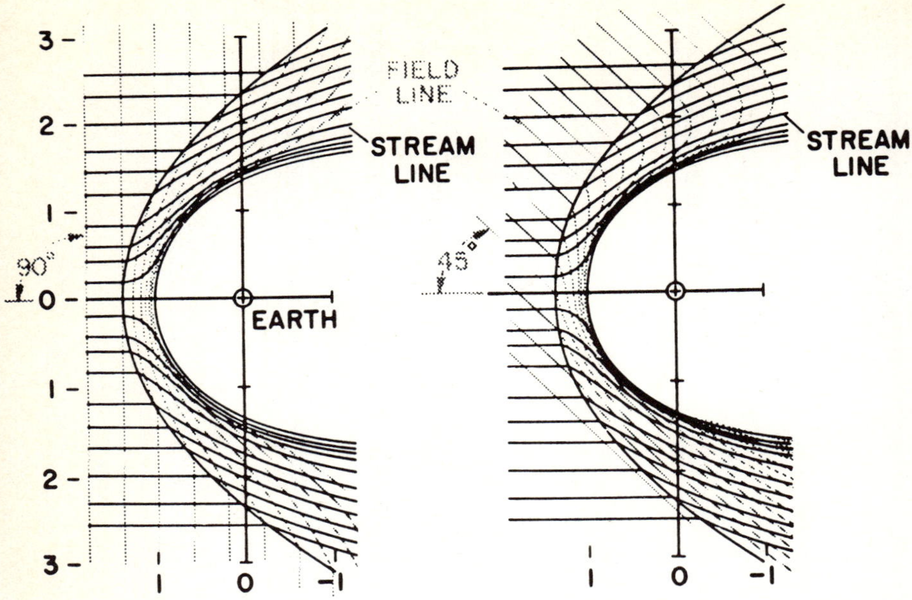

Fig. 6.6 Magnetic field lines in plane of free-stream velocity and magnetic-field vectors for supersonic flow past the magnetosphere; $M_\infty = 8.71$, $\gamma = 2$.

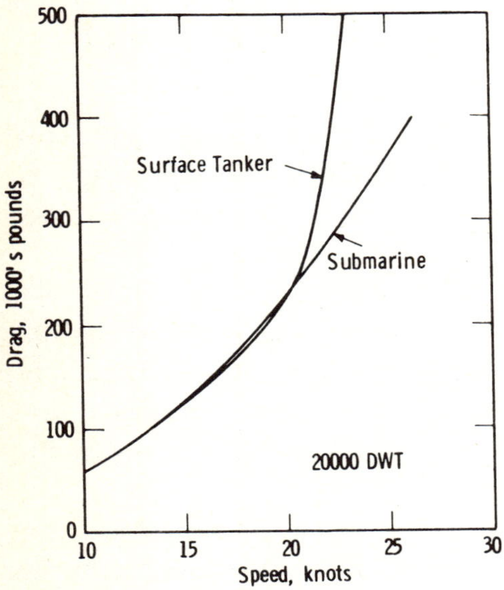

Fig. 6.7 Resistance of surface and submarine tankers of same dead weight.

for a given dead-weight tonnage. When conditions of speed and depth are met, submarines present a lower resistance than do surface ships; surface disturbances and formation of waves are eliminated.

Figure 6.7 shows this drag comparison for a fixed deadweight tonnage. The point of departure of the curves depends on the Froude number $u^2/g\Delta^{1/3}$; Δ is the submerged displacement, and it occurs at higher speeds for larger tankers.

Large cargo submarines are economically feasible only if powered by nuclear plants which will impose a greater initial cost than an equivalent surface ship. However, the size of these submarines is limited to the maximum draft that will permit their passage through the Panama and Suez Canals.

Magnetofluid-dynamic submarines were not an attractive alternative until large superconducting magnets became a possibility. They obtain their thrust from the Lorentz force produced by transverse electric and magnetic fields which pump the electrically conducting sea water through or past the submarine. The external field system will be discussed since a larger mass of fluid can be pumped by the fields and is the most efficient design. One disadvantage, however, is that the fields extend large distances from the hull and are detectable and may lift metal objects from the ocean floor. An example of the external field arrangement is shown in Fig. 6.8 for two electrodes that were designed to produce an electric field that is everywhere perpendicular to the magnetic field to obtain the maximum thrust.

The detailed analysis of the MFD submarine with $2\,k$ electrodes was provided by Stewart Way[6] and is repeated here. The geometry used in the analysis is shown in Fig. 6.9 for six electrodes but is applicable for $2\,k$ electrodes. The active length a of the electrodes is assumed to be at least four hull diameters and the flow along their length is assumed to have a uniform velocity and is two dimensional. The required electric and magnetic fields can be described in polar coordinates r and θ through the use of conjugate harmonic functions ϕ and ψ which are the real and imaginary parts of a function of a complex variable. They are

$$\mathbf{E} = -\frac{v_e}{\phi_e}\nabla\phi \qquad \mathbf{B} = -\frac{B_0 b}{k}\nabla\psi \tag{6.48}$$

and the functions

$$\phi = \frac{1}{2}\ln\frac{r^{2k} - 2r^k b^k \sin k\theta + b^{2k}}{r^{2k} + 2r^k b^k \sin k\theta + b^{2k}} \tag{6.49}$$

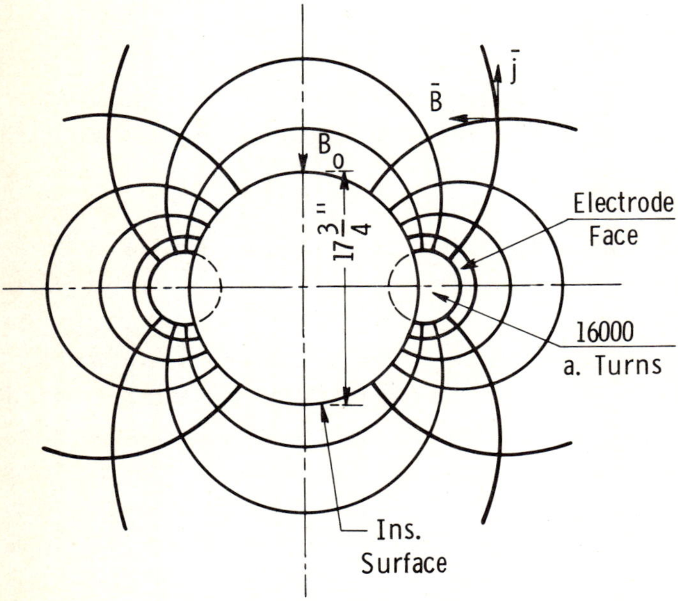

Fig. 6.8 Configuration of magnetic field and current flow pattern.

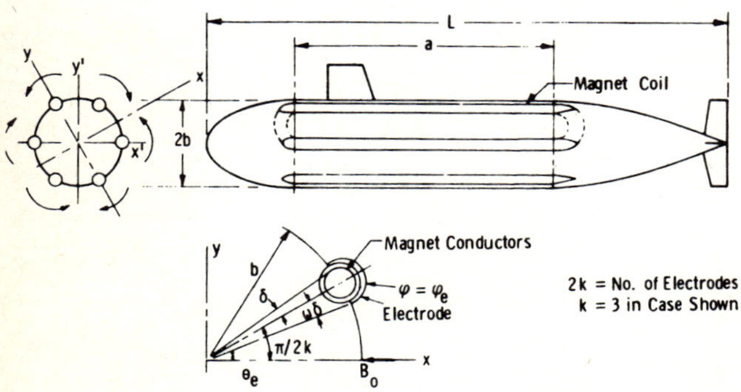

Fig. 6.9 Electromagnetic submarine configuration (schematic).

$$\psi = \tan^{-1}\left(\frac{r^k \sin k\theta - b^k}{r^k \cos k\theta}\right) - \tan^{-1}\left(\frac{r^k \sin k\theta + b^k}{r^k \cos k\theta}\right) \tag{6.50}$$

were obtained from

$$w = \phi + i\psi = \ln\left[\frac{(\xi/b)^k - i}{(\xi/b)^k + i}\right] \tag{6.51}$$

The magnitudes of the vectors **E** and **B** are then

$$E = -\frac{V_e h}{\phi_e} \qquad B = \frac{B_0 bh}{k} \tag{6.52}$$

and

$$h = \left|\frac{dw}{d\xi}\right| = |\operatorname{grad}\phi| = |\operatorname{grad}\psi| \tag{6.53}$$

ϕ_e is negative and is the value of ϕ at the electrode surface which is a surface of constant ϕ. V_e is one-half the voltage between the electrodes and B_0 is the magnetic induction at $r = b$ and $\theta = 0$.

The current density in the active region is

$$\mathbf{j} = \sigma(\mathbf{E} + \mathbf{u} \times \mathbf{B}) \tag{6.54}$$

and since $\mathbf{u} \times \mathbf{B}$ is in the direction opposite to $\mathbf{E}$, which produces $\mathbf{u}$ by the Lorentz force $\mathbf{j} \times \mathbf{B}$, it represents a loss. This expression may be simplified by defining

$$n = \frac{uB}{E} = -u\frac{B_0 b}{kV_e}\phi_e \tag{6.55}$$

and then the magnitude of the current density is

$$j = \sigma E(1 - n) = -\frac{V_e h\sigma}{\phi_e}(1 - n) \tag{6.56}$$

and n is a positive quantity less than unity. The magnitude of the Lorentz force per unit volume is then

$$jB = -V_e \frac{h^2 \sigma B_0 b(1 - n)}{\phi_e k} \tag{6.57}$$

The total force is obtained by integrating the Lorentz force over the 4 k sectors in the external region, that is,

$$F = 4ka \int_{\phi_e}^{0} \int_{-\pi/2}^{0} \frac{jB d\phi d\psi}{h^2} = -\frac{4\sigma B_0 ab(1 - n)}{\phi_e} \int_{\phi_e}^{0} \int_{-\pi/2}^{0} d\phi \, d\psi$$

$$= 2\pi\sigma B_0 V_e ab(1 - n) \tag{6.58}$$

The thrust power is defined as

$$P_T = Fu = 2\pi\sigma B_0 u V_e ab(1 - n) \tag{6.59}$$

and the current between two adjacent electrode is the integral of the current density along the x axis from the hull to ∞, that is,

$$I = a \int_{b}^{\infty} j_{y=0} dx = -\frac{\pi a}{2} \frac{\sigma V_e}{\phi_e} (1 - n) \tag{6.60}$$

where h in Eq. (6.56) was evaluated at $\theta = 0$. The total power required is

$$P_E = 4kV_e I = -2\pi ka\sigma \frac{V_e^2}{\phi_e} (1 - n) \tag{6.61}$$

and the propulsive efficiency is

$$\eta = \frac{P_T}{P_E} = -\frac{uB_0 b\phi_e}{kV_e} = n \tag{6.62}$$

This efficiency can be cast in a different form with the use of

$$2\pi\sigma B_0 V_e ab(1 - n) = \tfrac{1}{2}\rho u^2 S C_D \tag{6.63}$$

which requires the thrust produced by fields to equal the submarine's drag. C_D is the drag coefficient based on the wetted surface S. The efficiency is then

$$\eta = \frac{1}{1 - \alpha C_D k/2\xi \phi_e} \tag{6.64}$$

with

$$\xi = \frac{2a\sigma B_0{}^2}{\rho u} \qquad \alpha = \frac{S}{\pi b^2} \tag{6.65}$$

Since the largest thrust is desired, the magnetic field strength should be chosen as the largest or critical B_{cr} value for superconducting materials which will occur at the hull and $\theta = \pi/2k - \delta$, shown in Fig. 6.9. Thus,

$$B_0 = B_{cr} \sin k\delta \tag{6.66}$$

and the location where the electrode meets the hull is

$$\theta_e = \frac{\pi}{2k} - \omega\delta \tag{6.67}$$

ω is a constant that provides a small gap between the electrode and the magnetic coil for thermal insulation. The propulsive efficiency is now

$$\eta = \frac{1}{1 - \alpha k C_D/2\xi' \phi_e \sin^2 k\delta} \tag{6.68}$$

where

$$\xi' = \frac{2a\sigma B_{cr}{}^2}{\rho u}$$

Figure 6.10 is a plot of the efficiency as a function of C_D/ξ' for selected values of the geometric parameters. The efficiency is higher as the number of electrodes and the magnetic coil radius are decreased.

Way and Devlin[7] designed and constructed a model MFD submarine using the two-pole arrangement in Fig. 6.8, and as shown in Figs. 6.11 and 6.12. The design data are shown in Table 6.1.

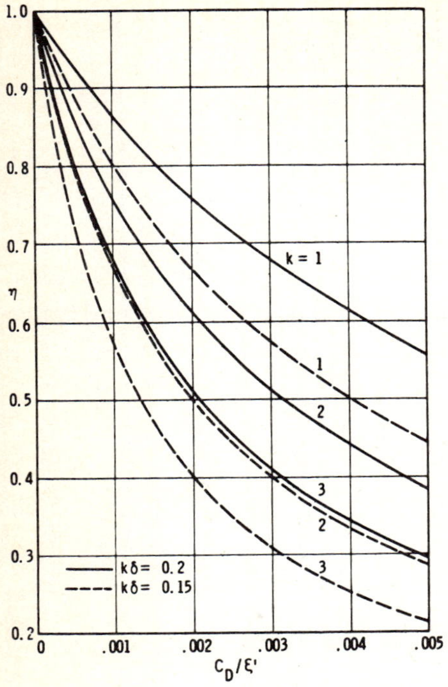

Fig. 6.10 Application to submarine tankers—propulsive efficiencies for $S/\pi b^2 = 27.6$ and $\omega = 1.1$.

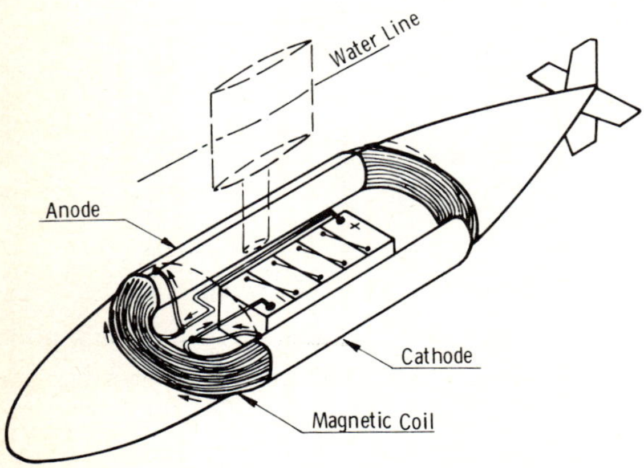

Fig. 6.11 Schematic view of model showing electrical connections.

Fig. 6.12 View of assembled model.

TABLE 6.1 Design data for model MFD submarine using two-pole arrangement.

Characteristic	Measurement
Length	3.05m (10 ft)
Displacement	408 kg (900 lb)
Diameter	0.45m (17.75 in.)
Active field length (a)	1.0m
Hull radius (b)	0.225m
Characteristic field (B_0)	0.015 weber/m^2
Assumed water conductivity in design (σ)	4 mho/m
Sea water density (ρ)	1030 kg/m^3
Design speed (u)	0.4 m/sec
Hull drag coefficient (C_D)	0.09 (based on cross section)
Electrode diameter	0.136m (5.4 in.)
Parameter (ϕ_e)	$-1.914 = \ln\epsilon/(2b + \epsilon)$
Interaction parameter (ξ)	$2a\sigma B_0{}^2/\rho u = 4.37 \times 10^{-6}$
Estimated drag (D)	1.18 newtons
Estimated thrust power (P_T) . . .	0.472 watt
Parameter ($K = E/uB$)	5,380
Magnet ampere turns	16,900
Electrode current (I)	91.3 amperes
Electrode height above hull outline	ϵ

The submarine was powered by five 217 amp-hour, 6-volt, lead-acid batteries which weighed 69 lb each and, when fully discharged in 20 minutes each, produced 378 watt-hours. The external and magnetic field circuits were in parallel and drew 110–120 amperes and 27 volts at the batteries. The center section was constructed from a welded 18-in.-diameter steel pipe with a 3/8-in. wall; the nose and tail sections were of reinforced epoxy fiberglass. During the tests, the electrode current was approximately 110–120 amperes instead of the design 91.3 amperes because the ocean had a higher conductivity of 4.2 mho/m and the fringing currents had been neglected in the design. Also the measured maximum magnetic field strength was higher at 0.0155 weber/m^2. The measured speeds ranged between 0.4 and 0.5 m/sec and successfully demonstrated the principle of the MFD submarine.

Magnetofluid-dynamic ejectors. A simple ejector uses a high-velocity primary jet to accelerate to a combined uniform exit velocity, a larger volume, lower velocity secondary jet. This process occurs over a distance in the flow direction called the *mixing distance*. The mixing is accomplished by the fluid's viscosity which diffuses away the initial vorticity between the primary and secondary jet. Since the mixing length is a function of the viscosity, it can't be controlled by external means. However, a magnetic field tends to suppress vorticity in an electrically conducting fluid and provides a possible method for controlling the mixing length externally. A schematic arrangement of an MFD ejector that was studied by Heiser[8] is shown in Fig. 6.13. The applied magnetic field induces currents in the flowing conducting fluid which produces Lorentz forces that retard the higher velocity jet and accelerate the secondary jet when the walls are electrically insulated. This latter effect will also produce a more uniform velocity and can be analyzed for a steady, inviscid, laminar flow of two jets of the same fluid when the magnetic field is uniform over the mixing length and the magnetic Reynolds number is very small. The analysis is considerably simplified by assuming that the velocities, currents, electric fields, and pressure are only functions of the axial coordinate x. Also, the exit static pressure is assumed to be equal to the stagnation pressure of the secondary jet at the inlet. Even though there is no wall

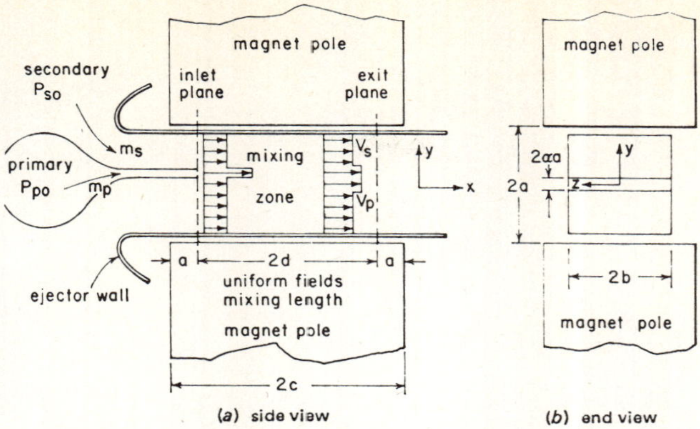

Fig. 6.13 Schematic representation of an MHD ejector.

function loss, by assumption, there will be pressure losses due to currents produced in the flow as it enters and leaves the magnetic field prior to the inlet plane and after the exit plane. Heiser accounted for these pressure losses by using the relation

$$\Delta p = (\beta \sigma b B^2) V = KV \tag{6.69}$$

which is applicable when the magnetic number squared, i.e., $R_m{}^2 = \sigma b B^2 / \rho V$ is less than 2. β is a constant that is equal to 0.11 for a square channel and 0.27 when channel height "a" is very small compared to the width "b." Two relations for the mass flow ratio are obtained with the momentum equations for both jets, i.e.,

$$p_{p0} - p_i = \frac{\rho V_{pi}{}^2}{2} + KV_{pi} \tag{6.70}$$

and

$$p_{s0} - p_i = \frac{\rho V_{si}{}^2}{2} + KV_{si} \tag{6.71}$$

When combined with the general momentum balance for the ejector between the inlet and exit planes

$$\dot{m}_p V_{pi} + \dot{m}_s V_{si} + p_i A = (\dot{m}_p + \dot{m}_s) V_e + p_{s0} A + KV_e A \tag{6.72}$$

and the continuity relations

$$\dot{m}_p = \alpha \rho A V_{pi} \tag{6.73}$$

$$\dot{m}_s = (1 - \alpha) \rho A V_{si} \tag{6.74}$$

and

$$\dot{m}_s + \dot{m}_p = \rho A V_e \tag{6.75}$$

give

$$\frac{1 - 2\alpha + 2\alpha^2}{2(1 - \alpha)^2}\left(\frac{\dot{m}_s}{\dot{m}_p}\right)^2 + 2\frac{\dot{m}_s}{\dot{m}_p} - \frac{1 - \alpha}{\alpha} = -\frac{KA}{\dot{m}_p}\left(\frac{2 - \alpha}{1 - \alpha}\frac{\dot{m}_s}{\dot{m}_p} + 1\right) \tag{6.76}$$

and

$$\Gamma\left(\frac{KA}{\dot{m}_p}\right)^2 + \left(\frac{1}{1 - \alpha}\frac{\dot{m}_s}{\dot{m}_p} - \frac{1}{\alpha}\right)\frac{KA}{\dot{m}_p} + \frac{1}{2}\left(\frac{1}{1 - \alpha}\frac{\dot{m}_s}{\dot{m}_p}\right)^2 = \frac{1}{2\alpha^2} \tag{6.77}$$

These last two equations must be solved simultaneously for the mass flow ratio $(\dot{m}_s/\dot{m}_p)$ and $KA/\dot{m}_p$ in terms of the constants α and Γ which are the ratio of the primary jet inlet area to the total flow area and

$$\Gamma = \frac{\rho(p_{p0} - p_{s0})}{K^2} \tag{6.78}$$

where p_{p0} and p_{s0} are the primary and secondary inlet stagnation pressures and p_i is the inlet static pressure.

The mixing length is obtained from the combined solution of the two momentum equations

$$\rho V_p \, dV_p = -dp - j_p B \, dx \tag{6.79}$$

$$\rho V_s \, dV_s = -dp - j_s B \, dx \tag{6.80}$$

and the Ohm laws

$$j_p = \sigma(E_z + V_p B) \tag{6.81}$$

$$j_s = \sigma(E_z + V_s B) \tag{6.82}$$

which gives

$$\ln\left(\frac{V_p}{V_e} - 1\right) + \frac{V_p}{V_e} + \left(\frac{1}{1 + \dot{m}_p/\dot{m}_s}\right)\left[\left(\frac{V_p}{V_e}\right)\left(1 + \frac{\dot{m}_s}{\dot{m}_p}\right) - 1\right]$$
$$- \left(\frac{1}{1 + \dot{m}_s/\dot{m}_p}\right) \ln\left[\frac{V_p}{V_e} - \left(\frac{1}{1 + \dot{m}_s/\dot{m}_p}\right)\right] = -\frac{\sigma B^2 x}{\rho V_e} \tag{6.83}$$

where the boundary condition $V_p/V_e = 1$ as $x \to \infty$ was used to evaluate the integration constant. The mixing length can now be computed for different mass flow ratios by solving Eq. (6.83) for the inlet value of $(\sigma B^2 x/\rho V_e)_i$ and subtracting it from the exit solution. The inlet velocity ratio is found by dividing Eqs. (6.73) and (6.75) to give

$$\left(\frac{V_p}{V_e}\right)_i = \frac{1}{\alpha(1 + \dot{m}_s/\dot{m}_p)} \tag{6.84}$$

and the exit velocity ratio is defined as

$$\left(\frac{V_p}{V_e}\right)_e - 1 = \frac{1}{10}\left[\left(\frac{V_p}{V_e}\right)_i - 1\right] \tag{6.85}$$

since the velocity across the ejector is then approximately uniform. The ratio of the mixing length and the channel width is then

$$\frac{d}{b} = \left(\frac{\beta N}{2}\right)\left(\frac{\dot{m}_p}{KA}\right)\left(1 + \frac{\dot{m}_s}{\dot{m}_p}\right) \tag{6.86}$$

where

$$N = \frac{2\sigma B^2 d}{\rho V_e} \qquad 2d = x_e - x_i$$

and is a function of α and Γ since the parameter $(\dot{m}_p/KA)$ and the mass flow ratio are also functions of α and Γ.

The pumping ratio and thrust augmentation parameter ϕ can be defined in terms of the ideal mass flow $\dot{m}_{p,\,\text{ideal}}$ which is the mass flow produced by exhausting the primary jet reversibly to the secondary reservoir pressure, i.e.,

$$\dot{m}_{p,\,\text{ideal}} = \alpha\rho A V_{p,\,\text{ideal}} = \alpha A[2\rho(p_{p0} - p_{s0})]^{1/2} \tag{6.87}$$

The pumping ratio is then

$$\frac{\dot{m}_s + \dot{m}_p}{\dot{m}_{p,\,\text{ideal}}} = \left(\frac{\dot{m}_p}{kA}\right)\left[\frac{1 + \dot{m}_s/\dot{m}_p}{\alpha(2\Gamma)^{1/2}}\right] \tag{6.88}$$

and the thrust augmentation parameter is

$$\phi = \frac{(\dot{m}_s + \dot{m}_p)V_e}{\dot{m}_p V_{p,\,\text{ideal}}} = \left(\frac{\dot{m}_p}{KA}\right)\frac{(1 + \dot{m}/\dot{m}_p)^2}{(2\Gamma)^{1/2}} \tag{6.89}$$

These analytical results were obtained for laminar, inviscid jets and are applicable when

$$1600N\left(\frac{d}{b}\right)^2 < \text{Re} < 6.25 \times 10^4 N \tag{6.90}$$

and where $\text{Re} = 2\rho V_e d/\mu$ is the Reynolds number. The lower limit must be exceeded for the flow to be considered inviscid and the upper limit must not be exceeded for the flow to remain laminar.

The numerical results are shown in Figs. 6.14–6.16 for the mixing length, pumping ratio, and the thrust augmentation. An increase in the magnetic field strength will reduce the parameter Γ, considerably reducing the mixing length. A comparable mixing length for a nonmagnetic ejector is 7 and thus, for square channels, can be significantly reduced for a Γ of 1,000. Results for the channel with $a \ll b$ can be obtained from

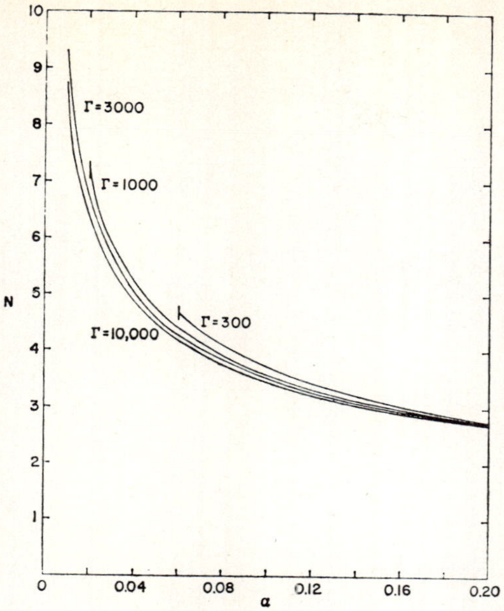

Fig. 6.14 Dimensionless mixing length as a function of α and Γ.

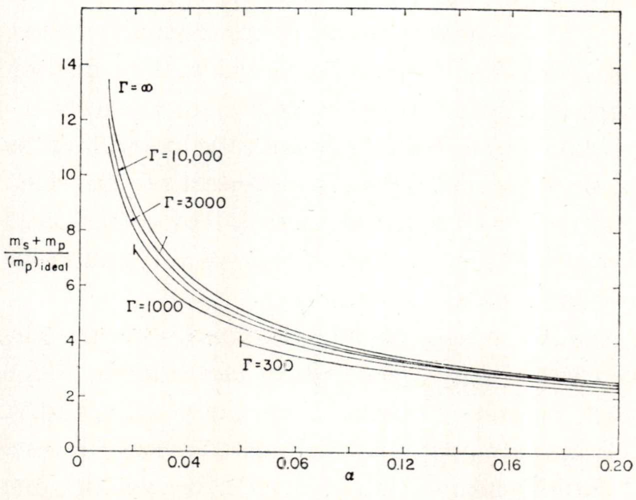

Fig. 6.15 Pumping ratio as a function of α and Γ.

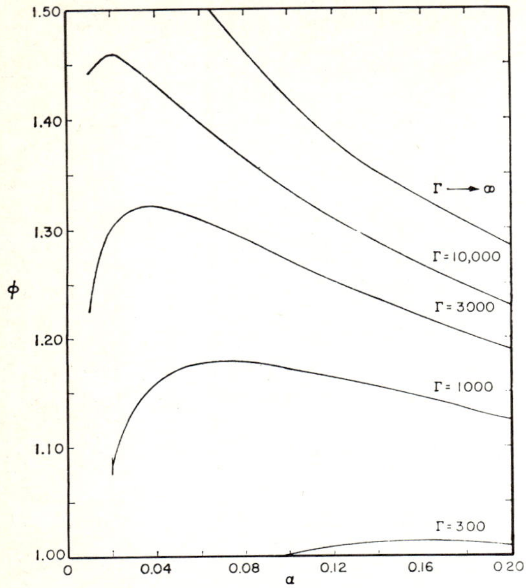

Fig. 6.16 Thrust augmentation as a function of a and Γ.

$$\left(\frac{d}{b}\right)_{a \ll b} = 2.4\left(\frac{d}{b}\right)_{a=b} \tag{6.91}$$

The reduction of the mixing length is gained at the expense of the pumping ratio and thrust augmentation which is apparent in the other figures where $\Gamma \to \infty$ corresponds to the nonmagnetic ejector. The pumping ratio loss for $\Gamma = 1{,}000$ is not as serious as the reduction in the thrust augmentation which is very sensitive to changes in Γ. However, if Γ is designed to be 3,000 or greater, the loss will not be too large and a modest reduction in the mixing length will be gained along with the added feature of external control. Any reduction in vorticity between the jets by the magnetic field will also reduce the mixing length to provide an additional benefit.

Magnetofluid-dynamic accelerators. Magnetofluid dynamic accelerators are part of a larger class called *plasma accelerators*, which employ electric and magnetic fields to produce their thrust. They are generally classed into continuous or pulsed devices with and without electrodes. Besides being electromagnetic devices, they are characterized by their performance capabilities which are:

Specific impulses . . . $I_{sp} = V_{Exhaust}/g$	greater than 1,000 sec
Thrust-to-weight ratios	less than unity

and are always operated supersonically so that the Lorentz force produces thrust. MFD accelerators have a potential performance capability less than 4,000 sec while chemical rockets are less than 400 sec and liquid hydrogen rockets produce less than 1,000 sec, and both the latter have thrust-to-weight ratios greater than unity. Obviously, plasma accelerators are useful as engines only in space and must be carried there by other type rockets. However, once in space, these plasma accelerators are superior to other engines for long missions because of their higher specific impulses. These higher specific impulses are produced through higher operating temperatures and direct energy addition by electric and magnetic fields. Chemical rockets are limited by the total energy available in the combustion gases. Higher operating temperatures are required to produce an electrically conducting gas but considerable energy is required to produce the disassociation and ionization which is lost since it remains in the exhaust gases. Also, higher temperatures produce a greater heat loss to the walls that can be reclaimed for preheating. Since the accelerator efficiency is the ratio of the thrust energy to the total energy input, the thermal energy in the exhaust gases must be minimized. This can be accomplished, in part, by seeding lower temperature gases with an easily ionized material such as cesium or potassium. High efficiencies are essential to reduce the weight of the electric and magnetic equipment and high specific impulses are required to reduce the total mass expended during long missions.

Magnetofluid-dynamic accelerators are designed with the "approximately" one-dimensional theory (discussed in Chap. 3) that assumes the gas obeys the perfect gas law. The electrical conductivity is a constant scalar quantity; the magnetic Reynolds number is small; and the electric and magnetic fields are arbitrary functions of the axial distance. Exact solutions of the equations are possible when friction and heat losses are disregarded and if any one of the variables is held constant, i.e., constant area, constant velocity, constant temperature, constant density, constant pressure, constant Mach number, and constant $p^{\epsilon-1}A^{\epsilon}$. Baker and Rogers[9] have provided extensive numerical results for these cases for selected inlet Mach numbers and specific heat ratios. In Chap. 3, we remarked that constant-flow property solutions are of limited interest to engineers, since comparison of different operating conditions required comparing channels with different geometries. However, such solutions are useful for selecting engine types in system feasibility studies and Marston[10] has condensed the many results into a general and compact form which will be presented here.

The approximately one-dimensional MFD equations, neglecting friction and heat-transfer losses, are:

$$\rho u \frac{du}{dx} + \frac{dp}{dx} = \sigma B^2 u(K - 1) \tag{6.92}$$

$$\frac{dc_p T}{dx} + u\frac{du}{dx} = \frac{\sigma B^2 u}{\rho} K(K - 1) \tag{6.93}$$

$$p = \rho R T = \left(\frac{\gamma - 1}{\gamma}\right)\rho c_p T \tag{6.94}$$

where $K = E/uB$ is the loading constant which is greater than unity for an accelerator; thus the induced Lorentz force represents a loss. The Maxwell equations are not included; since the electric and magnetic fields are arbitrary functions of x they need not be satisfied explicitly. However, if the magnetic field is taken as arbitrary, then the electric field function is determined by keeping K a constant. The required length of the electromagnetic interaction is determined from the interactor's parameter

$$L = \int_0^l \frac{\sigma B^2}{\rho_a u_a} dx \tag{6.95}$$

where the subscript a refers to the value at the channel entrance. Marston obtained the more general solution by requiring the ratio of the change in kinetic energy to the change in stagnation enthalpy to be constant through the accelerator rather than a particular flow property, i.e.,

$$\lambda = \frac{u^2/2RT_0 - u_a^2/2RT_0}{H_s - H_{s,a}} = \text{const} \tag{6.96}$$

where

$$H = \frac{c_p T}{RT_0} + \frac{u^2}{2RT_0} \tag{6.97}$$

the subscript zero refers to an assumed reference state and the subscript s to the stagnation value. The pressure-temperature relation that satisfies these equations has the form of the polytropic relation, i.e.,

$$\frac{p}{p_a} = \left(\frac{T}{T_a}\right)^{\bar{\mu}\frac{\gamma}{\gamma-1}} \tag{6.98}$$

where the parameter

$$\bar{\mu} = \left(\frac{\gamma - 1}{\gamma}\right)\frac{\ln(p_b/p_a)}{\ln(T_b/T_a)} = \frac{1 - \lambda K}{(1 - \lambda)K} \qquad (\lambda \neq 1)$$

equals unity for constant entropy. The regions of operation of the channel can now be represented on a map, Fig. 6.17, where λ and $\bar{\mu}$ are the coordinates and the constant velocity, constant temperature ($\lambda = 1$), constant pressure ($\lambda = 1/K$), and constant entropy ($\bar{\mu} = 1$) solutions are straight lines. The constant temperature, constant entropy, and $K \to \infty$ lines divide the map into the different regions of operation of the channels and each point in the map represents a channel operating condition. The pressure relation is used to integrate Eq. (6.93) and the result is used to evaluate the interaction parameter L to give:

$$L = \frac{\displaystyle\int_1^{T_b/T_a} \frac{(T/T_a)^{\bar{\mu}x/(\gamma-1)-1}\,d(T/T_a)}{\{(T/T_a - 1)[\lambda/(1-\lambda)] + (\gamma - 1)M_a^2/2\}^{1/2}}}{M_a[2(\gamma - 1)]^{1/2}(1 - \lambda)(K - 1)K} \qquad (\lambda \neq 1) \tag{6.99}$$

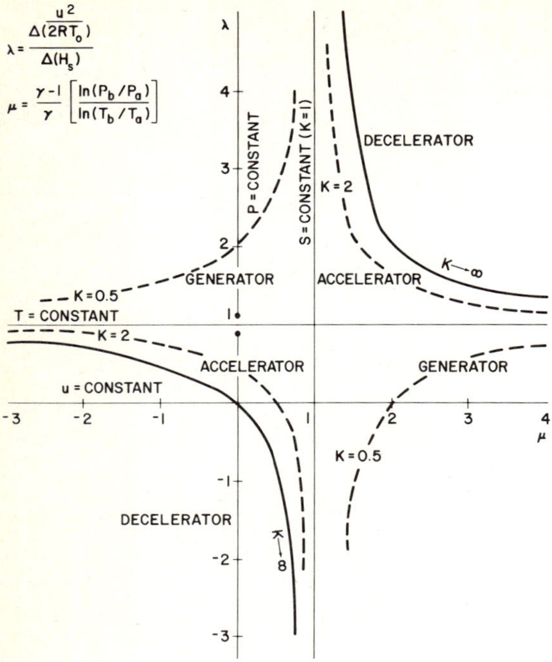

Fig. 6.17 One-dimensional perfect gas, MHD flow.

where the mass velocity is

$$\rho_a u_a = M_a p_a \left(\frac{\gamma}{RT_a}\right)^{1/2} \tag{6.100}$$

and the subscript b represents quantities evaluated at the channel exit. The interaction parameter is shown in Fig. 6.18 as a function of the entropy change and the enthalpy change through a constant pressure generator. This figure can be generally used since changes in the pressure ratio in the range $0.2 < p_b/p_a < 2.0$ produces only ±5-percent deviations in the entropy. Solution of the equations for a constant interaction parameter, while varying $\bar{\mu}$ and λ or T_b/T_a and the area ratio, are shown in Fig. 6.19 for three values of L and for M_a equal to unity. Significant velocity increases are possible by the proper selection of the pressure ratio and interaction parameter for a given constant area or constant temperature channel. The entropy change in the channel can then be estimated from Fig. 6.18.

These results were obtained by neglecting friction and heat transfer and an estimate of their effect can be obtained from Fig. 6.20 provided by Baker and Rogers for a constant area channel. They obtained a closed form solution of the equations including both friction and heat transfer by assuming that the friction factor was a constant fraction K of the Lorentz force and the heat transfer is from the gas to a cold wall $T_\omega/T \ll 1$. The performance is considerably degraded at the higher efficiencies by the friction, but when the heat transfer is included the performance is improved.

A survey of experimental MFD accelerators was provided by Friedel and Lackner[11] and Table 6.2 gives representative values obtained by the various investigators. The specific impulses are very low, that is, 10–30 sec, but MFD accelerators are in an early stage of development.

6.4.2 Wave Propagation. The successful landing of the lunar excursion module LEM depended on the accuracy of the information obtained from its Doppler and position radar. These radar signals must pass through the LEM's rocket exhaust plume which contains ionized sodium impurities that will attenuate and refract their signals and shift their phase velocity. The exhaust plume is a complex flow that varies from a continuum gas near the nozzle to a free molecule gas in the outer regions. However, Luzzi and Grunberg[12] simplified the analysis of the distortions of the radar signals by assuming constant

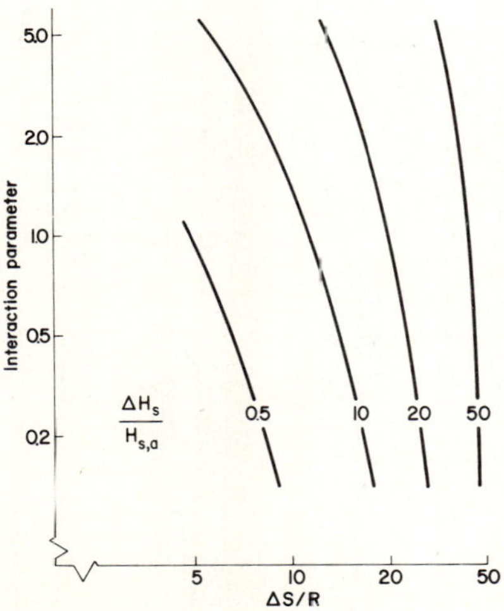

Fig. 6.18 Interaction parameter required for specified enthalpy and entropy increments at constant pressure ($M_a = 1$). Entropy increment varies from these curves approximately ±5% for a pressure ratio range 0.2 $< P_b/P_a <$ 2.0.

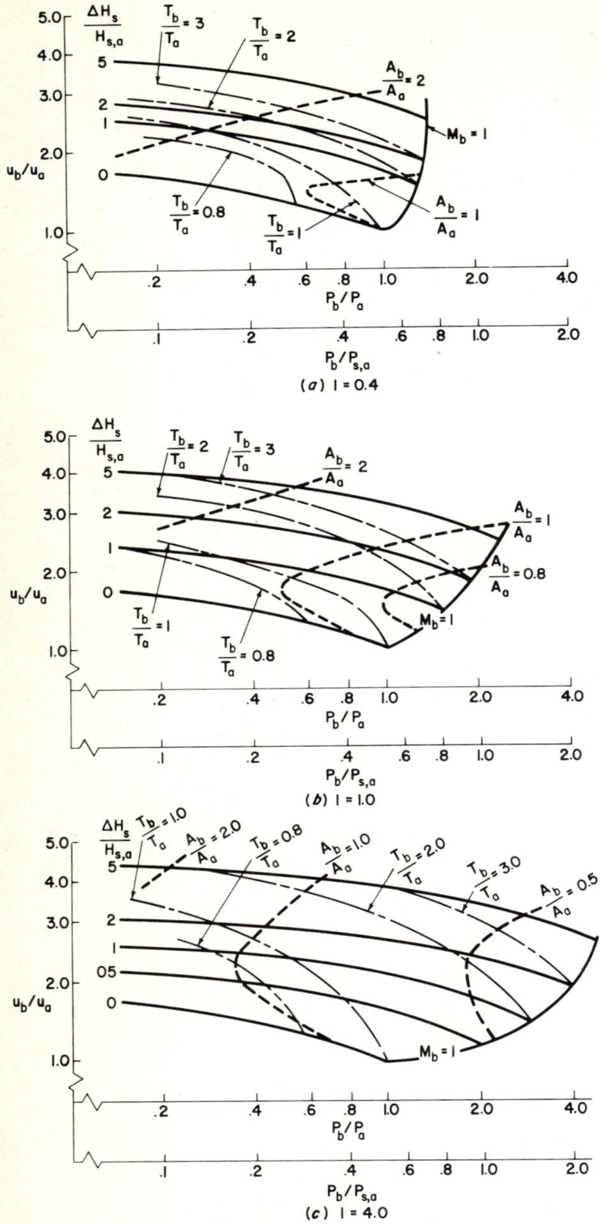

Fig. 6.19 Performance capability of a perfect gas, constant loading, constant γ, MHD accelerator ($\gamma = \frac{5}{3}$, $c_p/R = 2.5$, $M_a = 1$).

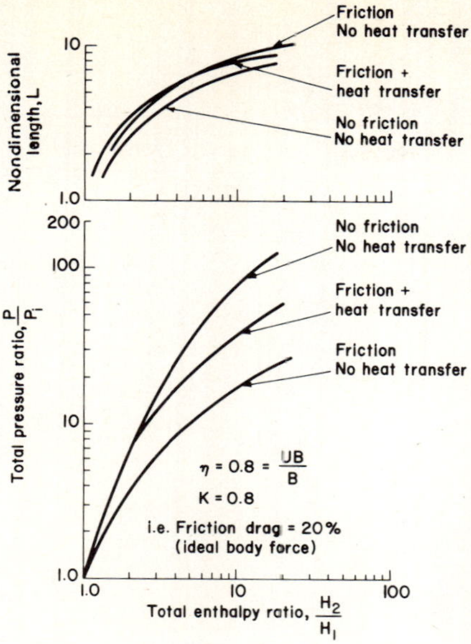

Fig. 6.20 Electromagnetic accelerator performance comparison, initial mach number, $(M_i) = 1.0$.

properties in the portion of the plume that produces the greatest effect. This selection was made after a study of the Mach number profiles in the plume which showed that the Doppler signals would pass through approximately 10 ft of the dense Mach 6 region. The properties were calculated with the use of the isentropic pressure and temperature relations along with the perfect gas law and with the electrical conductivity model provided by Drude.[13] The analysis of the wave distortion was then accomplished using Maxwell's equations and Ohm's law.

By assuming that the fields have a time harmonic form, i.e.,

$$\mathbf{A} = \mathbf{A} e^{i\omega t} \tag{6.101}$$

then the Maxwell equations and Ohm's law become

$$\nabla \times \mathbf{H} = \mathbf{J} + i\omega\epsilon_0 \mathbf{E} \tag{6.102}$$

$$\nabla \times \mathbf{E} = -i\omega\mu_0 \mathbf{H} \tag{6.103}$$

$$\nabla \cdot \mathbf{E} = \frac{\rho}{\epsilon_0} \tag{6.104}$$

TABLE 6.2 Parameters of Plasma Accelerators

Parameter	Investigator				
	Wood	Wood	Oates	Ragusa	Demetriades
	N_2	N_2	A	A	A
E,TH*	E	E	E,TH	E	E
type	Faraday				
$\rho\,[10^{-6}\,g\,cm^{-3}]$	6	8	160	5.4	2.3
p [atm]	0.75	0.08	1	0.05	8.10^{-3}
$T\,[10^3\,°K]$	5.4	7	3	4	1.4
$v\,[10^6\,cm\ \ sec^{-1}]$	0.2	0.3	0.1	0.15	0.3
$B\,[10^3 G]$	10	11	10	6	4.6
σ [mho cm^{-1}]	1	2	3.2	1	1
M	2	2.4	1	2.7	4
R_H	0.025	0.06	0.045	0.017	0.055
S	1	0.7	0.2	0.5	0.3
β	—	—	—	—	—
L [cm]	8.7	7	10	8	10
τ_{ion} [cm]	0.15	0.2	0.1	0.15	0.6
$\tau_{electr.}$[cm]	2.5×10^{-6}	3×10^{-6}	1.3×10^{-6}	1.8×10^{-6}	7×10^{-6}
λ [cm]	0.02	0.025	10^{-4}	3.10^{-4}	6×10^{-3}
τ_{tr}[sec]	4.3×10^{-5}	2×10^{-5}	10^{-4}	5.5×10^{-5}	3×10^{-5}
λ_D[cm]	1.5×10^{-7}	1.1×10^{-7}	$7{,}2.10^{-9}$	2.1×10^{-7}	5.4×10^{-8}

*E—experimentally tested device
TH—results of feasibility studies

$$\nabla \cdot \mu_0 H = 0 \tag{6.105}$$

and

$$\mathbf{J} = \sigma \mathbf{E} \tag{6.106}$$

where $i = \sqrt{-1}$ and the second term on the right-hand side of Eq. (6.102) is the displacement current. Now we will examine the behavior of a wave that is transverse to the direction of propagation by combining Eqs. (6.102), (6.103), and (6.106) to give

$$\nabla \times (\nabla \times \mathbf{E}) = -i\omega \nabla \times \mathbf{H} = -i\omega(\sigma + i\omega\epsilon)\mathbf{E} \tag{6.107}$$

or

$$\nabla^2 \mathrm{E} - (i\omega\sigma - \omega^2\epsilon)\mathrm{E} = 0 \tag{6.108}$$

since $\nabla(\nabla \cdot \mathrm{E}) = 0$. By defining the complex wave number as

$$k^2 = -(i\omega\sigma - \omega^2\epsilon) \tag{6.109}$$

then

$$\nabla^2 \mathrm{E} + k^2 \mathrm{E} = 0 \tag{6.110}$$

which describes the electric field wave that has the general form

$$\mathrm{E} = E_0 e^{i(\omega t - kz)} \tag{6.111}$$

and propagates in the z direction.

Since Eq. (6.109) contains the term $-i\omega\sigma$, then the conductivity produces a damping term which can be appreciable when the plasma's conductivity is large. Now when the conductivity and permittivity of the medium are zero then $\nabla \times \mathbf{H} = 0$ and there is no electromagnetic wave. However, the electric field may still have longitudinal oscillations at the plasma frequency defined as

$$\omega_p = e\sqrt{\frac{n_e}{m}} \tag{6.112}$$

where e is the unit electric charge, n_e the number density of the electrons, and m the electron mass. The required relations assumed for the electrical conductivity and permittivity are

$$\sigma = \frac{\nu \omega_p{}^2}{\nu^2 + \omega^2} \tag{6.113}$$

and

$$\epsilon = 1 - \frac{\omega_p{}^2}{\nu^2 + \omega^2} \tag{6.114}$$

where ν is the frequency of collision of the electrons with neutral particles in the plume.

The dispersion relation, i.e., relation between the frequency and wave numbers, can now be obtained by substituting Eqs. (6.113) and (6.114) into the relation for k^2:

$$k^2 = -i\omega\left[\frac{\nu\omega_p^2}{\nu^2 + \omega^2} + i\omega\left(1 - \frac{\omega_p^2}{\nu^2 + \omega^2}\right)\right] \tag{6.115}$$

or

$$k = \omega\left(1 - \frac{\omega_p^2}{\nu^2 + \omega^2} - i\,\frac{\omega_p^2}{\nu^2 + \omega^2}\,\frac{\nu}{\omega}\right)^{1/2} \tag{6.116}$$

Since the Doppler frequency, 13.5 KMC, is much larger than the plasma frequency, $\omega_p = 1.31 \times 10^9$, then

$$\frac{\omega_p^2}{\nu^2 + \omega^2} \ll 1 \qquad \text{and} \qquad \left(\frac{\omega_p^2}{\nu^2 + \omega^2}\right)\frac{\nu}{\omega} \ll 1$$

regardless of the value of the collision frequency ν. Thus, expanding Eq. (6.116) with the binomial theorem

$$k \simeq \omega\left[1 - \frac{1}{2}\,\frac{\omega_p^2}{\nu^2 + \omega^2} - \frac{i}{2}\left(\frac{\omega_p^2}{\nu^2 + \omega^2}\right)\frac{\nu}{\omega}\right] \tag{6.117}$$

and since $\nu = 1.32 \times 10^9$ is much smaller than the Doppler frequency, then

$$k \simeq \omega\left[1 - \frac{1}{2}\left(\frac{\omega_p}{\omega}\right)^2 - \frac{i}{2}\left(\frac{\omega_p}{\omega}\right)^2\frac{\nu}{\omega}\right] \tag{6.118}$$

is the complex dispersion relation which can be used in the general representation of the electric field wave. Another useful result is obtained from the definition of the index of refraction, i.e.,

$$N^2 = \frac{k_r^2}{\omega^2} \tag{6.119}$$

where k_r is the real part of the wave number. Using Eq. (6.118)

$$k_r^2 \simeq N^2\omega^2 = \left[1 - \frac{1}{2}\left(\frac{\omega_p}{\omega}\right)^2\right]\omega^2 \tag{6.120}$$

and thus

$$N = \left[1 - \frac{1}{2}\left(\frac{\omega_p}{\omega}\right)^2\right]^{1/2} \simeq 0.995 \tag{6.121}$$

which demonstrates that the radar beam is refracted upon passing through the plume. The wave phase velocity is then increased past the light speed C, i.e.,

$$V_\phi = \frac{C}{N} = \frac{3 \times 10^{10}}{0.995} = 3.015 \times 10^{10} \text{ cm/sec} \tag{6.122}$$

The depth of penetration of the wave is determined from the imaginary part of k

$$\delta = \frac{C}{|k_I|} = 2\left(\frac{\omega}{\omega_p}\right)^2 \frac{C}{\nu} = 4.6 \times 10^3 \text{ cm} \tag{6.123}$$

and is the distance required to damp the initial amplitude by a factor e^{-1}.

The numerical example demonstrated that the Doppler radar beam would be seriously affected and it was subsequently redirected.

6.4.3 Magnetofluid-Dynamic Lubrication. Chawla[14] has shown that when crossed magnetic and electric fields are applied to a slider bearing containing an electrically conducting fluid, the load carrying capacity can be increased considerably. His bearing was formed by positioning the slider at a small angle $\theta = \alpha$ to the bearing surface $\theta = 0$ in a cylindrical coordinate system (ζ, θ, z) fixed to the slider and with their intersection line at the z axis. The azimuthal magnetic field was assumed to be formed by a current along the z axis and the constant electric field in the z direction is produced between electrodes fastened to the slider ends which are far enough apart so that there are no changes in the z direction. The slider which is moving at a velocity u_0 and the bearing surface are assumed to be insulated; the length of the slider is l. The outlet is located a distance a from the z axis and h_1 and h_2 are the widths of the inlet and outlet. Since the angle α is small, then $h_2/\alpha = \tan\alpha$. The magnetic field, Ohm's law, continuity and momentum equations are then

$$\mathbf{H} = (0, H_0/\zeta, 0) \tag{6.124}$$

$$\mathbf{J} = \sigma(\mathbf{E} + \mu_e \mathbf{u} \times \mathbf{H}) \tag{6.125}$$

$$\nabla \cdot \mathbf{u} = 0 \tag{6.126}$$

$$0 = -\nabla p + \mu \nabla^2 \mathbf{u} + \mu_e \mathbf{J} \times \mathbf{H} \tag{6.127}$$

where the inertia terms have been omitted since the flow is approximately a Couette flow when α is small. Using Ohm's law, the momentum and continuity equations in the cylindrical coordinate system become

$$0 = -\frac{\partial p}{\partial \zeta} + \mu\left(\frac{\partial^2 u}{\partial \zeta^2} + \frac{1}{\zeta}\frac{\partial u}{\partial \zeta} + \frac{1}{\zeta}\frac{\partial^2 u}{\partial \theta^2} - \frac{u}{\zeta^2} - \frac{a}{\zeta^2}\frac{\partial v}{\partial \theta}\right) - \frac{\mu_e \sigma H_0}{\zeta}\left(E_z + \frac{\mu_e H_0}{\zeta} u\right) \tag{6.128}$$

$$0 = -\frac{1}{\zeta}\frac{\partial p}{\partial \theta} + \mu\left(\frac{\partial^2 v}{\partial \zeta^2} + \frac{1}{\zeta}\frac{\partial v}{\partial \zeta} + \frac{1}{\zeta^2}\frac{\partial^2 v}{\partial \theta^2} - \frac{v}{\zeta^2} + \frac{2}{\zeta^2}\frac{\partial u}{\partial \theta}\right) \tag{6.129}$$

$$\frac{\partial \zeta u}{\partial \zeta} + \frac{\partial v}{\partial \theta} = 0 \tag{6.130}$$

and the boundary conditions are

$$\begin{aligned} \theta &= 0 \quad & u &= -U_0 \quad & v &= 0 \\ \theta &= \alpha \quad & u &= 0 \quad & v &= 0 \\ \zeta &= a \quad & p &= 0 & & \\ \zeta &= a + l \quad & p &= 0 & & \end{aligned} \tag{6.131}$$

These equations are nondimensionalized by the relations

$$\begin{aligned}
u &= \bar{u}U_0 \qquad \zeta = a + lx = a(1 + bx)\\
v &= \frac{h_2}{l}\bar{v}U_0 \qquad \theta = \alpha y = \frac{h_2}{a}y\\
p &= \frac{ulU_0}{h_2^2}P \qquad E_z^* = \frac{E_z h_2}{U_0}\sqrt{\frac{\sigma}{\mu}}\\
M &= \frac{\mu_e H_0 h_2}{a}\sqrt{\frac{\sigma}{\mu}} \qquad b = \frac{l}{a} = \frac{h_1}{h_2} - 1
\end{aligned} \tag{6.132}$$

and can be simplified by neglecting second and higher powers of h_2/l which is small compared to l and a. The resulting equations are

$$0 = -\frac{\partial P}{\partial x} + \frac{1}{(1 + bx)^2}\frac{\partial^2 \bar{u}}{\partial y^2} - \frac{M}{(1 + bx)}\left(E_z^* + \frac{M\bar{u}}{1 + bx}\right) \tag{6.133}$$

$$0 = \frac{\partial P}{\partial y} \tag{6.134}$$

$$\frac{\partial}{\partial x}\int_0^1 \bar{u}(1 + bx)\,dy = 0 \tag{6.135}$$

and the boundary conditions are

$$\begin{aligned}
y &= 0 \qquad \bar{u} = -1 \qquad \bar{v} = 0\\
y &= 1 \qquad \bar{u} = 0 \qquad \bar{v} = 0\\
x &= 0 \qquad P = 0\\
x &= 1 \qquad P = 0
\end{aligned} \tag{6.136}$$

Chawla then solved these equations for the velocities and pressure which are

$$\begin{aligned}
\bar{u} &= \left(\frac{\cosh M}{\sinh M}\sinh My - \cosh My\right)\\
&+ \left(\cosh My - \frac{\cosh M \sinh My}{\sinh M} + \frac{\sinh My}{\sinh M} - 1\right)\left[\left(\frac{\cosh M - 1}{M\sinh M - 2\cosh M + 2}\right)\right.
\end{aligned} \tag{6.137}$$

$$\left(\frac{2(1+b)}{(2+b)(1+bx)} - 1\right) + \frac{E_z^*}{M}\frac{2(1+b)^2 \log(1+b)}{b(2+b)(1+bx)}\Bigg] \qquad (6.137)\ (\text{cont'd})$$

$$\bar{v} = \frac{b}{M \sinh M}\left\{[\cosh M - \cosh M(y-1)] + \frac{\cosh M - 1}{M \sinh M - 2\cosh M + 2}\right.$$

$$\left. \times [\cosh My + \cosh M - My \sinh M - 1 - \cosh M(y-1)]\right\} \qquad (6.138)$$

$$P = \left[\frac{M^2(\cosh M - 1)}{M \sinh M - 2\cosh M + 2}\right]\frac{bx(1-x)}{(2+b)(1+bx)^2}$$

$$+ E_z^* M\left\{\left(\frac{1+b}{b}\right)^2 \frac{\log(1+b)}{2+b}\left[1 - \frac{1}{(1+bx)^2}\right] - \frac{1}{b}\log(1+bx)\right\} \qquad (6.139)$$

The load carrying capacity per unit length of the slider bearing is then

$$W^* = \frac{Wh_2^2}{\mu U_0 l^2} = \int_0^1 P\,dx = \left[\frac{M^2(\cosh M - 1)}{M \sinh M - 2\cosh M + 2}\right]$$

$$\left\{\frac{1}{b}\left[\frac{1}{b}\log(1+b) - \frac{2}{2+b}\right]\right\} + \frac{E_z^* M}{b}\left[1 - \frac{2(1+b)}{b(2+b)}\log(1+b)\right] \qquad (6.140)$$

and the corresponding nonmagnetic capacity is

$$W_0^* = \frac{6}{b}\left[\frac{1}{b}\log(1+b) - \frac{2}{2+b}\right] \qquad (6.141)$$

The load carrying capacity is a function of the Hartmann number for a given b and E_z^* and demonstrates that it can be increased substantially by magnetic and electric fields. Results for the special cases of an open circuit ($I = 0$) and short circuit ($E_z^* = 0$) are shown in Table 6.3 which shows that in the open circuit case the load capacity increases significantly with the Hartmann number.

TABLE 6.3 Results for special open-circuit and short-circuit cases

M	W_{oc}/W_0		W_{sc}/W_0
	$b = 1$	$b = 2$	
0	1	1	1
1	1.13	1.14	1.02
2	1.57	1.51	1.06
3	2.17	2.05	1.14
4	2.98	2.72	1.24
5	3.93	3.41	1.36
6	4.99	4.14	1.48
7	6.18	4.95	1.63
8	7.38	5.71	1.78
9	8.79	6.60	1.93
10	10.35	7.39	2.08

6.4.4 Magnetofluid-Dynamic Thin Airfoil. The MFD-produced forces may be used to control the lift of airfoils and the following discussion of the thin airfoil was provided by Sears and Resler.[15] We shall consider the steady, incompressible, inviscid flow of highly conducting fluid past a lifting airfoil with no thickness and with the magnetic field parallel to the flow. The MFD equations are then:

$$\nabla \cdot \mathbf{q} = 0 \tag{6.142}$$

$$\mathbf{q} \cdot \nabla \mathbf{q} + \frac{1}{\rho} \nabla p = \frac{\mu_e}{\rho} \mathbf{J} \times \mathbf{H} \tag{6.143}$$

$$\mathbf{J} = \sigma(\mathbf{E} + \mu_e \mathbf{q} \times \mathbf{H}) \tag{6.144}$$

$$\mu_e \nabla \times \mathbf{H} = \mathbf{J} \tag{6.145}$$

$$\nabla \times \mathbf{E} = 0 \tag{6.146}$$

$$\nabla \cdot \mathbf{H} = 0 \tag{6.147}$$

The equations to be solved are obtained by substituting Eq. (6.145) into Eq. (6.143) to give the first relation

$$\mathbf{q} \cdot \nabla \mathbf{q} + \frac{1}{\rho} \nabla p = \frac{\mu_e}{\rho} \left[\mathbf{H} \cdot \nabla \mathbf{H} - \tfrac{1}{2} \nabla H^2 \right] \tag{6.148}$$

and by eliminating J and E from Ohm's law with Eqs. (6.142) and (6.145)–(6.147) to provide the second relation.

$$\mathbf{q} \cdot \nabla \mathbf{H} - \mathbf{H} \cdot \nabla \mathbf{q} = \frac{1}{\sigma \mu_e} \nabla^2 \mathbf{H} \tag{6.149}$$

We now assume that the flow and magnetic field are uniform except for small disturbances and are described by:

$$\begin{aligned} \mathbf{q} &= (U + u, v, 0) \quad \text{where} \quad u, v \ll U \\ \mathbf{H} &= (H_0 + h_x, h_y, 0) \quad \text{where} \quad h_x, h_y \ll H_0 \end{aligned} \tag{6.150}$$

where the ratios of the disturbances to the free stream values are assumed to be all the same order. Applying these assumptions to Eq. (6.148) and neglecting products of small quantities produces

$$U \frac{\partial \mathbf{v}}{\partial x} + \frac{1}{\rho} \nabla p = \frac{\mu_e}{\rho} \left(H_0 \frac{\partial \mathbf{h}}{\partial x} - H_0 \nabla h_x \right) \tag{6.151}$$

where $\mathbf{v}$ and $\mathbf{h}$ represent the vectors (u, v) and (h_x, h_y). The integrated x component of this equation is then

$$\rho U u + p = p_\infty \tag{6.152}$$

where p_∞ is the undisturbed static pressure. The y component can now be represented by

$$\Omega = \frac{H_0 \mu_e}{\rho U} \xi \tag{6.153}$$

where $-\rho U \partial u / \partial y$ was substituted for $\partial p / \partial y$ which was obtained from Eq. (6.152). The quantities Ω and ξ are curls of the vector $\mathbf{q}$ and $\mathbf{H}$ and are

$$\Omega = \frac{\partial v}{\partial x} - \frac{\partial u}{\partial y}$$

$$\xi = \frac{\partial h_y}{\partial x} - \frac{\partial h_x}{\partial y} = j_z \tag{6.154}$$

Application of the small perturbation assumptions to Eq. (6.149) gives

$$U\frac{d\mathrm{h}}{\partial x} - H_0\frac{\partial \mathrm{v}}{\partial x} = \frac{1}{\sigma\mu_e}\nabla^2 h \approx 0 \tag{6.155}$$

since the conductivity is very large. Equation (6.155) is then integrated to give

$$\frac{\mathrm{h}}{H_0} - \frac{\mathrm{v}}{U} = \mathrm{const}(y) = 0 \tag{6.156}$$

where the constant is evaluated at a very large y where the flow and magnetic field are undisturbed. This relation is our first result for a highly conducting fluid and demonstrates that the flow and magnetic field are disturbed in the same manner which should be expected for a small magnetic diffusivity $1/\mu_e\sigma$ since the magnetic field is frozen in the fluid. The second result follows from this since now

$$\Omega = \xi\frac{U}{H_0} \tag{6.157}$$

and then

$$\Omega = \frac{\mu_e H_0^2}{\rho U^2}\Omega \tag{6.158}$$

which is only correct for the special case that the flow velocity is equal to the Alfvén velocity $\sqrt{\mu_e/\rho}\,H_0$. Thus, the flow must be irrotational and the current density perturbation j_z in the flow is zero to the first order as a result of the perturbation assumptions. With these conclusions, the flow over the thin body reduces to the classical case of a nonconducting fluid. However, there is a current with an infinite density at the surface due to the discontinuous jump of the x component of the magnetic field perturbation across the airfoil. The lift of the thin airfoil is then altered by the resulting $\mu_e \mathbf{J}_s \times \mathbf{H}_0$ force. This surface current can be determined by applying the small perturbation assumptions to Ampère's law, Eq. (6.145), and integrating the result in the y direction across the thin airfoil, that is, $y = 0^-$ to $y = 0^+$. Thus

$$J_s(x) = -h_x(x, 0^+) + h_x(x, 0^-) \tag{6.159}$$

where J_s is the surface current per unit length and is in the negative z direction. The classical, irrotational small perturbation flow past a thin airfoil can be represented by a vortex sheet which produces a discontinuity in the u component of the velocity while keeping v continuous. Since the flow and magnetic field are disturbed in the same manner, then the discontinuity in u is the same as the discontinuity in h_x and both depend on the x coordinate. The lifting force can then be computed from

$$l(x) = [p(x, 0^-) - p(x, 0^+)] + \mu_e H_0 J_s \tag{6.160}$$

and using Eqs. (6.152) and (6.159) then

$$l(x) = -\rho U [u(x, 0^-) - u(x, 0^+)] + \mu_e H_0 [-h_x(x, 0^+) + h_x(x, 0^-)] \tag{6.161}$$

which can be simplified further with Eq. (6.156) to give

$$l(x) = \left(\rho U - \frac{\mu_e H_0^2}{U}\right)[u(x, 0^+) - u(x, 0^-)] \tag{6.162}$$

Since $u(x, 0^-)$ is in the negative direction on the 0^- side of the vortex sheet, Eq. (6.157) becomes

$$l(x) = 2\rho U u(x, 0^+)\left(1 - \frac{\mu_e H_0^2}{\rho U^2}\right) \tag{6.163}$$

where $\mu_e H_0^2/\rho U^2$ is the square of the reciprocal of the magnetic Mach number. The lift and moment coefficients for the highly conducting fluid can be represented in terms of the nonconducting results, that is,

$$C_l = C_{l_0}\left(1 - \frac{\mu_e H_0^2}{\rho U^2}\right)$$

and (6.164)

$$C_m = C_{m_0}\left(1 - \frac{\mu_e H_0^2}{\rho U^2}\right)$$

Equations (6.163) and (6.164) demonstrate that to *first order*, the lift and moment of a thin airfoil in a highly conducting fluid is decreased by an increase in the applied magnetic field strength when the flow is super-Alfvén. They are also less than their nonconducting fluid values. The lift and moment become negative for sub-Alfvén flow and an increase in the magnetic field strength increases their negative behavior. Thus the lift and moment of a thin airfoil in a highly conducting fluid can be controlled by applying a magnetic field of a strength sufficient to produce nonnegligible magnetic Mach numbers.

6.4.5 Fusion-Pinch Instability. The quest for a method to control nuclear fusion for electric power production began in the mid-1940s and has since stimulated one of the century's major research efforts. Success in this venture could provide the world with adequate power for billions of years. Bishop[16] has reviewed the United States controlled fusion program, called *Project Sherwood*, from 1951 to 1958. Most of the major concepts and fundamental problems were formulated during that period.

Energy is released in the fusion reaction by converting mass into energy when light nuclei are combined to produce heavier particles but with less total mass and is the source of energy in stars. Since the energy required to produce the fusion reaction is very large, the light elements will be stripped of their electrons; the nuclei will be positively charged and considerable impact velocities are required to overcome their repelling electrostatic fields. This requirement can be minimized by selecting elements with the lowest nuclear charges and the best choices are hydrogen and its two isotopes, deuterium and tritium. However, hydrogen has such a slow rate of reaction, that it is not of practical interest and tritium does not occur in nature. By comparison, deuterium can be extracted cheaply from sea water in the ratio of 1:6,000 parts of hydrogen. The two reactions that have been considered are:

$$\underset{\substack{(1\ \text{proton})\\(1\ \text{neutron})}}{\overset{(\text{Deuterium})}{\mathrm{D}}} + \underset{\substack{(1\ \text{proton})\\(2\ \text{neutrons})}}{\overset{(\text{Tritium})}{\mathrm{T}}} \longrightarrow \underset{\substack{(2\ \text{protons})\\(2\ \text{neutrons})}}{\mathrm{He}^4} + \underset{(1\ \text{neutron})}{\mathrm{n}} + 17.6\ \text{Mev} \tag{6.165}$$

$$\mathrm{D} + \mathrm{D} \begin{cases} \overset{\substack{(2\ \text{protons})\\(1\ \text{neutron})}}{\mathrm{He}^3} + \mathrm{n} + 3.2\ \text{Mev} \\ \underset{(1\ \text{proton})}{\mathrm{T}} + \mathrm{p} + 4.0\ \text{Mev} \end{cases} \simeq \text{probability} \tag{6.166}$$

and the D–D reaction is the most interesting since in the D–T reactions, the tritium must be manufactured. Both reactions require temperature of about 100,000°K or 1/10 of a million electron volts, Mev, to overcome the repelling electrostatic fields to permit the short-range nuclear forces to interact and cause the particles to coalesce.

In the D–T reaction, 80 percent of the energy released appears as kinetic energy of the neutrons whereas in the D–D reactions, only one-third of the energy released is transferred to the neutrons with two-thirds appearing as kinetic energy of the other reaction products. The energy can be extracted by slowing down the neutrons in a moderator and transferring the heat generated to a working fluid or in the D–D reaction by direct interaction of the energetic charged particles with a magnetic field. The fusion reaction produces only slightly more energy per unit mass than the fission reaction. However, the cost of the fuel is negligible for the D–D reaction and no radioactive waste material is produced. The reactor would still have to be shielded, though, since the high-energy neutrons would cause the surrounding materials to become radioactive. The reaction in a practical fusion device must be self-sustaining. It must operate at very high temperature levels since sufficient energy is required to continue to produce the reaction while energy is being removed and is lost by thermal radiation due to the collision of charged particles.

The sustaining temperature for the D–D reaction is $4 \times 10^8\,°\mathrm{K}$ and for the D–T reaction it is $4.5 \times 10^7\,°\mathrm{K}$. To keep the pressure reasonable at these temperatures, the density of particles must be less than 10^{18} particles/cc and must be greater than 10^{14} particles/cc to produce useful amounts of energy. Also, the particles must be confined long enough for the fusion reaction to occur since the particles have a much higher probability of scattering and must approach each other many times. The plasma must not be permitted to touch the walls since the energy lost would not permit the reaction to be self-sustaining. The walls would not be damaged by such a contact since the energy content is too low at the appropriate densities. These containment requirements are characterized by the product of the number density and the confinement time $n\tau$ which must be greater than 10^{16} for the D–D reaction. Example $n\tau$'s are $n = 10^{15}$ particles/cc and 10 sec and 10^{18} particles/cc with 0.01 sec. The D–T reaction requires $n\tau$ to be greater than 10^{14} and thus must be confined only 1/100 of the time of the D–D reaction with the same number densities.

Confining the plasma for the required $n\tau$ is the major problem that has scuttled all attempts to control the fusion reaction. The most promising approaches use magnetic fields to confine the plasma. However, instabilities in the plasma destroy the confinement. These instabilities are of macroscopic and microscopic origin and when one instability is eliminated, another type occurs with perplexing regularity. No completely stable configuration has been found.

One of the earliest confinement schemes that was theoretically and experimentally investigated is the self-pinch discharge. A high current density ring discharge is created in a doughnut-shaped container and the magnetic field produced by the current interacts with the current to produce a $\mathbf{J} \times \mathbf{B}$ force in the radially inward direction which pinches the discharge into a denser current filament. However, for thermonuclear conditions, the current would have to be approximately 10^6 amperes which is presently impossible to produce continuously. The pinch discharge is then, of necessity, an intermittent device.

An early analytical study predicted that the self-pinch discharge would develop a kink instability since the density of the magnetic lines is greater on the inside of the Ω-shaped kink which produces a force that rapidly pushes the discharge to the container walls. The kink instability was found in the first experimental device called the *perhapsatron* and it caused the plasma to be quenched in a few microseconds. The long wavelength kinks were found to be suppressed by surrounding the discharge tube with a conducting metal wall. The currents induced in the metal wall by a deviation of the discharge from the center line counteracts the displacement and suppresses the growth of the instability. The short wavelength kinks were not suppressed by this technique and an axial magnetic field was applied to form a "backbone" to maintain the discharge in the desired location. Rosenbluth provided the theory for the stability of the self-pinch discharge with an axial magnetic field and conducting walls which is thoroughly discussed in Chandrasekhar's book entitled "Plasma Physics."[17] The theory was experimentally verified in a linear discharge tube, and temperatures of $5 \times 10^6\,°K$ were produced without kink instabilities, but new instabilities were found. Rosenbluth's theory, as presented in Ref. 17, will be repeated here to demonstrate the application of MFD theory to a specific containment problem when continuum theory is applicable.

We will consider an infinitely long plasma cylinder on the axis of a perfectly conducting tube of radius R_0 with an applied axial magnetic field. The circular magnetic field is produced by the current sheet on the surface, $r = r_0$, between the infinitely conducting plasma, and the vacuum. The undisturbed magnetic fields in the plasma and vacuum can be represented in terms of the circular magnetic field on the surface of the plasma by

$$\text{(plasma)} \qquad \mathbf{B}_p^{\circ} = \alpha_p B_0 \mathbf{1}_z \tag{6.167}$$

$$\text{(vacuum)} \qquad \mathbf{B}_v^{\circ} = B_0 \frac{r_0}{r} \mathbf{1}_\phi + \alpha_v B_0 \mathbf{1}_z \tag{6.168}$$

where α_p and α_v are arbitrary constants that must satisfy the inequality $\alpha_p^2 \leq 1 + \alpha_v^2$. This inequality is obtained from the equation for pressure balance at the surface, that is,

$$p + \frac{|\mathbf{B}_p^{\circ}|^2}{2} = \frac{|\mathbf{B}_r^{\circ}|^2}{2} \quad \text{at } r = r_0 \tag{6.169}$$

Substituting Eqs. (6.167) and (6.168) into this relation gives

$$\frac{2}{B_0^2}\, p\Big|_{r=r_0} = 1 + \alpha_v^2 - \alpha_p^2 \tag{6.170}$$

and thus $\alpha_p^2 \leq 1 + \alpha_v^2$ since $p \geq 0$.

The MFD equations for a plasma with an infinite electric conductivity are:

$$\rho \frac{D\mathbf{u}}{Dt} = -\nabla p + (\nabla \times \mathbf{B}) \times \mathbf{B} \tag{6.171}$$

$$\frac{\partial \mathbf{B}}{\partial t} = \nabla \times (\mathbf{u} \times \mathbf{B}) \tag{6.172}$$

$$\frac{\partial \rho}{\partial t} + \nabla \cdot (\rho \mathbf{u}) = 0 \tag{6.173}$$

and

$$\frac{d(p\rho^{-\gamma})}{dt} = 0 \tag{6.174}$$

When the plasma is undisturbed $\mathbf{u} = 0$, ρ and p are uniform and $\mathbf{B}_p^\circ$ is given by Eq. (6.167). These equations may be linearized by the usual method of assuming small perturbations in the quantities

$$\begin{aligned} \mathbf{u} &= \mathbf{u} \\ p &= p + \delta p \\ \mathbf{B}_p &= \mathbf{B}_p^\circ + \delta \mathbf{B}_p \end{aligned} \tag{6.175}$$

and discarding products of perturbed quantities. Using this procedure reduces the equations to

$$\rho \frac{\partial \mathbf{u}}{\partial t} = -\nabla \delta p + (\nabla \times \delta \mathbf{B}_p) \times \mathbf{B}_p^\circ \tag{6.176}$$

$$\frac{\partial \delta \mathbf{B}_p}{\partial t} = \nabla \times (\mathbf{u} \times \mathbf{B}_p^\circ) \tag{6.177}$$

$$\frac{\partial \delta p}{\partial t} = -\rho \nabla \cdot \mathbf{u} \tag{6.178}$$

and

$$\delta p = \frac{\gamma p}{\rho} \delta \rho \qquad (6.179)$$

Since these equations are linear, we can examine the behavior of the normal modes of the disturbed quantities in the form

$$\mathcal{T} = f(r) \exp[i(\omega t \pm m\phi + kz)] \qquad (6.180)$$

because the modes won't interact. The frequency of the oscillation is ω, and m and k are the azimuthal and axial wave numbers.

Rosenbluth stated that stability of this system requires the normal component of the gradient of the perturbation in the kinetic plus magnetic pressure to act in the opposite direction to the displacement at the plasma boundary. An analytical relation for this criterion will be derived.

Let $\boldsymbol{\xi}$ represent the perturbed displacement of a fluid element, then

$$\mathbf{u} = \frac{\partial \boldsymbol{\xi}}{\partial t} \qquad (6.181)$$

and the equations become:

$$\frac{\partial^2 \xi}{\partial t^2} = -\nabla \delta p + (\nabla \times \delta \mathbf{B}_p) \times \mathbf{B}_p^\circ \qquad (6.182)$$

$$\delta \mathbf{B}_p = \nabla \times (\boldsymbol{\xi} \times \mathbf{B}_p^\circ) \qquad (6.183)$$

and

$$\delta \rho = -\rho \nabla \cdot \boldsymbol{\xi} \qquad (6.184)$$

where Eq. (6.183) was integrated with time. Now the marginal state occurs when $\partial/\partial t = 0$ and thus Eq. (6.182) becomes

$$\nabla \delta p = (\nabla \times \delta \mathbf{B}_p) \times \mathbf{B}_p^\circ \qquad (6.185)$$

and then

$$\mathbf{B}_p^\circ \cdot \nabla \delta p = 0 \tag{6.186}$$

and with relations for $\mathbf{B}_p^\circ$ and δp then

$$\alpha_p B_0 \frac{\partial \delta p}{\partial z} = \alpha_p B_0 ik \delta p = 0 \tag{6.187}$$

Since $\alpha p B_0 ik \neq 0$, then δp must be zero and then Eq. (6.185) becomes

$$(\nabla \times \delta \mathbf{B}_p) \times \mathbf{B}_p^\circ = 0 \tag{6.188}$$

and Eq. (6.184) is

$$\nabla \cdot \boldsymbol{\xi} = 0 \tag{6.189}$$

Now Eq. (6.183) can be expanded with the vector relations and since $\nabla \cdot \boldsymbol{\xi} = 0, \nabla \cdot \mathbf{B}_p = 0$, and $\mathbf{B}_p^\circ$ is constant the result is

$$\delta \mathbf{B}_p = (\mathbf{B}_p^\circ \cdot \nabla)\boldsymbol{\xi} = ik\alpha_p B_0 \boldsymbol{\xi} \tag{6.190}$$

which when substituted in Eq. (6.188) produces

$$(\nabla \times \boldsymbol{\xi}) \times \mathbf{B}_p^\circ = 0 \tag{6.191}$$

Since $\mathbf{B}_p^\circ$ is in the z direction, then the $(\nabla \times \boldsymbol{\xi})_{r,\phi}$ components must be zero and as a result of the form of the perturbation, the $(\nabla \times \boldsymbol{\xi})_z$ component is also zero. Thus,

$$\boldsymbol{\xi} = \nabla \phi \quad \text{and} \quad \nabla^2 \phi = 0 \tag{6.192}$$

since both the curl and divergence are zero. The solution of Laplace's equation for the scalar potential ϕ that is finite at the axis is

$$\phi = CI_m(kr) \exp[i(kz \pm m\phi)] \tag{6.193}$$

where $I_m(kr)$ is the modified Bessel function of the first kind and order m and C is a constant. The displacement is now

$$\boldsymbol{\xi} = C\left[kI_m'(kr)\mathbf{1}_r + \frac{im}{r}I_m(kr)\mathbf{1}_\phi + ikI_m(kr)\mathbf{1}_z\right]\exp[i(kz \pm m\phi)] \tag{6.194}$$

and since the boundary at $r = r_0$ is displaced as

$$r = r_0\{1 + \delta\exp[i(kz \pm m\phi)]\} \tag{6.195}$$

then

$$C = \frac{r_0\delta}{kI_m'(kr_0)} \tag{6.196}$$

and the equation for the magnetic field disturbance is

$$\delta\mathbf{B}_p = \frac{i\alpha_p B_0 r_0 \delta}{I_m'(kr_0)}\left[kI_m'(kr)\mathbf{1}_r + \frac{im}{r}I_m(kr)\mathbf{1}_\phi + ikI_m(kr)\mathbf{1}_z\right]\exp[i(kz \pm m\phi)] \tag{6.197}$$

To determine the stability criteria, we must calculate the difference between the disturbances in the combined gas pressure and magnetic pressure in the plasma and the magnetic pressure in the vacuum at the perturbed boundary. Thus, the pressure at the boundary in the perturbed plasma is

$$p_i = \frac{|\mathbf{B}_p^\circ + \delta p|^2}{2} + (p + \delta p) = \frac{|\mathbf{B}_p^\circ|^2}{2} + \mathbf{B}_p^\circ\cdot\delta\mathbf{B}_p + p \tag{6.198}$$

since $\delta p = 0$. The pressure at $r = r_0$ can be calculated with a Taylor expansion, that is,

$$\begin{aligned} P_i(r_0 + \delta r) &= \frac{|\mathbf{B}_p^\circ|^2}{2} + \mathbf{B}_p^\circ\cdot\delta\mathbf{B}_p + p + (\boldsymbol{\xi}\cdot\nabla)\left(\frac{|\mathbf{B}_p^\circ|^2}{2} + p\right) \\ &= \frac{|\mathbf{B}_p^\circ|^2}{2} + p + \mathbf{B}_p^\circ\cdot\delta\mathbf{B}_p \quad \text{at} \quad r = r_0 \end{aligned} \tag{6.199}$$

since $|\mathbf{B}_p^\circ|^2/2 + p$ is a constant. Thus, the perturbation in the combined pressure inside the plasma is:

$$\delta P_i(r_0) = \mathbf{B}_p^\circ \cdot \delta \mathbf{B}_p = \alpha_p B_0 \delta B_{p,z}$$

$$= -\alpha_p{}^2 B_0{}^2 k^2 r_0{}^2 \delta \frac{I_m(kr_0) \exp[i(kz \pm m\phi)]}{kr_0 I_m'(kr_0)} \qquad (6.200)$$

The magnetic field in the vacuum can be computed from a scalar potential since there are no currents there and $\nabla \times \mathbf{B} = 0$. Thus

$$\delta \mathbf{B}_v = \nabla \psi \quad \text{and} \quad \nabla^2 \psi = 0 \qquad (6.201)$$

since $\nabla \cdot \mathbf{B}$ is always zero. The solution to the Laplace equation in the perturbation form is

$$\psi = [AI_m(kr) + BK_m(kr)] \exp[i(kz \pm m\phi)] \qquad (6.202)$$

where $I_m(kr)$ and $K_m(kr)$ are modified Bessel functions of the first and second kinds, and A and B are constants to be determined from the boundary conditions. At $r = R_0$ the wall is perfectly conducting and then the radial component of $\delta \mathbf{B}_v$ is zero. Since

$$\delta \mathbf{B}_v = k\psi' \mathbf{1}_r + \frac{im}{r} \psi \mathbf{1}_\phi + ik\psi \mathbf{1}_z \qquad (6.203)$$

then

$$AI_m'(kR_0) + BK_m'(kR_0) = 0 \qquad (6.204)$$

The other boundary condition is obtained from the requirement that the normal component of the magnetic field must be continuous across the disturbed plasma surface, that is, $\mathbf{n} \cdot \mathbf{B}_p = \mathbf{n} \cdot \mathbf{B}_v$ or when linearized

$$\mathbf{1}_r \cdot \delta \mathbf{B}_p + \delta \mathbf{n} \cdot \mathbf{B}_p^\circ = \mathbf{1}_r \cdot \delta \mathbf{B}_r + \delta \mathbf{n} \cdot \mathbf{B}_v^\circ \quad \text{at} \quad r = r_0 \qquad (6.205)$$

where $\delta \mathbf{n}$ is the change in the unit normal vector due to the perturbation $\boldsymbol{\xi}$. The surface of the plasma moves with the plasma and thus

$$\frac{D\mathbf{n}}{Dt} = \mathbf{n} \times [\mathbf{n} \times \nabla(\mathbf{u} \cdot \mathbf{n})] = -\nabla(\mathbf{u} \cdot \mathbf{n}) + \mathbf{n}[\mathbf{n} \cdot \nabla(\mathbf{u} \cdot \mathbf{n})] \qquad (6.206)$$

and when linearized

$$\frac{\partial \delta \mathbf{n}}{\partial t} = -\nabla(\mathbf{n}^\circ \cdot \mathbf{u}) + [\mathbf{n}^\circ \cdot \nabla(\mathbf{n}^\circ \cdot \mathbf{u})]\mathbf{n}^\circ \tag{6.207}$$

Substituting for u and integrating with time gives

$$\delta \mathbf{n} = -\nabla \xi_r + \mathbf{1}_r(\mathbf{1}_r \cdot \nabla \xi_r) \tag{6.208}$$

Substituting this result and the expressions for the magnetic terms in Eq. (6.205) gives

$$\psi'(r = r_0) = \frac{iB_0\delta}{k}(\alpha_v k r_0 \pm m) \tag{6.209}$$

and the constants are evaluated as

$$A = \frac{iB_0(\alpha_v k r_0 \pm m)\delta/k}{I'_m(kr_0) - I'_m(kR_0)K'_m(kr_0)/K'_m(kR_0)} \qquad B = -A\,\frac{I'_m(kR_0)}{K'_m(kR_0)} \tag{6.210}$$

Now the change in the exterior pressure is:

$$\delta P_0 = \frac{\delta |\mathbf{B}_v|^2}{2} = \mathbf{B}_v^\circ \cdot \delta \mathbf{B}_v + \frac{\boldsymbol{\xi} \cdot \nabla |\mathbf{B}_v^\circ|^2}{2}$$

$$= B_0(\delta B_\phi + \alpha_v \delta B_z) - B_0{}^2\delta \exp[i(kz \pm m\phi)] \quad \text{at} \quad r = r_0 \tag{6.211}$$

Substituting for δB_ϕ and δB_z gives

$$\delta P_0(r_0) = B_0{}^2\delta \exp[i(kz \pm m\phi)] \left[-1 + (\alpha_v k r_0 \pm m)^2 \frac{G_m(kr_0)P_m(kr_0) - Q_m(kr_0)}{1 - G_m(kr_0)}\right] \tag{6.212}$$

where

$$P_m(kr_0) = \frac{I_m(kr_0)}{kr_0 I'_m(kr_0)} \qquad Q_m(kr_0) = \frac{K_m(kr_0)}{kr_0 K'_m(kr_0)}$$

and

$$G_m(kr_0) = \frac{K'_m(kR_0)I'_m(kr_0)}{K'_m(kr_0)I'_m(kR_0)} \tag{6.213}$$

The gradient of the pressure disturbance in the opposite direction of displacement must be

$$\frac{\delta P_0 - \delta P_i}{r - r_0} > 0 \tag{6.214}$$

for stability as stated by Rosenbluth. In terms of the relations for δP_0 and δP_i, the stability criterion is

$$\alpha_p{}^2 k^2 r_0{}^2 P_m(kr_0) + (m \pm \alpha_v k r_0)^2 \frac{G_m(kr_0)P_m(kr_0) - Q_m(kr_0)}{1 - G_m(kr_0)} > 1 \tag{6.215}$$

The regions of complete stability are shown in Fig. 6.21 for selected values of α_v that satisfy the required inequality. The results show that the largest tube diameter for stable operation is five times the plasma diameter.

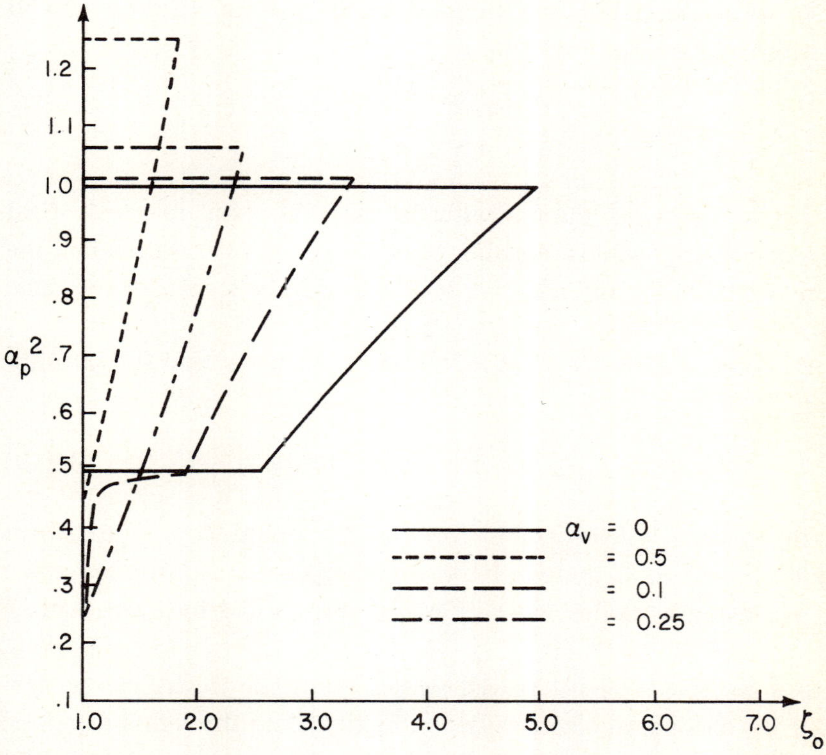

Fig. 6.21 Stability Diagram. A stable pinch occurs in the region bounded on the left by the $\alpha_p{}^2$-axis, on the right by the marginal curve for $m = -1$ and above by the equilibrium condition $\alpha_p{}^2 < 1 + \alpha_v{}^2$.

The following specific cases are also contained in the general results.

No axial magnetic field or conducting walls. Let $\alpha_v = \alpha_p = 0$ and $R \to \infty$, then $G \to \infty$ and the criteria for instability becomes:

$$m^2 Q(kr_0) + 1 > 0 \tag{6.216}$$

or

$$kr_0 + m^2 \frac{K_m(kr_0)}{K'_m(kr_0)} > 0 \tag{6.217}$$

Thus, the self-pinch without an axial magnetic field or conducting wall is unstable for the $m = 0$ or $m = 1$ modes. For large values of kr_0, $K_m(kr_0)/K'_m(kr)$ approaches -1 and then the plasma is stable for kr_0 values up to approximately m^2.

No axial magnetic field trapped in the plasma. Let $\alpha_p = 0$ and the criteria for stability becomes

$$(\alpha_v kr_0 \pm m)^2 \frac{G_m(kr_0)P_m(kr_0) - Q_m(kr_0)}{1 - G_m(kr_0)} > 1 \tag{6.218}$$

and the self-pinched plasma with an external, axial magnetic field and a conducting wall is unstable when $\alpha_v kr_0 - m = 0$. Also when the plasma is stable for any negative m, then it will be stable for the same positive m.

No conducting wall. Let $R \to \infty$ and then $G \to 0$ and the stability criteria is

$$\alpha_p^2 k^2 r_0^2 P_m(kr_0) - (m \pm \alpha_v kr_0)^2 Q_m(kr_0) > 1 \tag{6.219}$$

and the plasma will be stable for the $m = 0$ and $m > 1$ modes for proper choices of α_p and α_v. However, the plasma is unstable for the long wavelength modes when it is not surrounded by a conducting wall.

6.4.6 Magnetofluid-Dynamic Power Generation. Electric power can be generated by passing an electrically conducting fluid through a magnetic field and by connecting electrodes, located in the walls pointed at by $\pm \mathbf{V} \times \mathbf{B}$, to an external load. The power produced is the scalar product of the load voltage and the $\mathbf{V} \times \mathbf{B}$ current and is dc power when the flow and magnetic field are both steady and ac power when either are oscillated. Since energy must be added to the fluid to provide the flow velocity, then the most suitable method is to add thermal energy to an ionizable gas. Unfortunately, extremely high temperatures are required to obtain a practical range of conductivity. This requirement can be lowered by adding an easily ionized seed material to the gas and then the temperature will be compatible with usable refractory materials. The resulting temperatures are also sufficiently high to provide a sizeable increase in the Carnot efficiency over steam power plants. However, the conductivity of the seeded gas will be lost if large temperature drops are permitted to occur through the generator channel. This requires the MFD generator to be a topping unit in a system that uses its exhaust gas to further produce power by more conventional methods. The hoped for efficiency of such a system would be about 80 percent as compared to 55 percent for steam turbine systems.

The following analysis of the dc MFD generator was provided by Sutton.[18] The current generated is given by Ohm's law

$$J = \sigma(E + UB) = \sigma(1 - K)UB \tag{6.220}$$

where the working fluid is assumed to be grossly neutral. The electric field E is that imposed by the external circuit at the electrodes and $K = -E/UB$ is the generator coefficient. The power generated per unit volume is then

$$P = -EJ = \sigma K(1 - K)U^2B^2 \tag{6.221}$$

and reaches its maximum when $K = 0.5$. This condition occurs when the external resistance equals the internal resistance which is demonstrated by the following calculation. Let the local voltage be

$$V = KUBh \tag{6.222}$$

where h is the channel width and let the external load current per unit length be

$$I = \sigma(1 - K)UBb \tag{6.223}$$

where b is the electrode height. The external load resistance per unit length is then

$$R_L = \frac{V}{I} = \frac{Kh}{(1 - K)\sigma b} = \frac{K}{1 - K} R_i \tag{6.224}$$

where $R_i = h/\sigma b$ is the effective internal resistance. The generator coefficient can then be written as

$$K = \frac{1}{1 + R_i/R_L} \tag{6.225}$$

and equals 1/2 when $R_L = R_i$.

The equations for the power generated contains the electrical conductivity and the mean velocity U and both can vary along the channel depending on its shape. The performance of an MFD generator can then be studied by examining various channel shapes with the quasi-one-dimensional theory of Chap. 3. The equations in the present form are

$$\frac{d(\rho UA)}{dx} = 0 \tag{6.226}$$

$$\rho U \frac{dU}{dx} + \frac{dp}{dx} = \sigma UB^2(K - 1) \tag{6.227}$$

$$\rho U \frac{d(C_p T + U^2/2)}{dx} = \sigma U^2 B^2 K(K - 1) \tag{6.228}$$

$$p = \rho RT \tag{6.229}$$

and

$$\frac{\sigma}{\sigma_0} = \left(\frac{T}{T_0}\right)^{\omega} \tag{6.230}$$

These equations have been solved by assuming a constant temperature, constant velocity, constant pressure, constant density, or a constant area. The length required to extract a given power, when $K = 1/2$, for each case is shown in Fig. 6.22. The minimum

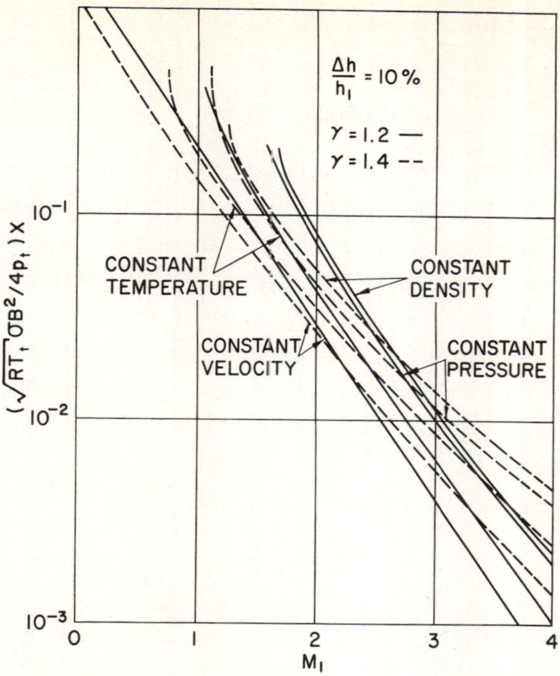

Fig. 6.22 Generator length required for different channels, $K = 0.5$.

length occurs for the constant velocity case, which will be discussed in detail. However, the constant area channel, not shown in the figure, is the simplest to construct.

The solution of these equations provide the following relations required for a detailed study of the constant velocity channel. The power generated by the entire channel is simply the mass flow rate times the static enthalpy change through the channel, that is,

$$P_G = \rho_1 U_1 A_1 C_p (T_1 - T_2) \tag{6.231}$$

where T_1 and T_2 are the entrance and exit static temperatures. The static temperature change along the channel is given by

$$\frac{T}{T_1} = \left\{1 + \left[\frac{\gamma - \omega K(\gamma - 1)}{K(\gamma - 1)}\right]\left[\frac{\sigma_1 U B^2 K(K - 1)\rho_1 C_p}{T_1}\right] x\right\}^{\frac{K(\gamma - 1)}{\gamma - \omega K(\gamma - 1)}} \tag{6.232}$$

and for a given inlet and exit temperatures the required length is

$$x \sim \frac{p_1}{[\gamma - \omega K(\gamma - 1)]\sigma U B^2 (1 - K)} \tag{6.233}$$

This relation shows that the lower the pressure, the shorter the length; and the stronger the magnetic field, the shorter the length. However, these methods of shortening the length are limited, since both will produce a tensor character in the electrical conductivity which has been assumed to be a scalar quantity. A further decrease in the pressure lengthens the channel and a further increase in the magnetic field produces no additional shortening. The tensor character of the conductivity will be discussed in Chap. 7. Also, the closer the generator parameter is to unity, the longer the channel.

The pressure drop occurring through the channel is

$$\frac{p_2}{p_1} = \left(\frac{T_2}{T_1}\right)^{\frac{\gamma}{K(\gamma - 1)}} = \left\{1 + \frac{\sigma_1 U B^2 x}{\rho_1 T_1 C_p (K - 1)}\left[\frac{\omega + \gamma(2 - \omega)}{\gamma - 1}\right]\right\}^{\frac{2\gamma}{\omega + \gamma(2 - \omega)}} \tag{6.234}$$

and the area increase is

$$\frac{A_2}{A_1} = \left(\frac{T_2}{T_1}\right)^{1 - \gamma/K(\gamma - 1)} \tag{6.235}$$

The entropy increase is

$$\frac{S_2 - S_1}{R} = (1 - K)\ln\left(\frac{p_1}{p_2}\right) \tag{6.236}$$

and the closer K is to unity, the smaller the entropy increase but the longer the channel. In an actual channel, there are also skin friction losses which increase with length and then there is a length for which the entropy increase is a minimum.

The efficiency of the channel can be defined as the ratio of the power produced by the channel to the isentropic power which can be computed with Eqs. (6.231) and (6.234) by setting $K = 1$, that is,

$$P_{Gi} = \rho_1 U_1 A_1 C_p T_1\left[1 - \left(\frac{T_2}{T_1}\right)_i\right] = \rho_1 U_1 A_1 C_p T_1\left[1 - \left(\frac{p_2}{p_1}\right)^{(\gamma - 1)/\gamma}\right] \tag{6.237}$$

and then

$$\eta = \frac{P_G}{P_{Gi}} = \frac{(p_2/p_1)^{K(\gamma-1)/\gamma} - 1}{(p_2/p_1)^{(\gamma-1)/\gamma} - 1} \tag{6.238}$$

Figure 6.23 demonstrates that the efficiency increases with the generator coefficient and hence the length for a fixed pressure ratio and the length decreases as the pressure ratio is decreased for a fixed efficiency. However, in the design of a generator for a given efficiency, the material costs and weight will determine the choice of the generator coefficient.

Since the power produced is directly proportional to the constant velocity, the largest value would seem desirable. However, the energy required to produce the flow is limited by the temperature of the heat source and large increases in the velocity are accompanied by large static temperature drops which in turn lowers the electrical conductivity. The static pressure also drops which increases the electrical conductivity. The optimum Mach number for the maximum power production from the gas is in the high subsonic

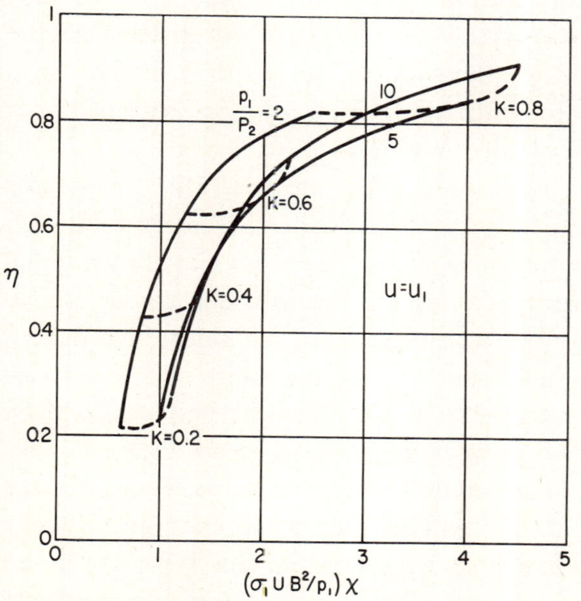

Fig. 6.23 Generator efficiency and length for various pressure ratios and generator coefficients.

range. Now all the energy in the gas, in the temperature range that provides an adequate conductivity, is not available for producing net power. In addition to the pressure losses previously mentioned, energy is required to excite the electromagnet; heat is lost to the water that cools the generator walls, and power is lost due to reverse currents through the gas between the high potential anode and low potential cathode at the entrance and exit of the generator passage. Studies of these losses have shown that the fraction of the power required to excite the electromagnet decreases as the generated power increases and a 500-megawatt generator would require 8 percent. Generator passages must be cooled when the working gas is about 5,000°F, but the bulk temperature should not be reduced more than a few percent to maintain a practical conductivity. The energy lost to the cooling water will then be small. The power lost due to reverse currents at the inlet and exit of the electrodes can amount to one-third of the power produced for a short channel and one-fourth for a long channel. Several methods of reducing these end losses have been proposed, but the best method is to extend the magnetic field upstream and downstream with an exponential decay. This field shading also increases the power produced. However, these gains are at the expense of the larger pressure drop through the generator, which limits the efficiency to less than 60 percent for a channel with a length-to-width ratio of 1 and 80 percent for a length-to-width ratio of 6. See Fig. 6.24.

Feasibility studies have considered various MFD generator systems and a typical large plant would produce 500 megawatts with a combined MFD-steam turbine system. The MFD passage would be constructed from refractory oxides and the working gas would be combustion products seeded with 1 percent, by weight, potassium. The gas temperature would enter the MFD passage at 2,250 ft/sec and 4,900°R, and leave at 4,200°R for a conductivity range of 15 mho/m to 5 mho/m. The magnet would be a liquid cooled copper coil electromagnet producing 20,000 to 30,000 gauss. The generator coefficient K would be 0.8. A typical open-cycle plant arrangement is shown in Fig. 6.25. In a practical system of this type, the ac–dc inverter cost per kilowatt would exceed the turbine generator cost and be about 30 percent of the total plant cost. Also, the power lost

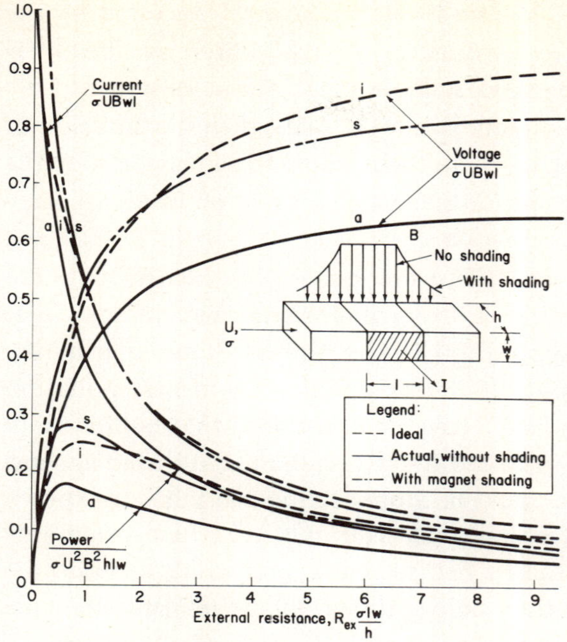

Fig. 6.24 Operating characteristics of a steady dc MHD generator, for $h = l$.

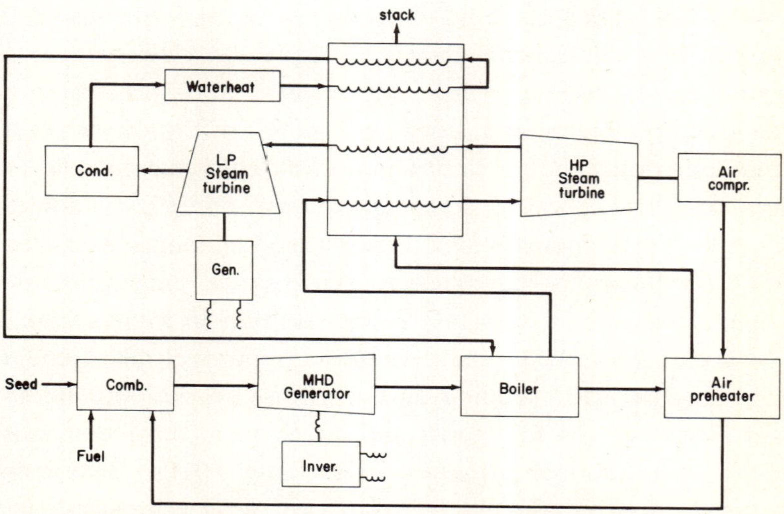

Fig. 6.25 Typical MFD plant.

in the inverters would be about 4 percent of that generated by the MFD channel. The direct production of ac power by the MFD generator would eliminate these undesirable features. However, if the power is to be transmitted more than 300 miles, then transmission of high voltage dc power will counterbalance the disadvantages of the inverters at the receiving station since ac transmission losses are large at such distances.

The unique features of MFD power generators, i.e., high system efficiency and rapid starting, are very important characteristics that should stimulate their application in the United States due to the rapidly approaching energy limitation and environmental pollution problems. Economic studies[19] have demonstrated that combustion-driven MFD generators up to a 100-megawatt capacity used as short period reserve generators could be more economical than any known alternative when operated less than 100 hr/year. These generators could be started in less than a half-second which would provide the required power until other conventional units are put on line. Aircraft gas turbine units require 5 min for starting, steam turbine plants require 15 min, and hydroelectric plants can be started in less than 2 min. High efficiency MFD topping plants for base power could reduce thermal pollution at the site, since, for example, if a station is required to produce 2,000 megawatts at 30-percent efficiency, 6,666 megawatts must be produced, and 4,666 megawatts exhausted to the environment whereas a 60-percent efficient station would produce 3,333 megawatts and reject 1,333 megawatts—a significant difference for the environment. Since the MFD plant would require seeding the gas to obtain useful electrical conductivity levels and the high operating temperatures will produce oxides of nitrogen, then these commercially valuable contaminants would be recovered, producing a very clean smokestack gas—again, an environmental benefit. A very interesting addition to these systems would be a coal gasification plant where its byproduct, char, could be burned for the MFD topping unit. Coal gasification will be an important source of gas for homes and industry since the demands will outstrip our national resources in the late 1970s. Economic studies[20] have been conducted on combined MFD generators and nuclear power plants. However, their state of development is not as advanced as combustion systems, and they suffer from plasma instabilities occurring as a consequence of the low temperature of the reactor exhaust gas. This low temperature limits the efficiency of the station and increases the thermal pollution as compared to combustion systems. Promising methods for controlling the instabilities are evolving and, if successful, would eliminate the need for costly development programs for high temperature nuclear reactors.

REFERENCES

1. L. I. Sedov, "Similarity and Dimensional Methods in Mechanics," Academic Press, Inc., New York, 1959.
2. C. P. Yu and P. S. Lykoudis, Some Simple Theoretical Examples of Explosions in Magnetic Polytropic Stars, *Purdue University A & ES* 63-8, August, 1963.
3. C. F. Kennel and H. E. Petshek, Van Allen Belt Plasma Physics, *AVCO-Everett Res. Lab. Rept.* 259, December, 1966.
4. John R. Spreiter, A. L. Summers and A. Y. Alksne, Hydromagnetic Flow Around the Magnetosphere, *Planet. Space Sci.*, vol. 14, pp. 223–253, 1966.
5. J. R. Spreiter and W. P. Jones, *J. Geophys. Res.*, vol. 68, p. 3555, 1963.
6. S. Way, Electromagnetic Propulsion for Cargo Submarines, *AIAA/Sname Advance Marine Vehicles Meeting*, Paper no. 67–363, Norfolk, Virginia, May 22–24, 1967.
7. S. Way and C. Devlin, Construction and Testing of a Model Electromagnetically Propelled Submarine, *Proc. Eng. Aspects of Magnetohydrodynamics*, Stanford University, California, 1967.
8. W. H. Heiser, Analysis of a Magnetohydrodynamic Ejector, Paper no. 65-APM-25, Transactions of the ASME, *J. Appl. Mech.*, 1965.
9. J. Baker and T. Rogers, One-Dimensional Magnetogasdynamic Channel Flow with Constant Conversion Efficiency, *Marquardt Corp. Rept.* MR 20, 146, September, 1961.
10. Charles H. Marston, MHD Accelerator Performance for Specified Interaction Parameter, *AIAA J.*, vol. 4, no. 11, November, 1966.
11. H. Friedel and K. Lackner, Magnetohydrodynamic Propulsion, *Inst. Theoret. Phys. Sci. Rept.* 29, University of Innsbruck, Austria, December, 1965.
12. T. Luzzi and R. Grunberg, Radar Penetration of the LEM Rocket Exhaust Gases, *Grumman Res. Dept. Memo* RM-236, July, 1964.

13. J. Jackson, "Classical Electrodynamics," John Wiley & Sons, Inc., New York, 1962.
14. S. S. Chawla, The Magnetohydrodynamic Inclined Slider Bearing, *Japan. J. Appl. Phys.*, vol. 5, no. 3, March, 1966.
15. W. R. Sears and E. L. Resler, Theory of Thin Airfoils in Fluids of High Electrical Conductivity, *J. Fluid Mech.*, vol. 5, pp. 257–273, 1959.
16. A. S. Bishop, "Project Sherwood," Addison-Wesley Publishing Company, Inc., Reading, Mass., 1958.
17. S. Chandrasekhar, "Plasma Physics," University of Chicago Press, Chicago, Ill., 1960.
18. George W. Sutton, Design Considerations of a Steady DC Magnetohydrodynamic Electrical Power Generator, *Gen. Elec. Rept.* R59SD432, 1959.
19. R. Bunde and H. Muntenbruch, The Combustion MHD Generator as a Reserve Plant in Electricity Supply Systems, 12th Symposium of Engineering Aspects of MHD, March 27–29, 1972, Argonne National Laboratory, Argonne, Ill.
20. T. Bohn, K. Grawatch, P. Komarek, and G. Noack, Applied Magnetohydrodynamics MHD-Nuclear Power Plants, Julich Nuclear Research Establishment, Inc., Institute for Applied Physics *ARGAS* 1970, no. 5, pp. 1–231, Germany.

Chapter 7

ADVANCED TOPICS OF MAGNETOFLUID DYNAMICS

7.1 INTRODUCTION

The treatment of magnetofluid dynamics in the previous chapters of this book may be referred to as the *classical magnetofluid dynamics*. The basic assumptions in these analyses are:

A generalized Ohm's law is used instead of the exact differential equations of electric current density

The fluid has a scalar electrical conductivity

The temperatures of all species in the electrically conducting fluid are the same

Such a system of equations gives good results for the flow of electrically conducting liquid as well as an ionized gas if the strength of the magnetic field is not too large and the density of the gas is not too small.

The current trend for the application of magnetofluid dynamics is toward a strong magnetic field (so that the influence of electromagnetic force is noticeable) and toward a low density of the gas (such as in space flight and in nuclear fusion research). Under these conditions, the Hall current and ion slip will be important and we should not assume that the electrical conductivity of the gas is a scalar quantity. In Secs. 7.2 and 7.3, we discuss the magnetofluid dynamics for a gas with tensor electrical conductivity. First, we extend the ordinary generalized Ohm's law by including the Hall current and ion slip (Sec. 7.2) and then we discuss some simple flow problems such as quasi-one-dimensional flow and flow in channel for gas with Hall current and ion slip. It is interesting to see that the flow pattern with Hall current and ion slip may differ appreciably from that in the case of scalar electrical conductivity. We shall also discuss the criterion under which the results of scalar electrical conductivity are a good first approximation of the results of tensor electrical conductivity.

When the degree of ionization of the gas is large and the density of the gas is low, gas-dynamic forces will affect the electrical current density distribution; the temperature of electrons may differ appreciably from that of heavy particles such as ions or neutral particles. We have to improve our fundamental equations of magnetofluid dynamics so that these phenomena should be taken into account. Such improvements may be derived from the multifluid theory of magnetofluid dynamics. The complete theory of multifluid theory of magnetofluid dynamics is too complicated for the present book. We shall only discuss the essential features of such a theory (Sec. 7.4) without going into the detailed computations; we will offer some examples of multifluid theory to show some interesting new results which cannot be obtained in the classical theory of magnetofluid dynamics (Sec. 7.5).

When the density of the gas is too low, the continuum equations of gas dynamics will not be a good approximation to the actual conditions. It is better to use the kinetic theory of gases. The similar situation is true for plasma or ionized gas. In Sec. 7.6, we shall discuss briefly the rarefied plasma dynamics based on the kinetic theory. We shall not go into the details of the theory but we shall discuss what new results would occur in the rarefied ionized gas.

Another treatment of rarefied plasma dynamics is the particle motion of a plasma. In Sec. 7.7, we shall discuss some basic results of particle motion of plasma.

Finally when the temperature is very high and the density is very low, the thermal radiation will be of the same order of magnitude as the heat conduction or heat convection. In Sec. 7.8, we discuss the criterion when thermal radiation effects will be important as well as some well-known thermal radiation effects in magnetofluid dynamics.

Since the complete analysis of these advanced topics should be discussed in a more advanced textbook than this one, we shall not give detailed derivation of the results but simply state the main results of these topics. However some important references will be given so that the readers may pursue further these subjects.

7.2 TENSOR ELECTRICAL CONDUCTIVITY, HALL CURRENT, AND ION SLIP

The electrical current density J represents the relative motion of charged particles in a fluid. The equation of electric current density may be derived from the diffusion velocities of the charged particles.[1,2] The major forces on charged particles are electromagnetic forces. If we consider only the electromagnetic forces, we may obtain the generalized Ohm's law. However, the deduction from the diffusion velocities of charged particles, is more complicated than the generalized Ohm's law of Eq. (2.122) because of the fact: When we apply an electric field E, there will be an electrical current in the direction of E. If the magnetic field H is perpendicular to E, there will be an electromagnetic force $J \times B$ which is perpendicular to both E and H. Such a force will cause the charged particles to move in the direction perpendicular to both E and H. We have a new component of electric current density in the direction perpendicular to both E and H which is known as *Hall current* and which was first discovered by Hall. Furthermore, the masses of ions and electrons are different. For the same electromagnetic force, the motion of ions is different from that of electrons. Usually, the diffusion velocity of electrons is much larger than that of ions. As a first approximation, we may consider that the electric current density is determined mainly by the diffusion velocity of the electrons in a plasma. However, when the electromagnetic force is very large (such as in the case of very strong magnetic field), the diffusion velocity of ions may not be negligible. If we include the diffusion velocity of ions as well as that of electrons, we have the phenomena of ion slip. When we include the Hall current, ion slip, and collisions between electrons and neutral particles, the generalized Ohm's law should be written in the following form:[2]

$$\sigma \mathbf{E}_u = A_1 \mathbf{i} + A_2 \mathbf{i} \times \mathbf{B} + A_3(\mathbf{i} \times \mathbf{B}) \times \mathbf{B} \tag{7.1}$$

where σ is the scalar electric conductivity, $\mathbf{E}_u = \mathbf{E} + \mathbf{q} \times \mathbf{B}$ and the coefficients A_1, A_2 and A_3 are functions of the magnitude of magnetic induction B, electron number density and friction coefficients or collision frequencies between various species in the plasma. In general, the effect of collision between electrons and neutral particles is much smaller than that between ions and neutral particles. Equation (7.2) may be written as follows:

$$\sigma \mathbf{E}_u = \mathbf{i} + \zeta\sigma(\mathbf{i} \times \mathbf{B}) - \frac{\beta_i \beta_e}{B^2}(\mathbf{i} \times \mathbf{B}) \times \mathbf{B} \tag{7.2}$$

where

$$\sigma = \frac{n_e{}^2 e^2}{K_{ei}} = \text{scalar electrical conductivity} \tag{7.3}$$

$K_{ei} = m_e f$ = friction coefficient between ions and electrons with f as the collision frequency between ions and electrons.

$$\beta_e = \frac{e n_e B}{K_{ei}} = \frac{\omega_c}{f} = WB\left(\frac{\rho_0}{\rho}\right) = \text{Hall current factor} \tag{7.4}$$

$$\omega_c = \frac{e n_e B}{m_e} = \text{cyclotron frequency of electron} \tag{7.5}$$

with m_e as the mass of an electron.

$$\zeta = \frac{1}{n_e e} = \text{Hall factor} \tag{7.6}$$

$$\beta_i = \frac{e n_e B}{(1 + n_e/n_a)K_{ai}} = \text{ion slip factor} \tag{7.7}$$

with n_e as the electron number density, n_a as the neutral particle number density and K_{ai} as the friction coefficient between ions and neutral particles.

If we solve for the conduction current i from Eq. (7.2), we have

$$\mathbf{i} = \frac{1 + \beta_i \beta_e + \beta_e^2}{(1 + \beta_i \beta_e)^2 + \beta_e^2} \sigma \mathbf{E}_u^{\parallel} + \frac{(1 + \beta_i \beta_e)\sigma \mathbf{E}_u^{\perp}}{(1 + \beta_i \beta_e)^2 + \beta_e^2} + \frac{\sigma \beta_e}{(1 + \beta_i \beta_e)^2 + \beta_e^2} \left(\frac{\mathbf{B}}{B} \times \mathbf{E}_u \right) = \sigma_T \cdot \mathbf{E}_u \tag{7.8}$$

where σ_T is the tensor electrical conductivity defined in Eq. (7.8) and the electrical field $\mathbf{E}_u = \mathbf{E}_u^{\parallel} + \mathbf{E}_u^{\perp}$ has two components, that is,

$$\mathbf{E}_u^{\parallel} = \frac{\mathbf{E}_u \cdot \mathbf{B}}{B^2} \mathbf{B} = \text{component of } \mathbf{E}_u \text{ parallel to } \mathbf{B}$$

$$\mathbf{E}_u^{\perp} = \frac{\mathbf{B} \times (\mathbf{E}_u \times \mathbf{B})}{B^2} = \text{component of } \mathbf{E}_u \text{ perpendicular to } \mathbf{B}$$

From Eq. (7.8) we see that the electrical conduction current i has three components: one is parallel to B or $\mathbf{E}_u^{\parallel}$; the other is perpendicular to B but parallel to $\mathbf{E}_u^{\perp}$ and the third (Hall current) is perpendicular to both B and $\mathbf{E}_u$. The electrical conductivity in the direction parallel to B is larger than those perpendicular to B. If the ion slip factor is negligible, the electrical conductivity in the direction parallel to B is simply the scalar electrical conductivity σ which has been used in our previous chapters. The electrical conductivity in the direction of $\mathbf{E}_u^{\perp}$ when the ion slip factor is negligible is

$$\sigma^{\perp} = \frac{\sigma}{1 + \beta_e^2} = \frac{\sigma}{1 + (\omega_c/f)^2} \tag{7.9}$$

It is evident that $\sigma^{\perp}$ is always smaller than σ. When the ratio of cyclotron frequency of electron ω_c is negligibly small in comparison to collision frequency f, $\sigma^{\perp}$ is equal to σ. In other words, if the magnetic induction is not large and the density of the plasma is not small such that $(\omega_c/f) \ll 1$, $\sigma^{\perp} \simeq \sigma$.

The electrical conductivity of the Hall current is the factor of the third term in Eq. (7.8) and when ion slip factor is negligible, we have the electrical conductivity of the Hall current as

$$\sigma_b = \beta_e \sigma^{\perp} = \frac{\omega_c}{f} \sigma^{\perp} \tag{7.10}$$

Hall current is negligible when $(\omega_c/f) \ll 1$. Hence if $\beta_e = (\omega_c/f) \ll 1$, Eq. (7.8) reduces to the ordinary generalized Ohm's law which we have been using in the previous chapters, i.e.,

$$\mathbf{i} = \sigma \mathbf{E}_u \tag{7.11}$$

The Hall current factor is proportional to B and is a function of density and temperature of the ionized gas. We may express β_e as a function of T and ρ. Figure 7.1 shows an example of such variations.[3] The scalar electrical conductivity σ is also a function of T and ρ which is also shown in Fig. 7.1.

The ion slip factor β_i is usually smaller than the Hall current factor β_e because the friction coefficient K_{ei} is proportional to the mass of electron m_e while the friction coefficient K_{ei} is proportional to the mass of neutral particle m_a and m_a is much larger than m_e.

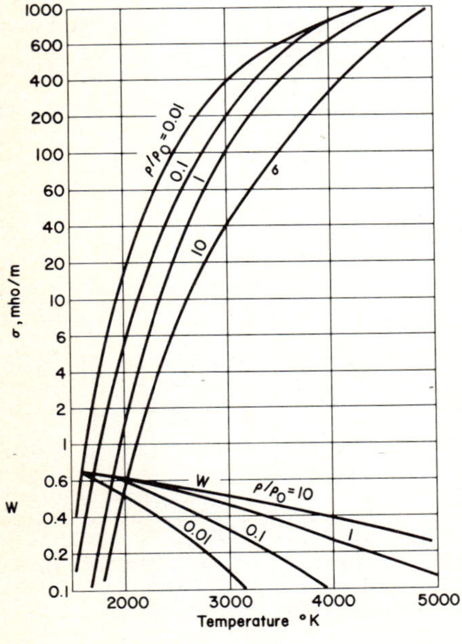

Fig. 7.1 Tensor conductivity of argon seeded with 1% potassium as function of density and temperature. ρ_0 = density of air in standard atmosphere at sea level. (*Taken from R. J. Rosa, Physical Principles of MHD Power Generation, AVCO-Everett Res. Lab. Rept. 69, 1960.*)

7.3 SOME FLOW PROBLEMS WITH TENSOR ELECTRICAL CONDUCTIVITY

In order to show the effect of tensor electrical conductivity on the flow pattern of magnetogas dynamics, we reexamined two simple flow problems (discussed in previous chapters) for the case of scalar electrical conductivity.

7.3.1 MHD Channel Flow with Tensor Electrical Conductivity. We reexamine the generalized Hartmann flow (Chap. 4) with Hall current. We consider again the steady flow of a viscous incompressible fluid of constant properties through a rectangular channel. Two opposite faces of the channel are insulated walls situated at $y = \pm L$ while the other two, at $z \to \pm\infty$, are electrodes. All the variables are functions of y only, except the pressure p. Since the Hall current is included, a transverse electromagnetic force will be produced, thus we have a transverse flow velocity component in addition to the axial velocity component. Furthermore, the variation of the flow velocity in the channel will affect the induced electrical field and then the electrical current. Hence the conditions of this problem give the following expression for various variables:

$$u = Uu^*(y^*) \quad v = 0 \quad w = Uw^*(y^*) \quad E_x = E_{x0} \quad E_y = E_{y0}$$
$$E_z = E_{z0} \quad H_x = H_0 H_x^*(y^*) \quad H_y = H_{y0} \ (H_y^* = 1) \quad H_z = H_0 H_z^*(y^*)$$
$$p = \rho U^2 p^*(x^*, y^*, z^*) = \rho U^2[x^* p_x^* + z^* p_z^* + p_1^*(y^*)] \tag{7.12}$$

where U is a reference velocity and all lengths are in terms of L. The asterisk refers to nondimensional quantities; subscript 0 refers to a constant value of the quantity. The pressure gradients p_x and p_z are constants. With Hall current, we introduce the z-wise pressure gradient p_z, the z component of velocity w, and the z component of magnetic field H_z. Our fundamental equations for the five unknowns of our problem, i.e., u^*, w^*, H_x^*, H_z^* and p_1^*, are as follows:

$$\frac{1}{R_e}\frac{d^2u^*}{dy^2} + R_H \frac{dH_x^*}{dy^*} = p_x^* \tag{7.13}$$

$$\frac{1}{R_e}\frac{d^2w^*}{dy^{*2}} + R_H \frac{dH_z^*}{dy^*} = p_z^* \tag{7.14}$$

$$(1+\beta_e{}^2)\frac{d^2H_x^*}{dy^{*2}}+R_\sigma\frac{du^*}{dy^*}+\beta_e R_\sigma\frac{dw^*}{dy^*}=0 \tag{7.15}$$

$$(1+\beta_e{}^2)\frac{d^2H_z^*}{dy^{*2}}+R_\sigma\frac{dw^*}{dy^*}-\beta_e R_\sigma\frac{du^*}{dy^*}=0 \tag{7.16}$$

$$\frac{dp_1^*}{dy^*}=-\frac{R_H}{2}(H_x^{*2}+H_z^{*2}) \tag{7.17}$$

where $R_e = UL/\nu$ is the Reynolds number, $R_\sigma = UL/\nu_H$ is the magnetic Reynolds number, $R_H = B_0H_0/\rho U^2$ is the magnetic pressure number and $\beta_e = en_eB_0/K_{ei}$ is the Hall parameter.

In solving Eqs. (7.13) to (7.17), it is convenient to introduce the following complex variables.

$$V = u^* - iw^* \qquad H = H_x^* - iH_z^* \quad \text{and} \quad p_s = p_x^* - ip_z^* \tag{7.18}$$

In terms of the complex variables of (7.18), Eqs. (7.13) and (7.14) give

$$\frac{1}{R_e}\frac{d^2V}{dy^{*2}}+R_H\frac{dH}{dy^*}=p_s \tag{7.19}$$

and Eqs. (7.15) and (7.16) give

$$\frac{1}{R_\sigma^*}\frac{d^2H}{dy^{*2}}+\frac{dV}{dy^*}=0 \tag{7.20}$$

where

$$R_\sigma^* = \frac{R_\sigma}{1-i\beta_e} = \text{complex magnetic Reynolds number} \tag{7.21}$$

Equations (7.19) and (7.20) are identical to those equations for the case of scalar conductivity except that the variables and parameters are complex quantities only. Eliminating H from Eqs. (7.19) and (7.20), we have

$$\frac{d^3V}{dy^{*3}}-R_b^*\frac{dV}{dy^*}=0 \tag{7.22}$$

where

$$R_b^* = (R_e R_H R_\sigma^*)^{1/2} = \text{complex Hartmann number} \tag{7.23}$$

The general solution of Eq. (7.22) is

$$V = A_1 \cosh(R_b^* y^*) + B_1 \sinh(R_b^* y^*) + C_1 \tag{7.24}$$

where A_1, B_1, and C_1 are arbitrary constants to be determined by the boundary conditions. For instance, if we consider the Poiseuille flow, the boundary conditions are

$$\begin{aligned} \text{At } y^* &= \pm 1 \qquad u^* = w^* = 0 \qquad H_x^* = H_y^* = 0 \\ y^* &= L \qquad p_1 = p_0 = \text{const} \end{aligned} \tag{7.25}$$

where both the walls are assumed to be insulated walls. The velocity distributions of u^* and w^* are shown in Fig. 7.2 in which the axial velocity distribution u^* is similar to the case of $\beta_e = 0$ but the transverse velocity distribution is sinusoidal without any net flow in the z direction. If $\beta_e = 0$, w^* is identical to zero.

After V is obtained, we may easily calculate H from Eq. (7.20) and p_1 from Eq. (7.17). (See Prob. 7.3.)

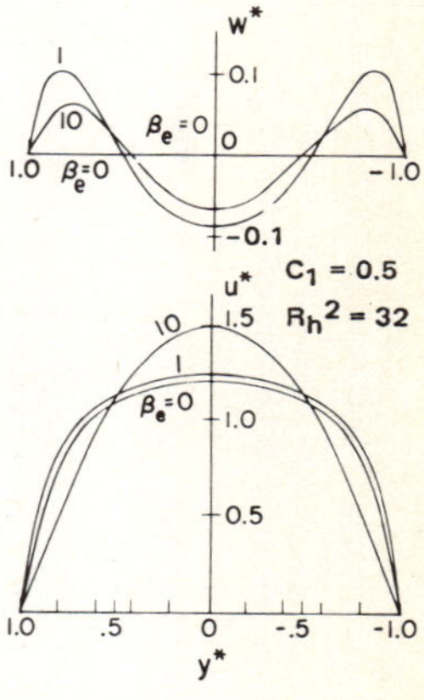

Fig. 7.2 Velocity distributions in an MHD channel with Hall current.

7.3.2 Quasi-One-Dimensional Flow with Tensor Electrical Conductivity. With Hall current and ion slip, the flow field is essentially two dimensional. The longitudinal Hall current gives a transverse body force, which in turn leads to transverse gradients in pressure, velocity, density, etc., as we have shown in Sec. 7.3.1. For engineering application, the quasi-one-dimensional analysis is preferable for its simplicity. Strictly speaking, we should derive the fundamental equations by averaging process described in Chap. 3 for the quasi-one-dimensional flow with tensor electrical conductivity. Since the quasi-one-dimensional analysis gives only a rough first approximation, we shall use the gas-dynamic approximation only in this section which gives some general effects of tensor electrical conductivity.

Let us consider the cartesian coordinate system x, y, z such that x is the axis of the channel. Hence there is only one x-component velocity u different from zero (in the sense of averaging over the cross section). The applied magnetic induction **B** is in the z direction, i.e., $B_x = B_y = 0$, $B_z = B$. Notice that these values of B are also in the sense of averaging process. The electrical field E has both x and y components, i.e., E_x and E_y, but the z component of electric field is assumed to be zero, i.e., $\overline{E}_z = 0$. Equation (7.8) in matrix notation for the present case becomes:

$$\begin{bmatrix} i_z \\ i_y \\ i_x \end{bmatrix} = \frac{\sigma}{(1+\beta_i\beta_e)^2+\beta_e^2} \begin{bmatrix} (1+\beta_i\beta_e+\beta_e^2) & 0 & 0 \\ 0 & (1+\beta_i\beta_e) & +\beta_e \\ 0 & -\beta_e & (1+\beta_i\beta_e) \end{bmatrix} \times \begin{bmatrix} E_z \\ E_y - UB \\ E_x \end{bmatrix} \tag{7.26}$$

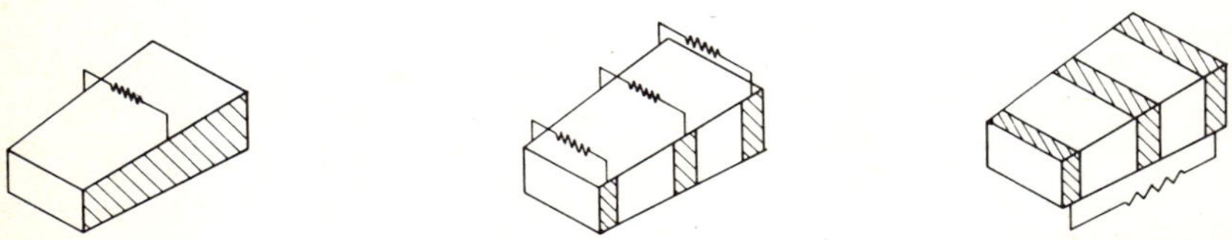

(*a*) Continuous electrode generator (*b*) Segmented electrode generator (*c*) Hall generator

Fig. 7.3 Various types of MFD generators.

We now apply the Eq. (7.26) for three types of MGD generators with constant velocity U as follows: (Fig. 7.3).

Continuous electrode generator (Fig. 7.3*a*). This is the simplest type but should be used only under the conditions of $\beta_e < 1$. When β_e approaches and exceeds unity, the electric Hall field will cause adverse effect on the pressure gradient of the channel. In this case the longitudinal electric field E_x is zero, and transverse electric current is

$$i_y = \frac{(1 + \beta_i\beta_e) UB (K - 1)}{(1 + \beta_i\beta_e)^2 + \beta_e^2} \tag{7.27}$$

where

$$K = \frac{E_y}{UB} \tag{7.28}$$

K is known as load factor and for generator of electric power K lies between 0 and 1. The power generated per unit volume P is

$$P = -\mathbf{E} \cdot \mathbf{i} = -E_y i_y = \frac{K(1 - K) U^2 B^2 (1 + \beta_i\beta_e)}{(1 + \beta_i\beta_e)^2 + \beta_e^2} \tag{7.29}$$

Evidently, when the Hall effect is predominant (large β_e) this generator has relatively low performance. In this case, there is a downstream longitudinal electric current i_x.

Generator with segmented electrodes (Fig. 7.3*b*). To improve the MGD generator at high values of β_e, we should interrupt the x-wise current i_x by using segmented electrodes. For this type of generator, we have $i_x = 0$ but $E_x \neq 0$ and the power generated is

$$P = \frac{U^2 B^2 K (1 - K)}{1 + \beta_i\beta_e} \tag{7.30}$$

This type of generator has the advantage of high power generated but the disadvantage of a complex loading circuit.

Hall generator (Fig. 7.3*c*). This is a modification of the segmented electrode generator but uses the axial current and voltage for electric power extraction. It is attractive because of the single load circuit but it requires a large β_e. In this case we have $E_y = 0$ and the power generated is then

$$P = -E_x i_x = \frac{K_H (1 - K_H) UB}{(1 + \beta_i\beta_e)^2 + \beta_e^2} \tag{7.31}$$

where

$$K_H = -\frac{E_x(1 + \beta_i\beta_e)}{\beta_e UB} \tag{7.32}$$

K_H is the load factor for Hall current.

7.4 MULTIFLUID THEORY OF MAGNETOFLUID DYNAMICS

We consider a plasma as a mixture of N species which consists of ions, electrons, and neutral particles. From the macroscopic point of view, it is sufficient to use the following variables to describe the gas-dynamic properties of each species:

T_s = (kinetic) temperature of sth species
p_s = pressure of sth species
$\rho_s = m_s n_s$ = (mass) density of sth species
u_s^i = ith component of the flow velocity vector of sth species

where $s = 1,2,—,N$ and $i = 1,2$, or 3. The number density of sth species is n_s and the mass of a particle of sth species is m_s.

Besides these gas-dynamic variables, we have to know the electromagnetic field strengths which are the same for all species, i.e.,

E^i = ith component of electric field strength
H^i = ith component of magnetic field strength

In multifluid theory, we deal with $6N + 6$ variables (T_s, p_s, n_s, u_s^i, E^i, and H^i). All properties such as mass m_s, electrical charge e_s, etc., are assumed to be known. In order to investigate these $6N + 6$ variables, we must find $6N + 6$ relations connecting these variables which are $6N$ gas-dynamic equations and 6 electromagnetic field equations.

The 6 electromagnetic field equations are the well known Maxwell equations of electromagnetic field which were derived in Chap. 2. The $6N$ gas-dynamic equations may be derived from the Boltzmann equations of the kinetic theory of gases.[1, 2] For engineering problems, we may assume that the mean free paths of all species are small, and a set of equations similar to those obtained in Chap. 2 for single fluid theory will be obtained. We shall give only some simple cases of the equations for multifluid theory in Sec. 7.5. Readers who are interested in the complete equations should read Ref. 1.

One of the interesting points is to find the relation between the approach of multifluid theory and the classical theory discussed in Chap. 2. In the classical theory, one discusses the gross variables of the plasma as a whole. The relations between these gross variables and the partial variables of multifluid theory are as follows:
Pressure of the mixture:

$$p = \sum_{s=1}^{N} p_s \tag{7.33}$$

Number density of the mixture:

$$n = \sum_{s=1}^{N} n_s \tag{7.34}$$

Density of the mixture:

$$\rho = mn = \sum_{s=1}^{N} m_s n_s = \sum_{s=1}^{N} \rho_s \tag{7.35}$$

where m is the mean mass of the mixture which is a function of the composition of the mixture.
Temperature of the mixture:

$$T = \frac{1}{n} \sum_{s=1}^{N} n_s T_s \tag{7.36}$$

ith Velocity component of the mixture:

$$u^i = \frac{1}{\rho} \sum_{s=1}^{N} \rho_s u_s^i \tag{7.37}$$

ith component of diffusion velocity of the mixture:

$$w_s^i = u_s^i - u^i \tag{7.38}$$

From the definition of the flow velocity u^i and diffusion velocity w_s^i, we have

$$\sum_{s=1}^{N} \rho_s w_s^i = 0 \tag{7.39}$$

Excess electric charge:

$$\rho_e = \sum_{s=1}^{N} n_s e_s = \sum_{s=1}^{N} \rho_{es} \tag{7.40}$$

ith Component of electrical current density:

$$J^i = \sum_{s=1}^{N} \rho_{es} u_s^i = \sum_{s=1}^{N} \rho_{es} w_s^i + u^i \sum_{s=1}^{N} \rho_{es} = i^i + \rho_e u^i \tag{7.41}$$

where i (i^i) is the electric conduction current density and $u^i \rho_e$ is the electric convection current density. Even though ρ_e and J^i or i^i are electrical quantities, they are due to gas-dynamic variables n_s and u_s^i.

There are two ways to describe the flow of a plasma by multifluid theory:

We may use the partial variables (T_s, p_s, n_s, u_s^i, E^i, H^i) to describe the flow field of a plasma where $s = 1,2,—,N$. In this way, we do not need the gross variables such as J^i, ρ_e, etc. because we may calculate the gross variables if we know all the partial variables.

We may use the gross variables (T, p, ρ, u^i, E^i, H^i) *and some of the partial variables* (T_r, p_r, n_r, w_r^i) where $r = 1,2,—,N-1$. In this manner, we may replace one of w_r^i by the electric current density J^i, one of n_r by the excess electrical charge ρ_e and the other n_r by the corresponding concentration $c_r = n_r/n$ or mass concentration $k_r = \rho_r/\rho$.

In principle, the above two approaches are identical as long as we consider $6N + 6$ variables. But for easy application of the multifluid theory to flow problems, the first approach (partial variables) is preferable because the internal energy of the mixture as a whole and the viscous stresses of the mixture as a whole depend on the diffusion phenomena and they are very complicated and difficult to analyze accurately. Hence in the next section, we shall use this approach to investigate the wave motion in a plasma.

It is easier to find the relation between the multifluid theory and the classical single fluid theory of Chap. 2 from the second approach (gross variables and some of the partial variables). To show this relation, we consider a case of partially ionized monatomic gas which consists of atoms, singly charged ions, and electrons, i.e., $N = 3$. According to the second approach, we should use the following 24 variables to describe the flow field of such a gas:

$$T,\ p,\ \rho,\ u^i,\ E^i,\ H^i,\ \rho_e,\ J^i,\ T_1,\ p_1,\ T_2,\ p_2,\ \rho_2,\ w_2^{\ i} \tag{7.42}$$

where subscript 1 refers to the value of electrons and subscript 2 refers to the values of ions.

For single fluid theory of magnetofluid dynamics, we use the first 16 variables of Eq. (7.42), i.e., T, p, ρ, u^i, E^i, H^i, ρ_e, J^i. The reduction of the 24 variables to 16 variables is based on the following approximations:

1. We assume that the temperature of all species are equal, i.e.,

$$T = T_1 = T_2 \tag{7.43}$$

Hence we do not have to consider the variables T_1 and T_2. This approximation is reasonably good if the masses of all species in a mixture of gases are approximately equal. Thus it is a good approximation for neutral air. But for a partially ionized air, the electrons are much lighter than ions and neutral particles. The temperature of electrons T_1 may differ considerably from that of the ions or atoms. At high temperatures, it is advisable to consider at least two temperatures in the flow field of a plasma: one is the electron temperature T_1 and the other is the temperature of heavy particles $T_2 = T_a$.

2. Usually we are not interested in the partial pressures p_1 and p_2 in most engineering problems. Furthermore, if we know the temperature and density of a species, we can easily calculate the partial pressure from the equation of state. Hence we do not need to analyze the partial pressure in our calculation of the flow variables.

3. If the masses of all species in a mixture are approximately equal, the diffusion velocities w_r^i are usually small. We may either neglect the diffusion velocities completely or use some simple formulas for the diffusion velocities in terms of diffusion coefficients. If the diffusion velocities are completely neglected, the concentrations of all species in a mixture of gases remain constant. As a result, we need to consider the gross variables p, ρ, T, u^i only in the gas-dynamic variables. This is the reason why the single fluid theory gives good results to a mixture of neutral gases such as ordinary air. However in a plasma, the electrons are much lighter than all the other heavy particles and we cannot neglect the diffusion velocity of electrons w_1^i. If the diffusion velocity of the ions w_2^i is still negligible, the electrical current density J^i is proportional to w_1^i. The diffusion phenomena of electrons may be expressed in terms of $J^i(w_1^i)$ and $\rho_e(\rho_1)$. This gives the results of classical magnetogas dynamics in which 16 variables are used instead of 24 variables for a partially ionized gas.

The fundamental equation for J^i may be derived from the equation of diffusion velocities of multifluid theory. In general such an equation for J^i depends on both the electromagnetic variables and the gas-dynamic variables and is very complicated. Only under some simplified assumptions can the generalized Ohm's law or the Ohm's law with Hall current and ions slip be obtained (see Ref. 2).

7.5 WAVE MOTIONS BASED ON MULTIFLUID THEORY[4]

Now we reexamine the waves of small amplitude in magnetogas dynamics (Chap. 5, Sec. 5.2) by the multifluid theory. For simplicity, we consider the case of fully ionized plasma in which we have only singly charged ions and electrons. Thus $N = 2$. As in Chap.

5, we assume that originally the plasma is at rest under an external magnetic field $\mathbf{H}_0$, i.e.,

$$\mathbf{H}_0 = \mathbf{i}H_x + \mathbf{j}H_y + \mathbf{k}\cdot 0 \tag{7.44}$$

where H_x and H_y are constants which may or may not be zero. In the undisturbed state of the field, there is neither electric field nor electric current nor excess electric charge. When the plasma is perturbed by a small disturbance, we shall have in the resultant motion the following quantities:

$$\begin{aligned} \mathbf{q}_s &= \mathbf{i}u_s(x,t) + \mathbf{j}v_s(x,t) + \mathbf{k}w_s(x,t) \\ p_s &= p_0 + p_s'(x,t); \quad n_s = n_0 + n_s'(x,t) \quad T = T_0 + T_s'(x,t) \\ \mathbf{E} &= \mathbf{i}E_x(x,t) + \mathbf{j}E_y(x,t) + \mathbf{k}E_z(x,t) \\ \mathbf{H} &= \mathbf{H}_0 + \mathbf{h} = \mathbf{H}_0 + \mathbf{i}h_x(x,t) + \mathbf{j}h_y(x,t) + \mathbf{k}h_z(x,t) \end{aligned} \tag{7.45}$$

where $s = 1$ or 2, and the subscript 1 refers to the values of electrons and subscript 2 refers to the values of ions. The prime refers to the perturbed values and subscript 0 refers to the undisturbed value. We assume that all the perturbed quantities are small so that the second and higher order terms of these quantities are negligible. For simplicity, we assume that all perturbed quantities are functions of one spatial coordinate x and time t only. Thus, we discuss wave propagation in the direction of the x axis.

We make the following assumptions in our fundamental equations:

Both ions and electrons may be considered as inviscid and non-heat-conducting gas.

Perfect gas law may be used for all species.

The interaction forces between ions and electrons are proportional to the difference between their mean velocities:

$$\mathbf{F}_{12} = K_{12}(\mathbf{q}_1 - \mathbf{q}_2) = -\mathbf{F}_{21} \tag{7.46}$$

where $K_{12} = K_{21}$ is the friction coefficient of the plasma, i.e., K_{ie} in Eq. (7.3).

The linearized fundamental equation for the two-fluid theory for the present case are as follows:

Maxwell's equations for electromagnetic fields:

$$\frac{\partial \mu_e h_x}{\partial t} = 0 \tag{7.47a}$$

$$\frac{\partial \mu_e h_y}{\partial t} = \frac{\partial E_z}{\partial x} \tag{7.47b}$$

$$\frac{\partial \mu_e h_z}{\partial t} = -\frac{\partial E_y}{\partial x} \tag{7.47c}$$

$$\frac{\partial \epsilon E_x}{\partial t} + en_0(u_2 - u_1) = 0 \tag{7.47d}$$

$$\frac{\partial \epsilon E_y}{\partial t} + en_0(v_2 - v_1) = -\frac{\partial h_z}{\partial x} \tag{7.47e}$$

$$\frac{\partial \epsilon E_z}{\partial t} + en_0(w_2 - w_1) = \frac{\partial h_y}{\partial x} \tag{7.47f}$$

Equations of state for sth species:

$$\frac{p'_s}{p_0} = \frac{n'_s}{n_0} + \frac{T'_s}{T_0} \tag{7.47g}$$

Equation of continuity of sth species:

$$\frac{\partial n'_s}{\partial t} + n_0 \frac{\partial u_s}{\partial x} = 0 \tag{7.47h}$$

where we assume that there is no source term.

Equations of motion of sth species:

$$m_s n_0 \frac{\partial u_s}{\partial t} = -\frac{\partial p'_s}{\partial x} + e_s n_0 E_x - e_s n_0 B_y w_s + K_{st}(u_s - u_t) \tag{7.47i}$$

$$m_s n_0 \frac{\partial v_s}{\partial t} = e_s n_0 E_y + e_s n_0 B_x w_s + K_{st}(v_s - v_t) \tag{7.47j}$$

$$m_s n_0 \frac{\partial w_s}{\partial t} = e_s n_0 E_z + e_s n_0 (u_s B_y - v_s B_x) + K_{st}(w_s - w_t) \tag{7.47k}$$

where $e_1 = -e, e_2 = e, \mathbf{B}_0 = \mu_e \mathbf{H}_0, s, t = 1$ or 2.

Energy equation for sth species:

$$m_s n_0 c_{ps} \frac{\partial T'_s}{\partial t} = \frac{\partial p'_s}{\partial t} \tag{7.47l}$$

where $m_s c_{ps} = (5/2)R$.

Among the 18 perturbed quantities, only h_x is independent of all other quantities and is given by Eq. (7.47a) alone. Since the divergence of $\mathbf{h}$ is zero, we conclude that h_x = constant which may be put equal to zero. All the other 17 perturbed quantities are interrelated by Eqs. (7.47b) to (7.47l). However there are four basic wave modes which interact with the external magnetic field. The interaction depends on the orientation of the magnetic field with respect to the direction of wave propagation which will be discussed later.

The dispersion relations may be obtained in a similar manner as in the classical case of Chap. 5. In other words, we may assume that all the perturbed quantities are proportional to the exponential form of Eq. (5.2) and find the relations between the wave number λ and the frequency ω from Eqs. (7.47a) through (7.47l).

7.5.1 Basic Waves. If there is no external magnetic field, $\mathbf{H}_0 = 0$, the 17 perturbed quantities may be separated into three independent groups as follows:

The first group consists of v_1, v_2, E_y, and h_z which represents a transverse wave.

The second group consists of w_1, w_2, E_z, and h_y, which represents another transverse wave.

The third group consists of u_1, u_2, p'_1, p'_2, n'_1, n'_2, T'_1, T'_2, and E_x which represents two longitudinal (sound) waves.

Transverse basic waves. Since there is no external magnetic field, we cannot distinguish y from z, hence the dispersion relation of the two transverse waves must be the same. The dispersion relation for the transverse waves of the first or second group is

$$c^2\lambda^2 = \omega^2 - \omega_e{}^2\left(\frac{1 - iK^*_{12}}{1 + K^{*2}_{12}}\right) \tag{7.48}$$

where c is the velocity of light

$$\omega_e = e\left(\frac{n_0}{\epsilon m_1}\right)^{1/2} = \text{electron plasma frequency} \tag{7.49}$$

and

$$K_{12}^* = \frac{K_{12}}{m_1 \omega n_0} \tag{7.50}$$

For an ideal plasma, $K_{12} = 0$, Eq. (7.48) becomes

$$V = \frac{\omega}{\lambda} = \frac{c}{(1 - \omega_e^2/\omega^2)^{1/2}} \tag{7.51}$$

Equation (7.51) is the well-known result of elementary theory of plasma oscillation and shows that there is a large damping for low frequency waves ($\omega < \omega_e$) which is the essential mechanism of radio blackout during the reentry of a space vehicle. Such a blackout may be avoided by applying an external magnetic field as will be discussed in Sec. 7.5.2.

In vacuum, $n_0 = 0$ and $\omega_e = 0$, Eq. (7.51) shows that $V = c$ and that electromagnetic wave is propagated at a speed of light c.

The effect of electrical conductivity σ or K_{12} is to introduce a damping on the transverse wave.

Longitudinal wave. The dispersion equation gives two roots of λ^2 for the longitudinal waves of the third group as follows:

$$\lambda^2 = \frac{\omega^2}{2a_2^2}\left\{\left[\left(1 - 2\frac{\omega_i^2}{\omega^2}\right) + 2iK_{12}^*\frac{m_1}{m_2}\right]\right.$$

$$\left.\pm\left[\left[\left(1 - 2\frac{\omega_i^2}{\omega^2}\right) + 2iK_{12}^*\frac{m_1}{m_2}\right]^2 - 4\frac{a_2^2}{a_1^2}\left(i - \frac{\omega_e^2}{\omega^2}\right) + iK_{12}^*\right]^{1/2}\right\} \tag{7.52}$$

where

$$\omega_i = e\left(\frac{n_0}{\epsilon m_2}\right)^{1/2} = \text{ion plasma frequency} \tag{7.53}$$

$$a_s = \left(\frac{\gamma_s p_0}{m_s n_0}\right)^{1/2} = \text{sound speed of } s\text{th species} \tag{7.54}$$

Equation 7.52) gives two longitudinal waves: one is known as *ion sound wave* (+sign) and the other, as *electron sound wave* (–sign).

The ion sound wave of an ideal plasma is propagated at a speed V_2 which is

$$V_2 = \frac{\omega}{\lambda_2} = \frac{a_2}{\sqrt{2}} \left\{ \left(1 - 2\frac{\omega_i^2}{\omega^2}\right) + \left[\left(1 - 2\frac{\omega_i^2}{\omega^2}\right)^2 - 4\frac{a_2^2}{a_1^2}\left(1 - \frac{\omega_e^2}{\omega^2}\right)\right]^{1/2} \right\}^{-1/2} \tag{7.55}$$

The value of V_2, (see Fig. 7.4) is always real and this wave is always an undamped wave with a finite speed of propagation. At very low frequency ($\omega \to 0$), V_2 is equal to the sound speed of the plasma as a whole, i.e., $V_2 = \sqrt{2}\, a_2 = a$ which agrees with the value of single-fluid theory of Chap. 5. As ω increases, V_2 decreases continuously. At very high frequency ($\omega \to \infty$), V_2 approaches a_2, the ion sound speed as a limit.

The electron sound wave of an ideal plasma is propagated at a speed V_1 which is

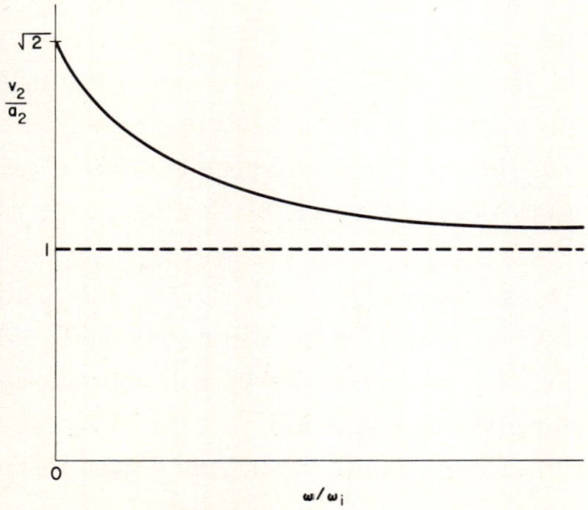

Fig. 7.4 Speed of propagation of ion sound wave in an ideal plasma.

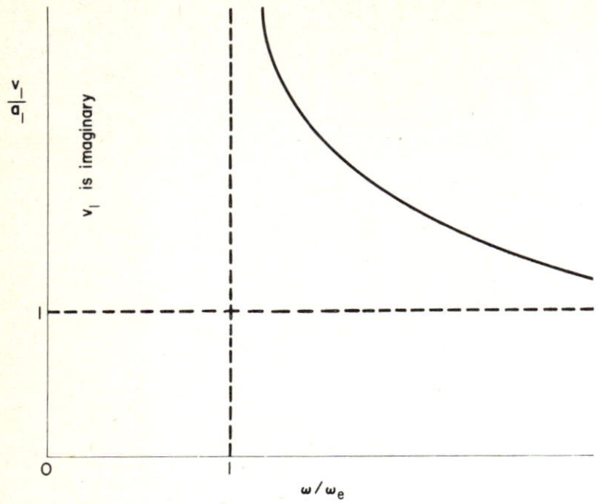

Fig. 7.5 Speed of propagation of electron sound wave in an ideal plasma.

$$V_1 = \frac{\omega}{\lambda_1} = \frac{a_2}{\sqrt{2}} \left\{ \left(1 - 2\frac{\omega_i^2}{\omega^2}\right) - \left[\left(1 - 2\frac{\omega_i^2}{\omega^2}\right)^2 - 4\frac{a_2^2}{a_1^2}\left(1 - \frac{\omega_e^2}{\omega^2}\right)\right]^2 \right\}^{-1/2} \tag{7.56}$$

As long as $\omega < \omega_e$, V_1 is pure imaginary which represents a damped exponential wave. As $\omega > \omega_e$, V_1 is real and it represents an undamped wave (see Fig. 7.5). The value of V_1 decreases from $V_1 \to \infty$ at $\omega = \omega_e$ to $V_1 = a_1$ at $\omega \to \infty$.

In the classical single-fluid theory of Chap. 5, we have only one mode of sound wave in an ideal plasma with a speed of propagation $V = a$ which is independent of the applied frequency ω. Hence the classical single fluid theory gives only the result of the ion mode at very low frequency.

7.5.2 Waves under Longitudinal External Magnetic Field. When $H_x \neq 0$ and $H_y = 0$, the basic longitudinal waves [Eq. (7.52)] will not be affected by the magnetic field but the two transverse waves will interact so that two new modes of transverse waves will result.

The dispersion relation for the interacted transverse waves in an ideal plasma[4] is

$$A_0\lambda^4 + 2A_2\lambda^2 + A_4 = 0 \tag{7.57}$$

where

$$A_0 = \left(1 - \frac{\omega_{x1}^2}{\omega^2}\right)\left(1 - \frac{\omega_{x2}^2}{\omega^2}\right)$$

$$A_2 = \frac{\omega_e^2}{c^2}\left[\left(1 - \frac{\omega^2}{\omega_e^2}\right) - \frac{\omega_{x1}\omega_{x2}}{\omega^2}\left(1 - \frac{\omega^2}{\omega_i^2} + \frac{\omega_{x1}\omega_{x2}}{\omega_e^2}\right)\right] \qquad (7.58)$$

$$A_4 = \frac{\omega_e^4}{c^4}\left[\left(1 - \frac{\omega^2}{\omega_e^2} + \frac{\omega_{x1}\omega_{x2}}{\omega^2}\right)^2 - \frac{\omega_{x1}\omega_{x2}\omega^2}{\omega_i^2\omega_e^2}\right]$$

$$\omega_{xs} = \frac{eB_x}{m_s} = \text{cyclotron frequency of } s\text{th species} \qquad (7.59)$$

Equation (7.57) gives two modes of transverse waves which are represented by the following roots of

$$\lambda_1^2 = \frac{-A_2 + \sqrt{A_2^2 - A_0A_4}}{A_0} = \text{extraordinary wave} \qquad (7.60)$$

$$\lambda_2^2 = \frac{-A_2 - \sqrt{A_2^2 - A_0A_4}}{A_0} = \text{ordinary wave} \qquad (7.61)$$

When there is no applied magnetic field, $\omega_{xs} = 0$, we have

$$\lambda_1^2 = \lambda_2^2 = \frac{\omega^2}{c^2}\left(1 - \frac{\omega_e^2}{\omega^2}\right) \qquad (7.62)$$

It is the result of the basic transverse waves.

When there is an applied longitudinal magnetic field and when the applied frequency ω is very small, we have

$$\lambda_1^2 = \lambda_2^2 = \frac{\omega^2}{V_x^2}\left(1 + \frac{V_x^2}{c^2}\right) = \lambda_m^2 \qquad (7.63)$$

At very low frequencies, the two modes of the transverse waves propagate at the same speed V, that is,

$$V = \frac{\omega}{\lambda_m} = \frac{V_x}{(1 + V_x^2/c^2)^{1/2}} \qquad (7.64)$$

where

$$V_x = H_x\left(\frac{\mu_e}{\rho}\right)^{1/2} = H_x\left(\frac{\mu_e}{m_2 n_0}\right)^{1/2} = \text{speed of Alfvén's wave under } H_x \tag{7.65}$$

If $V_x \ll c$, we have $V = V_x$ which is the well-known result of single-fluid theory, Alfvén's wave. Without an external magnetic field, Eq. (7.62) shows that there is no undamped wave when $\omega < \omega_e$, but with $H_x \neq 0$, undamped waves exist at low frequencies. Thus the applied magnetic field may improve the radio communication in a plasma and may avoid the blackout during the reentry of a space vehicle.

When the applied frequency ω is very large, $\omega \to \infty$, Eqs. (7.60) and (7.61) give

$$\lambda_1^2 = \lambda_2^2 = \frac{\omega^2}{c^2} \tag{7.66}$$

Thus at very high frequencies, the basic waves are decoupled and each is propagated at a speed of light as if there is no external applied magnetic field.

At intermediate frequencies, these two modes of transverse waves propagate at different speeds and one or both may become damped exponential waves (see Fig. 7.6). For $\omega < \omega_{x2}$, these two modes are both undamped waves. For $\omega_{x1} > \omega > \omega_{x2}$, the extraordinary wave λ_2 remains an undamped wave but the ordinary wave λ_1 becomes a damped exponential wave. This region is usually referred to as the *whistler mode* of the extraordinary wave and the group velocity $\partial\omega/\partial\lambda$ increases with the frequency ω. When $\omega > \omega_{x2}$, the extraordinary wave λ_2 will first change into a damped exponential wave and then revert to an undamped wave at very high frequencies. Of course, the ordinary wave will revert to an undamped wave at a lower frequency than that for an extraordinary wave. For strong applied magnetic fields, at most one of the transverse waves will be a damped wave but for weak applied magnetic fields, in a certain intermediate frequency range, both transverse waves may become damped waves simultaneously.

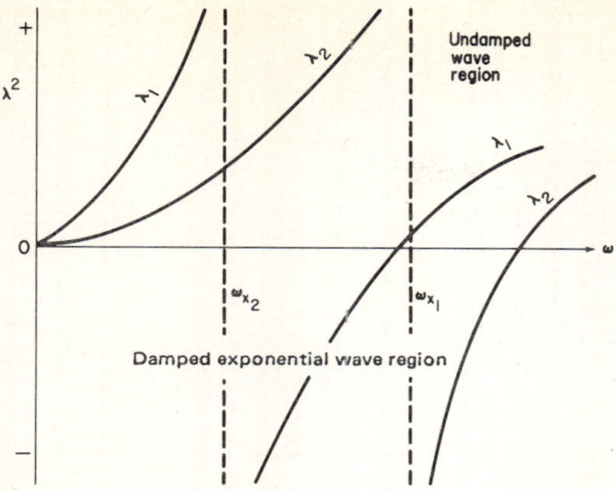

Fig. 7.6 Dispersion relations of interacted transverse waves under longitudinal magnetic field H_x.

At $\omega = \omega_{x2}$, the speed of propagation of the extraordinary wave V_2 is $\sqrt{2}\,V_x$ and that of the ordinary wave V_1 is zero. Hence the extraordinary wave is connected mainly with the ions because V_2 depends mainly on the mass of the ions.

At $\omega = \omega_{x1}$, we have $V_1 = c/(1 - \omega_e^2/2\omega_{x_1}^2)^{1/2}$ and $V_2 = 0$. Hence, the ordinary wave is connected mainly with the electrons because V_1 depends mainly on the mass of electrons. At other frequencies, there are interactions between ions and electrons and thus the speeds of propagation of these two transverse waves change with frequencies and one or both modes may change to damped waves.

The friction coefficient or electrical conductivity will introduce damping to these transverse waves and may change their speeds of propagation.

7.5.3 Waves under Transverse External Magnetic Field. When $H_x = 0$ and $H_y \neq 0$, the basic transverse wave of v_1, v_2, E_y, and h_z will not be affected by the magnetic field but the other basic transverse wave (w_1, w_2, E_z, h_y) and the two basic longitudinal waves will interact so that three new transverse-longitudinal waves will result.

The dispersion relation for these transverse-longitudinal waves in an ideal plasma is

$$\lambda^6 + S_4\lambda^4 + S_2\lambda^2 + S_0 = 0 \qquad (7.67)$$

where

$$S_4 = \frac{\omega^2}{c^2}\left(\frac{\omega_e^2}{c^2} - 1\right) + \frac{\omega^2}{a_2^2}\left(2\frac{\omega_i^2}{\omega^2} - 1 + \frac{\omega_{y1}\omega_{y2}}{\omega^2}\right)$$

$$S_2 = \frac{\omega^2}{c^2}\frac{\omega^2}{a_2^2}\left(2\frac{\omega_i^2}{\omega^2} - 1\right)\left(\frac{\omega_e^2}{\omega^2} - 1 + \frac{\omega_{y1}\omega_{y2}}{\omega^2}\right)$$

$$+ \frac{\omega^4}{a_1^2 a_2^2}\left[\left(1 - \frac{\omega_e^2}{\omega^2}\right) + \frac{\omega_{y2}^2}{\omega^2}\left(\frac{\omega_i^2}{\omega^2} + \frac{\omega_{y2}^2}{\omega^2} - 1\right)\right] \tag{7.68}$$

$$S_0 = -\frac{\omega^2}{c^2}\frac{\omega^4}{a_1^2 a_2^2}\left[\left(\frac{\omega_e^2}{\omega^2} - 1\right)^2 + \frac{\omega_{y1}^2}{\omega^2}\left(2\frac{\omega_i^2}{\omega^2} + \frac{\omega_{y2}^2}{\omega^2} - 1\right)\right]$$

$$\omega_{ys} = \frac{eB_y}{m_s} = y\text{-wise cyclotron frequency of } s\text{th species}$$

Equation (7.67) gives three roots of λ^2 which represent three different modes of the transverse-longitudinal waves (Fig. 7.7). When the applied frequency is very low ($\omega \to 0$), only one root λ_1^2 is positive. Hence, we have one undamped wave and two damped waves at low applied frequencies. The speed of propagation of this

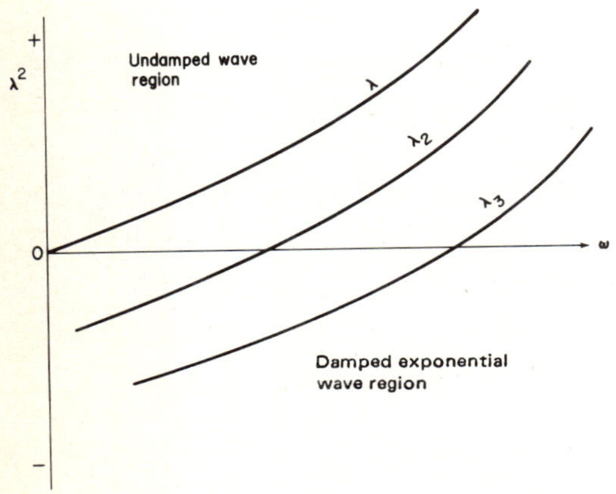

Fig. 7.7 Dispersion relations of transverse-longitudinal waves in an ideal plasma under transverse magnetic field H_y.

undamped wave is

$$V_1 = \frac{\omega}{\lambda_1} = \left[\frac{a^2 + V_y^2}{1 + (V_y^2/c^2)}\right]^{1/2} \tag{7.69}$$

where $V_y = H_y(\mu_e/m_2 n_0)^{1/2}$ is the y-wise speed of propagation of Alfvén's wave.

When $V_y \ll c$, Eq. (7.69) becomes

$$V_1 = \sqrt{a^2 + V_y^2} \tag{7.70}$$

Equation (7.70) is the well-known result of single-fluid theory of magnetogas dynamics [cf. Eq. (5.32)].

When we increase the magnetic field strength, the speed of propagation of this mode of wave at low frequencies increases, but its value is less than that given by Eq. (7.70). In the limit H_y, V_1 approaches the velocity of light c.

For very high frequencies, $\omega \to \infty$, the three modes of waves will be decoupled and they become the three basic modes.

As the frequency ω increases from zero, the first mode λ_1 is always an undamped wave but its speed of propagation changes. The second mode λ_2 will first change from the damped wave into an undamped wave and then the third mode λ_3 will change from damped wave into undamped wave. At very high frequencies, all the three modes are undamped waves. The critical values of frequency ω_c at which the second and the third modes change from damped wave into undamped wave depend on the magnitude of the applied magnetic field H_y. For weak magnetic fields, these critical values of frequency ω_c are close to the electron plasma frequency ω_e, i.e., at $\omega_2 = \omega_e - \Delta\omega$ the second mode changes into undamped wave and at $\omega_3 = \omega_e + \Delta\omega$, the third mode changes into undamped wave. For strong applied magnetic field, the critical frequencies are the corresponding cyclotron frequencies i.e., $\omega_2 = \omega_{y2}$ and $\omega_3 = \omega_{y1}$.

7.5.4 Waves under a Magnetic Field of Arbitrary Orientation. When both H_x and H_y are different from zero, all four basic waves interact so that four new waves will result. The dispersion relation for these waves may be written as

$$C_0\lambda^8 + C_1\lambda^6 + C_2\lambda^4 + C_3\lambda^2 + C_4 = 0 \tag{7.71}$$

where the coefficient C_n are complicated functions of the plasma and the cyclotron frequencies of ions and electrons, sound speed of ion and electrons, and friction coefficient. For an ideal plasma, $C_0 = A_0$ of Eq. (7.58). The dispersion relations of (7.71) in an ideal plasma are shown in Fig. 7.8. At low frequencies ($\omega \to 0$), only three of the four roots, λ^2, are positive which represent three undamped waves. For small applied magnetic field, these three undamped waves at very low frequencies reduce to the well-known results of single-fluid theory of magnetogas dynamics (Chap. 5, Sec. 5.5), the fast wave, slow wave, and transverse wave. As the frequency increases, the results of two-fluid theory differ from those of the classical single-fluid theory. One of the modes, corresponding to the fast wave, is always an undamped wave but its speed of propagation V_1 changes with frequency. At very high frequency V_1 tends to be a_2. The second mode λ_2 is a modified ordinary transverse wave and it changes into a damped wave at $\omega = \omega_{x2}$. At high frequency, it reverts into an undamped wave again and finally as $\omega \to \infty$, V_2 tends to be c. The third mode λ_3 is a modified extraordinary transverse wave and it changes into a damped wave at $\omega = \omega_{x1}$. At high frequency, it also reverts into an undamped wave and finally as $\omega \to \infty$, $V_3 \to c$. The

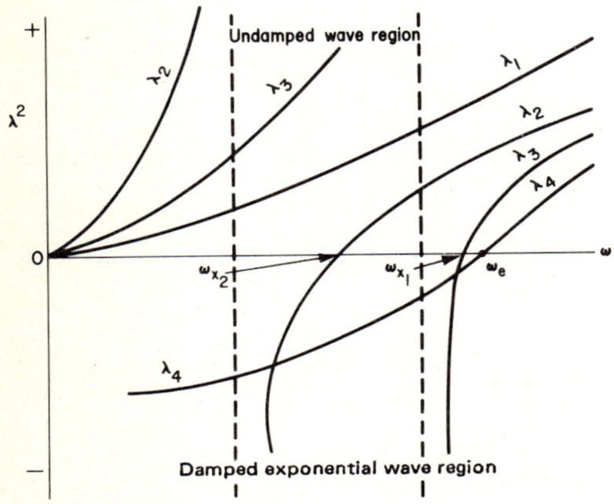

Fig. 7.8 Dispersion relations of magnetogas-dynamic waves in an ideal plasma under arbitrary applied magnetic fields H_x and H_y.

fourth wave is a modified electron sound wave. As long as $\omega < \omega_e$ it is a damped wave. When $\omega > \omega_e$, it changes into an undamped wave. As $\omega \to \infty$, $V_4 \to a_1$.

In conclusion, we find that the single-fluid theory may be a good approximation to the two-fluid theory when the magnetic field is not strong and when the applied frequency is very small. In general, single-fluid theory will not be a good approximation to the two-fluid theory. The two-fluid theory will give much more information concerning the flow problem of a fully ionized plasma which cannot be obtained by classical single-fluid theory.

The two-fluid theory may be extended into multifluid theory in a straightforward manner but the algebra of the determinental equation of λ^2 will be very complicated because more modes of waves will occur. For instance, if we consider the partially ionized monatomic gas, we should consider three species: ions, electrons, and atoms. An additional longitudinal wave, the sound wave of the atoms, will be introduced into this system.[5]

When we study flow problems by multifluid theory, one of the main interesting points is that the temperatures of all species are not equal, especially the electron temperature may differ considerably from those of heavy particles. As a result, the heat transfer will be greatly influenced by the electron temperature.[6]

7.6 RAREFIED PLASMA DYNAMICS, KINETIC THEORY

All of the previous discussions in this book are based on the assumption that the mean free path of the gas is small and that the gas may be considered as a continuum whether it is a single fluid or a mixture of gases. Some modifications should be made when the mean free path of the gas is not small. A more accurate analysis of the flow of a gas or a plasma is the kinetic theory of a gas.[7, 8] In this section, we are going to discuss a few essential points of kinetic theory of a plasma and its relations with the treatments discussed in this book.

A gas or a plasma is composed of a large number of particles which are generally called *molecules*. The term "molecules," in a general sense, includes ordinary molecules as well as atoms, ions, and electrons. Thus the molecules of a gas may be all of a kind or of several kinds. The number of molecules of a gas is usually much larger than the number of kinds of molecules. If we could investigate the detailed motion of all molecules in a gas flow, a complete description of the gas flow may be obtained. However, it is not possible to do so because in the first place we do not know the detailed conditions of all the gas molecules at a prescribed initial instant and in the second place the number of molecules of a gas is so large that it is not possible at the present time to calculate the motions of all the molecules even if we know the initial conditions of all the molecules. Hence we can only conveniently consider the statistical average of the motion of the molecules of a gas. It is known as the *kinetic theory* of gases.

In the kinetic theory of gas,[1] we describe the motion of the molecules of a kind by means of a molecular distribution function of this kind of molecule $F_s(\mathbf{r}, \mathbf{q}_m, t)$ which represents the expectation of the number of molecules of sth species per unit volume at the position $\mathbf{r}$ and time t within the molecular velocity range $\mathbf{q}_m$ and $\mathbf{q}_m + d\mathbf{q}_m$, i.e.,

$$dn_s = F_s(\mathbf{r}, \mathbf{q}_m, t)\, d\mathbf{q}_m \tag{7.72}$$

The average number density of sth species at $\mathbf{r}$ and t is then

$$n_s = \iiint F_s(\mathbf{r}, \mathbf{q}_m, t)\, d\mathbf{q}_m \tag{7.73}$$

where $d\mathbf{q}_m$ means the integral over the whole velocity plane, that is, $dq_m = d\xi d\eta d\zeta$ and ξ, η, and ζ are the x, y, and z component of the molecular velocity $\mathbf{q}_m$.

The average of any physical quantity Q of the molecules is

$$\overline{Q}_s = \iiint Q f_s \, d\xi \, d\eta \, d\zeta \tag{7.74}$$

where $F_s = nf_s$. For instance, the average x-wise component of velocity of the gas flow of sth species is

$$u_s = \iiint \xi f_s \, d\xi \, d\eta \, d\zeta \tag{7.75}$$

7.6.1 The Macroscopic Quantities. If we know the distribution function F (for simplicity, we drop the subscript s from now on and it is understood that the distribution function and the corresponding results may be applied to a gas of one kind or any one species in a mixture), we may calculate all the macroscopic quantities used in the continuum theory which we have discussed previously in this book. We introduce a random or peculiar velocity $\mathbf{c}$ with component c_i of the molecule as follows:

$$q_{mi} = q_i + c_i \tag{7.76}$$

where q_i is the ith component of the flow velocity $\mathbf{q}$ such as u_s of Eq. (7.75). By the definition of Eq. (7.74), we have the following relations for the variables used in the macroscopic treatment previously discussed in this book in terms of the molecular distribution function F.

The density. The density ρ is

$$\rho = mn = m \int F d\mathbf{c} \tag{7.77}$$

The flow velocity. The flow velocity $\mathbf{q}$ of component q_i has the relation

$$u_i = \frac{1}{n} \int q_{mi} \, F d\mathbf{c} = \bar{q}_{mi} \tag{7.78}$$

where bar refers to the average value of a quantity according to Eq. (7.74).

The pressure tensor. The pressure tensor p_{ij} and the pressure p are respectively

$$p_{ij} = mn\overline{c_i c_j} = m\iint c_i c_j F\, dc \tag{7.79}$$

and

$$p = \tfrac{1}{3}(p_{11} + p_{22} + p_{33}) \tag{7.80}$$

The viscous stress tensor with the ij component τ_{ij} is defined by the relation

$$\tau_{ij} = -p_{ij} + p\delta_{ij} \tag{7.81}$$

where $\delta_{ij} = 1$ if $i = j$ and $\delta_{ij} = 0$ if $i \neq j$.

The kinetic temperature. The kinetic temperature T is defined as

$$T = \frac{m}{3kn}\overline{c^2} = \frac{m}{3k}\int (c_1^2 + c_2^2 + c_3^2) F\, dc \tag{7.82}$$

where k is the Boltzmann constant.

The energy flux. The third moment of the peculiar velocity is

$$S_{ijk} = nm\overline{c_i c_j c_k} = m\int c_i c_j c_k F\, dc \tag{7.83}$$

The heat flux due to conduction is then

$$Q_{ci} = S_{i11} + S_{i22} + S_{i33} = m\int (c^2)\, c_i F\, dc \tag{7.84}$$

The distribution function F is governed by the Boltzmann equation which is

$$\frac{\partial F}{\partial t} + q_{mi}\frac{\partial F}{\partial x_i} + \frac{P_i}{m}\frac{\partial F}{\partial q_{mi}} = \frac{\delta F}{\delta t} \tag{7.85}$$

The left side of Eq. (7.85) represents the time rate of change of the distribution function F and P_i is the ith component of the force acting on a molecule. The first term is the time rate of change of F, the second term is the change in space and the third term is the change due to velocity variation by the body force P_i. The right-side terms are known as the *collision term* which represent the collision or encounter of the molecules considered with other molecules in the flow field. One of the main objects of the kinetic theory of gases is to find the proper expression for the collision terms.[7,8] For the discussion of the collision terms, see Refs. 7 and 8.

We may derive the fundamental equations of continuum theory from the Boltzmann equation by integration over the entire velocity space with a proper multiplication factor and using the definition of the continuum theory variables given by Eqs. (7.77) to (7.84). For instance, if we multiply Eq. (7.85) by m_s and integrate the resulting equation with respect to c_s, we have

$$\frac{\partial \rho_s}{\partial t} + \frac{\partial \rho_s q_{si}}{\partial x_i} = \int m_s \left(\frac{\delta F_s}{\delta t}\right) dc_s = \sigma(m_s n_s) \tag{7.86}$$

Equation (7.86) is the equation of continuity of sth species in the multifluid theory where σ is the source term. If there is no source term, Eq. (7.86) is the ordinary equation of continuity (2.196).

Similarly, if we multiply the Boltzmann equation by $m_s c_{si}$, we have the equation of motion. In general, the pressure tensor should be governed by a differential equation which may be obtained by multiplying the Boltzmann equation by mc_ic_j and integrating the resulting equation. Only when the mean free path of the gas is small may we approximate the differential equation by the simple Navier-Stokes relation of Eq. (2.234). Hence in the rarefied gas, we shall have rather complicated relations between stress and strain. Similarly, only when the mean free path is small, we may use the Fourier relation for the heat conduction flux and in general, we should have a rather complicated relation between the temperature and heat flux.

7.6.2 The Knudsen Number. It is advisable to divide the gas flow into various regions according to their degree of rarefication. The parameter which may be used to define these regions is known as the *Knudsen number* K which is defined as

$$K = \bar{l}L \tag{7.87}$$

where $\bar{l}$ is the mean free path of the gas and L is the characteristic length of the flow field. When the Knudsen number is not negligibly small, the flow should be considered as a rarefied gas flow. When the Knudsen number is negligibly small, the fluid may be considered as a continuous medium. From the kinetic theory of gas, the mean free path of a neutral gas is defined approximately by the following formula:

$$\nu = \frac{1}{2}v\bar{l} \tag{7.88}$$

where ν is the coefficient of kinematic viscosity and v is the mean molecular velocity which is related with the sound speed a of the gas by the relation

$$a = v\left(\frac{\pi\gamma}{8}\right)^{1/2} \tag{7.89}$$

where γ is the ratio of the specific heats. The Knudsen number is then

$$K = 1.26\sqrt{\gamma}\,\frac{M}{R_e} \tag{7.90}$$

where $M = U/a$ is the Mach number of the flow field, U is a typical velocity of the flow field and $R_e = UL/\nu$ is the Reynolds number of the flow field. Equation (7.90) shows that the Knudsen number becomes important when M is large and R_e is small.

For a plasma, besides the mean free path $\bar{l}$, there are two other characteristic lengths which also have influence on the rarefied effects of the corresponding flow problems. These are the Debye length and the Larmor radius.

The Debye length L_D is obtained from the condition that the electrical potential energy is equal to the thermal energy and it is given by the formula:

$$LD = \left(\frac{kT\epsilon}{ne^2}\right)^{1/2} \tag{7.91}$$

where k is the Boltzmann constant, n is the number density of the plasma, T is the absolute temperature of the plasma, e is the absolute electric charge, and ϵ is the inductive capacity. The nondimensional parameter

$$K_D = \frac{L_D}{L} \tag{7.92}$$

may be called the *electric Knudsen number* which is a measure of the distance over which the excess electric charge may be different appreciably from zero.

The Larmor radius L_L may be defined as

$$L_L = \frac{a}{\omega_c} = \frac{am}{eB} \tag{7.93}$$

where ω_c is the cyclotron frequency of the charged particle. The Larmor radius is a measure of the radius of the helical path of a charged particle of mass m and charge e in a magnetic induction B. We use the sound speed a to represent the particle velocity of the plasma because a is of the same order of the average molecular velocity of the plasma. The nondimensional parameter

$$K_L = \frac{L_L}{L} \tag{7.94}$$

may be called *magnetic Knudsen number* which shows the rarefied effects of a plasma under the influence of the magnetic field.

7.6.3 The Flow Region. We may divide the flow region according to the three Knudsen numbers given above. For simplicity, it is sufficient to consider the ordinary Knudsen number K alone. We may divide the flow into four regions as follows:

Continuum flow region. When the Knudsen number is very ..all, say $K < 0.01$, the gas or plasma may be considered as a continuum and the results discussed previously in this book are applicable.

Slip flow region. When the Knudsen number is not negligibly small but it is still small, say $0.01 \leq K \leq 0.1$, the flow is in the slip flow region. In this case, the fluid still behaves as a continuum medium and the fundamental equations given previously in this book may be still used to describe the flow field but the no-slip boundary conditions do not hold true. There will be a slip of velocity on the solid boundary and a jump in temperature on the solid boundary due to the rarefied effects even though the viscous and the heat-conducting effects are included.

Transition region. When the Knudsen number is large, say $0.1 \leq K \leq 3$, the flow is in a transition region in which we may not use the Navier-Stokes relations for the viscous stress-strain relation nor the Fourier relation for the heat conduction flux and the temperature gradient. The flow depends greatly on the discrete character of the fluid. Our present knowledge of the flow problem in the transition region is very limited.

Free molecule regime. When the Knudsen number is very large, say $K > 3$, the gas particles should be considered to be free in motion without the influence of other particles. The flow problem becomes again rather simple. In this case, the collisions between the molecules are much less than the collisions between the molecules and the surface of the body in the flow field. As a result, we may neglect completely the collisions between the particles and consider only the interaction between the gas molecules and the surface of the body in the flow field. In this case, we may assume the molecular distribution function is the maxwellian distribution function F_0, that is,

$$F_0 = n\left(\frac{m}{2\pi kT}\right)^{3/2} \exp\left[-\frac{m(\mathbf{q}_m - \mathbf{q})^2}{2kT}\right] \tag{7.95}$$

The maxwellian distribution function F_0 represents a uniform and steady state of a gas of mean velocity q_m and an absolute temperature T, and a number density n. In free molecular flow, we may use this maxwellian molecular distribution to find the change of velocity and temperature after the interaction of the gas molecules with the solid surface.

If we include both the electric and magnetic Knudsen numbers, we may have some flow regions in which the electric and magnetic field become very important. (See Chap. 14 of Ref. 1.)

In the continuum flow region, usually the actual molecular distribution function does not differ much from the maxwellian distribution. Hence we may write

$$F = F_0(1 + F') \tag{7.96}$$

where $|F'| \ll 1$. Some perturbation technique may be used to solve for F'. This is the way which has been extensively used to determine the transport coefficients of a gas or a plasma such as the coefficient of viscosity, coefficient of heat conductivity, and electrical conduction from the Boltzmann equation.

7.7 PARTICLE MOTIONS OF A PLASMA[9]

Even though the Boltzmann equation describes the motion of a highly rarefied gas as well as a dense gas, it is sometimes convenient to describe the motion of a highly rarefied gas by considering the motion of individual particles, particularly for the charged particles whose motion is largely dominated by the electromagnetic forces. Such an analysis has been widely used for highly rarefied plasma. For a charged particle, ion or electron, the equation of motion under the electromagnetic force is

$$m_s \frac{d\mathbf{q}_s}{dt} = e_s(\mathbf{E} + \mathbf{q}_s \times \mathbf{B}) \tag{7.97}$$

where m_s is the mass of a particle of sth species in the plasma, $\mathbf{q}_s$ is its velocity, and e_s, its electric charge. Equation (7.97) should be solved with the Maxwell equations for electromagnetic fields. Without going into the detailed analysis, Eq. (7.97) gives many interesting results. We shall consider a few simple cases from Eq. (7.97) alone.

onstant electric field. E = constant and no magnetic field 0. In this case, Eq. (7.97) becomes

$$\frac{d\mathbf{q}_s}{dt} = \frac{e_s\mathbf{E}}{m_s} \tag{7.98}$$

The particle has a constant acceleration in the direction of the electric field E.

No electric field. E = 0 and constant magnetic field B = constant. In this case, it is convenient to use cylindrical coordinates r, θ, z with z in the direction of B. The final equilibrium condition will be the case that the particle has no radial velocity component and has only the axial and tangential component. The axial velocity component q_z is determined by the initial condition of the particle and is unaffected by the magnetic field. The tangential velocity component q_θ is determined by the equilibrium condition between the centrifugal force and electromagnetic force, i.e.,

$$m_s \frac{q_\theta^2}{r_1} = e_s q_\theta B \tag{7.99}$$

where r_1 is the radius of the circle where the particle moves in the equilibrium condition. Eq. (7.99) may be written as

$$\omega_c = \frac{q_\theta}{r_1} = \frac{e_s B}{m_s} \tag{7.100}$$

where ω_c is the cyclotron frequency or Larmor frequency of the charged particle m_s (as has been discussed before). If q_z is zero, the particle moves on a circle of radius r_1 which is known as *Larmor radius*; while if q_z is different from zero, the particle will move in a helical path with radius r_1 of the helix. In our previous discussion, we assumed that $q_\theta = a$ and obtained Eq. (7.93).

Constant electric field, constant magnetic field, and E is perpendicular to B. In this case, it is convenient to define a new velocity of the particle v_s such that

$$\mathbf{v}_s = \mathbf{q}_s - \frac{\mathbf{E} \times \mathbf{B}}{B^2} = \mathbf{q}_s - \mathbf{q}_d \tag{7.101}$$

where

$$q_d = \frac{\mathbf{E} \times \mathbf{B}}{B^2} = \mathbf{i}_d \frac{E}{B} = \text{drift velocity} \tag{7.102}$$

where $\mathbf{i}_d$ is the unit vector in the direction of $\mathbf{E} \times \mathbf{B}$.

The velocity $\mathbf{v}_s$ satisfies the following equation

$$m_s \frac{d\mathbf{v}_s}{dt} = e_s \mathbf{v}_s \times \mathbf{B} \tag{7.103}$$

The velocity $\mathbf{v}_s$ is independent of the electric field $\mathbf{E}$.

If the electric field is not perpendicular to $\mathbf{B}$, we may resolve the electric field $\mathbf{E}$ into two components: one is perpendicular to B, i.e., $\mathbf{E}_\perp$, and the other is parallel to B, i.e., $\mathbf{E}_\parallel$. We use $\mathbf{E}_\perp$ in Eq. (7.102) for the drift velocity and $\mathbf{E}_\parallel$ gives the axial acceleration of the particle along the direction of $\mathbf{B}$.

If both $\mathbf{E}$ *and* $\mathbf{B}$ *are not constant.* In this case the motion of the particle is very complicated. However, if the variations of the electric and magnetic fields are small, the motion of the particle will be qualitatively the same as we have obtained above. Some of the interesting problems of a plasma may be brought to view by studying the particle motion in slowly varying fields such as stability of a plasma confined by magnetic fields.

In Eq. (7.97), we assume that the particle is subjected to an electromagnetic force only. We may generalize the motion of the particle by considering other forces such as gravitational force. In general the gravitational force has similar effect as the electric field on a charged particle.

7.8 THERMAL RADIATION EFFECTS OF MAGNETOFLUID DYNAMICS

In our previous study of heat-transfer problem, we considered only two modes of heat transfer, i.e., heat conduction and heat

,tion, while neglecting thermal radiation. In a gas or a plasma very high temperature and low density, the thermal radiation / become an important mode of heat transfer in comparison with heat conduction and heat convection. Hence, in many advanced engineering problems such as aerospace problems and fusion problems, we should consider the thermal radiation effects simultaneously with those of heat conduction and heat convection. The thermal radiation effects include the radiation stresses, the radiation energy density, and the heat flux of thermal radiation.

We may either consider the thermal radiation as a stream of photons or as electromagnetic waves. When we consider the thermal radiation as a stream of photons, we should use the relativistic Boltzmann equation to study the motion of these particles.[10] When we consider the thermal radiation as electromagnetic waves, we may use the geometrical optics to study the behavior of thermal radiation from the macroscopic point of view. The results of these two approaches are the same (see Ref. 10). In this section, we consider only the case of continuum theory and we shall use the treatment of geometrical optics. The thermal radiation may be expressed in terms of a specific intensity I_ν which is defined as follows:

$$I_\nu = \lim_{d\sigma_0,\, d\omega,\, dt,\, d\nu \to 0} \left(\frac{dE_\nu}{d\sigma_0 \cdot \cos\theta \cdot d\omega \cdot dt \cdot d\nu} \right) \tag{7.104}$$

where I_ν is a function of time, spatial coordinates, direction and the frequency of the wave ν. The amount of radiative energy flowing through the area $d\sigma_0$ in the frequency range ν and $\nu + d\nu$ in the direction L which makes an angle θ with the normal of the area $d\sigma_0$ within a solid angle $d\omega$ in the time interval dt is dE_ν. The total amount of energy radiated over the whole spectrum is

$$dE = \int_0^\infty \left(\frac{dE_\nu}{d\nu} \right) d\nu = I \cos\theta \, d\omega \, d\sigma_0 \, dt \tag{7.105}$$

where

$$I = \int_0^\infty I_\nu \, d\nu = \text{integrated intensity of thermal radiation} \qquad (7.106)$$

7.8.1 Thermal Radiation on the Flow Field. If the specific intensity I_ν is known, we may easily calculate the effects of thermal radiation on the flow field.

Flux of heat energy by radiation. The flux of heat energy by radiation is $q_R{}^i$, that is,

$$q_R{}^i = \int_{4\pi} I n^i \, d\omega \qquad (7.107)$$

where n^i is the directional cosine of the radiation ray with respect to the ith axis. We should add the divergence of this radiative heat flux in the energy equation, i.e.,

$$Q_R = \nabla \cdot \mathbf{q}_R \qquad (7.108)$$

In general, Q_R is a differentio-integral expression and the energy equation is a differentio-integral equation.

Energy density of radiation E_R. The energy density of radiation within the frequency range ν and $\nu + d\nu$ is

$$u_\nu = \frac{1}{c} \int_{4\pi} I_\nu \, d\omega \qquad (7.109)$$

and the energy density of radiation for the whole spectrum is then

$$E_R = \int_0^\infty u_\nu \, d\nu = \frac{1}{c} \int_{4\pi} I \, d\omega \qquad (7.110)$$

This radiative energy density should be added to the total energy of the gas and it behaves similar to the internal energy of the gas.

Stress tensor of radiation. The ijth component of stress tensor of radiation is

$$\tau_R{}^{ij} = -\frac{1}{c}\int_{4\pi} I n^i n^j d\omega \tag{7.111}$$

may define a radiation pressure p_R as follows:

$$p_R = -\frac{1}{3}(\tau_R{}^{11} + \tau_R{}^{22} + \tau_R{}^{33}) = \frac{1}{3c}\int_{4\pi} I d\omega = \frac{1}{3}E_R \tag{7.112}$$

Notice that we consider the gas as consisting of one species in the above formulas. If the gas consists of more than one species, we should have similar expressions for each species. For simplicity, in this section we consider the case for a single species only and the result is applicable to the mixture as a whole.

7.8.2 Specific Intensity of Radiation. In a macroscopic treatment, the specific intensity of radiation may be expressed in terms of two overall coefficients: one is the absorption coefficient of radiation k_ν and the other is the emission coefficient of radiation j_ν.

Absorption coefficient k_ν of radiation. The loss of specific intensity along the ray of radiation over a distance ds is

$$dI = -\rho k_\nu \; I_\nu \; ds \tag{7.113}$$

The integration of Eq. (7.113) gives

$$I_\nu(s) = I_\nu(s_0)\exp\left(-\int_{s_0}^{s} \rho k_\nu \, ds\right) = I(s_0)\exp(-\tau_\nu) \tag{7.114}$$

where τ_ν is known as the *optical thickness* of the layer $(s - s_0)$ and

$$L_R = \frac{1}{\rho k_\nu} = \text{mean free path of radiation} \tag{7.115}$$

The absorption coefficient k_ν is a function of the temperature and density of the medium as well as the frequency ν. In general k_ν consists of two parts: one is the true absorption and the other is that due to scattering. In the macroscopic theory of radiation gas dynamics, we assume that k_ν is a known function of temperature and density and frequency ν. The determination of k_ν can be made by microscopic theory or experiment. The absorption coefficient in radiation gas dynamics has a similar position as the other transport coefficients such as viscosity and thermal conductivity in ordinary gas dynamics. It may be considered as a new transport coefficient.

Emission coefficient of radiation j_ν. The radiation energy emitted from a mass dm is

$$dE_e = j_\nu \, dm \, d\nu \, d\omega \, dt \tag{7.116}$$

If we have both the absorption coefficient and the emission coefficient of the medium, the conservation of radiative energy gives the radiative transfer equation which governs the specific intensity of radiation, i.e.,

$$dE_0 - dE_i = dE_e + dE_a - dE_t \tag{7.117}$$

The difference between the outgoing radiative energy dE_0 and the incoming radiative energy dE_i must be equal to the sum of the energy emitted and energy absorbed minus the net change of radiative energy in the volume with time dE_t. Equation (7.117) in terms of k_ν and j_ν is

$$\frac{1}{c}\frac{\partial I_\nu}{\partial t} + n^i \frac{\partial I_\nu}{\partial_x{}^i} = \rho k_\nu (J_\nu - I_\nu) \tag{7.118}$$

where

$$J_\nu = \frac{j_\nu}{k_\nu} = \text{source function of radiation} \tag{7.119}$$

One of the most difficult problems in radiation gas dynamics is to determine the source function of radiation J_ν. At the present stage of investigation, we usually assume local thermodynamic equilibrium which may be considered as a first approximation of the actual case.

Under complete thermodynamic equilibrium, the specific intensity of thermal radiation is given by the Planck radiation function B_ν which is also known as the *black body radiation function* and which is

$$B_\nu(\nu, T) = \frac{2h\nu^3}{c^2} \frac{1}{\exp(h\nu/kT) - 1} \tag{7.120}$$

where h is the Planck constant. The properties of B_ν had been known before Planck found the correct expression [Eq. (7.120)]. For instance, Kirchhoff knew that under thermodynamic equilibrium

$$I_\nu = \frac{j_\nu}{k_\nu} = B_\nu(T) \tag{7.121}$$

The function $B_\nu(T)$ must be a function of temperature and independent of the material. Wien found the displacement law:

$$\frac{u_\lambda}{T^5} = G\left(\frac{c}{\lambda T}\right) \tag{7.122}$$

where $u_\lambda d\lambda = u_\nu d\nu$ and λ is the wavelength. Rayleigh-Jean found that at low frequency:

$$B_\nu(\nu, T) = \frac{1}{c^2} 8\pi\nu^2 kT \tag{7.123}$$

Of course, Eq. (7.123) is not valid for high frequency which causes the ultraviolet catastroph of Rayleigh-Jean law.

When we assume that the gas is in local thermodynamic equilibrium, i.e., the emission is determined by the local temperature, Eq. (7.118) becomes

$$\frac{1}{c}\frac{\partial I_\nu}{\partial t} + \frac{\partial I_\nu}{\partial s} = \rho k'_\nu (B_\nu - I_\nu) \tag{7.124}$$

where

$$k'_\nu = k_\nu \left[1 - \exp\left(-\frac{h\nu}{kT}\right)\right] \tag{7.125}$$

The reduction of the absorption coefficient is due to induced emission.

7.8.3 Essential Features of Effects of Thermal Radiation. In general we have to solve the radiative transfer equation (7.118) or (7.124) with the other fundamental equations of gas dynamics, including the radiation terms. In order to get some essential features of the effects of thermal radiation, some approximations have been used. They are discussed in some detail in Ref. 10. Here we list a few most common ones:

Optical thick medium. When the mean free path of radiation is very small in comparison with the dimension of the flow field, the solution of Eq. (7.124) may be expressed as follows:

$$I_\nu = B_\nu - L_{R\nu}\left(n^i \frac{\partial I_\nu}{\partial_x^{\,i}}\right) + O(L_{R\nu}{}^2) \tag{7.126}$$

where $L_{R\nu} = 1/\rho k'_\nu$. If we neglect higher order terms of $L_{R\nu}{}^2$, the radiation terms in gas dynamic equations can be easily evaluated, i.e.,

$$E_R = a_R T^4 = 3p_R \tag{7.127}$$

where

$$\begin{aligned} a_R &= 7.67 \times 10^{-15} \text{ erg. cm}^{-3}\,{}^\circ\text{K}^{-4} \\ &= \text{Stefan-Boltzmann constant for radiation} \end{aligned} \tag{7.128}$$

$$\mathbf{q}_R = D_R \nabla E_R \tag{7.129}$$

and

$$D_R = \frac{c}{3\rho K_R} = \text{Rosseland diffusion coefficient of radiation} \tag{7.130}$$

and

$$K_R = \frac{\displaystyle\int_0^\infty \frac{\partial B_\nu}{\partial T}\, d\nu}{\displaystyle\int_0^\infty \frac{1}{k'_\nu}\frac{\partial B_\nu}{\partial T}\, d\nu} = \text{Rosseland mean absorption coefficient} \tag{7.131}$$

For the optically thick medium, the differential-integral equations are reduced to differential equations. The fundamental equations are essentially the same as the ordinary Navier-Stokes equations of a compressible fluid with a few more terms.

One-dimensional approximation. If the mean free path of radiation is not small, we have to use more terms in the series expansion of Eq. (7.126). It is questionable how many terms we should use and it is not known about the convergence of this series. Hence it is better to solve the integro-differential equations. In many practical problems such as boundary-layer flows, the gradient of temperature in one direction is much larger than those in the other directions. Hence we may assume that the specific intensity of radiation I_ν is essentially a function of the predominant coordinates, say y, i.e., $I_\nu(y, \theta)$. If we define the optical thickness in terms of y, i.e.,

$$\tau_\nu = \int_0^y \rho k'_\nu \, dy \tag{7.132}$$

we may carry out the integration of solid angle θ and ϕ. Hence, the integral expressions of the radiative terms will be simplified. (See Ref. 10).

Gray gas approximation. The coefficient of absorption k'_ν is in general a function of frequency ν. As a result, the integral expression of radiation terms may not be integrated analytically. Gray gas approximation has been used by assuming that the absorption coefficient is independent of frequency. Actually we use some mean value of the absorption for the overall flow phenomena. The Rosseland mean absorption coefficient [Eq. (7.131)] is the one for optically thick medium. Planck mean absorption coefficient K_P is the one used for finite mean free path of radiation which is defined as

$$K_P = \frac{\int_0^\infty k'_\nu B_\nu \, d\nu}{\int_0^\infty B_\nu \, d\nu} \tag{7.133}$$

Optically thin medium. In many gas-dynamic problems, the mean free path is large in comparison with the flow dimension. Furthermore, the effect of thermal radiation is not too large. For a first approximation, we may use the optically thin approximation that the absorption of the gas can be neglected. In this case, we usually consider only the radiative heat flux and neglect the radiation stresses and radiation energy density.

Under the one-dimensional approximation and gray gas approximation, we may show that when the optical thickness is very large, we have the radiative heat flux as

$$q_R = e_g \sigma T^4 \tag{7.134}$$

where

$$e_g = 2\rho K_P L = \text{emissivity of a gas layer of thickness } L \tag{7.135}$$

and

$$\begin{aligned} \sigma &= a_R \frac{c}{4} = 5.68 \times 10^{-5} \text{ erg. cm}^2 \text{ sec. }^\circ\text{K}^{-4} \\ &= \text{Stefan-Boltzmann constant for radiative flux.} \end{aligned} \tag{7.136}$$

7.8.4 Effects of Radiation in Magnetogas Dynamics. *Wave motion in magnetogas dynamics including thermal radiation effects.* Now we reconsider the wave of small amplitude in magnetogas dynamics discussed in Chap. 5. For an ideal plasma the coefficients of viscosity, heat conductivity and magnetic diffusivity are zero and we should also assume that the diffusion coefficient of thermal radiation is zero. For an ideal plasma, the influence of thermal radiation is due to radiation stresses and radiation energy density and the only effect is that the sound speed is increased by the radiation effect. With thermal radiation, the sound speed of the plasma is

$$C_R = a_0 \sqrt{\frac{1 + 20\,[(\gamma - 1)/\gamma]R_p + 16\,[(\gamma - 1)/\gamma]R_p{}^2}{1 + 12\,(\gamma - 1)R_p}} \tag{7.137}$$

If we use the sound speed of radiation C_R to replace the ordinary sound speed a_0, the formulas for the fast and slow waves remain unchanged. The factor R_p is the ratio of radiation pressure to gas pressure, i.e.,

$$R_p = \frac{a_R T_0^3}{3R\rho_0} \tag{7.138}$$

The main effect of thermal radiation is on the temperature and it affects the velocity distribution very little. For instance, consider a normal shock including the thermal radiation effects. If the effective Mach number in front, the shock $M_{e1} = u_1/C_{R1}$ is very large, no matter what kind the gas, the effective ratio of the specific heats γ has the value of 4/3. Hence the ratio of the velocity behind the shock u_2 to the velocity in front of the shock u_1 is 1/6 instead of 1/4 if γ is 5/3 for monatomic gas. But the jump in temperature with the radiation effect will be proportional to the square root of Mach number M_{e1} instead of proportional to square of Mach number without the radiation effect, i.e.,

$$\frac{T_2}{T_1} = 1.033\, M_{e1}{}^{1/2} \tag{7.139}$$

Plane Couette flow in radiation magnetogas dynamics. Our second example is to consider an ionized gas flowing between two parallel plates: one of the plates is at rest and the other has a uniform motion with the velocity U which has been discussed in Chap. 4, Sec. 4.2. Assume that the transport coefficients are constants, then the velocity and magnetic field distributions are not affected by the thermal radiation when the radiation stresses and energy density are neglected. We may use the results of the velocity and magnetic field distributions obtained in Chap. 4 to calculate the temperature distribution from the energy equation which is now

$$\frac{1}{(\gamma - 1)P_r}\frac{dT}{dy} + M^2\mu\frac{d}{dy}\left(\frac{1}{2}u^2\right) + R_e R_E R_H M^2 H_x + R_{p0} R_F R_e q_R(T) = \text{constant} \tag{7.140}$$

where R_{p0} is the reference radiation pressure number given by Eq.

(7.138) and R_F is known as radiation flux number which is

$$R_F = \frac{cL}{UL_R} \tag{7.141}$$

and the rest are nondimensional parameters with their usual meanings.

In order to solve Eq. (7.140), we have to know the expression of the radiative flux q_R. Two expressions have been used for q_R: One is the optically thick approximation which is

$$q_R = \frac{4}{\gamma} T^3 \frac{dT}{dy} \tag{7.142}$$

and the other is the modified optically thin approximation. In Eq. (7.134), we neglect the absorption completely which is not a very good approximation if the mean free path of radiation is not too large. Hence we shall retain the first-order term of absorption and the corresponding radiative flux is

$$q_R(T) = \frac{3}{2\gamma}\left[(T_1^4 + 1)\tau_a + \int_{\tau_a}^{\tau_{a2}} T^4(t)\,dt - \int_0^{\tau_a} T^4(t)\,dt\right] \tag{7.143}$$

where T_1 is the nondimensional temperature of the upper wall with respect to the reference temperature and

$$\tau_a = \tau\frac{L_R}{L} \qquad \tau = \int_0^{y^*} \rho k' dy^* \tag{7.144}$$

and y^* is the dimensional coordinate.

The integration of the energy equation with q_R given by Eq. (7.142) is

$$A[T^4 - 1 - y(T_1^4 - 1)] + \frac{\gamma}{(\gamma - 1)P_r}[T - 1 - y(T_1 - 1)]$$
$$= \gamma M^2\left(\frac{R_E}{R_e}\right)R_b^2\left[y\int_0^1 H_x dy - \int_0^y H_x dy\right] + \tfrac{1}{2}\gamma M^2(y - u^2) \tag{7.145}$$

where $A = R_{p0}R_F R_e$ and with the expression of q_R by Eq. (7.143) is

$$\frac{3}{2}A\left[\frac{1}{2}(T_1^2 + 1)(y^2 - y) + \int_0^y I_R\,dy - y\int_0^1 I_R\,dy\right]$$
$$+ \frac{\gamma}{(\gamma - 1)P_r}[T - 1 - y(T_1 - 1)] = \frac{1}{2}\gamma M^2(y - u^2)$$
$$+ \gamma M^2\left(\frac{R_E}{R_e}\right)R_b^2\left[y\int_0^1 H_x dy - \int_0^y H_x dy\right] \quad (7.146)$$

where

$$I_R = \int_y^1 T^4(y)\,dy - \int_0^y T^4(y)\,dy \quad (7.147)$$

Figure 7.9 shows some typical temperature distributions with and without thermal radiation effects. Without thermal radiation effect, $A = 0$, the temperature in the flow field is enormously high for the case of large Hartmann number R_b and high electric field number R_E. With radiation effects considered, the maximum temperature in the flow field drops a great deal. For the optically thick medium approximation, the formula overestimates the effect of radiation a little. It is better to use the integral expression. Due to the effect of

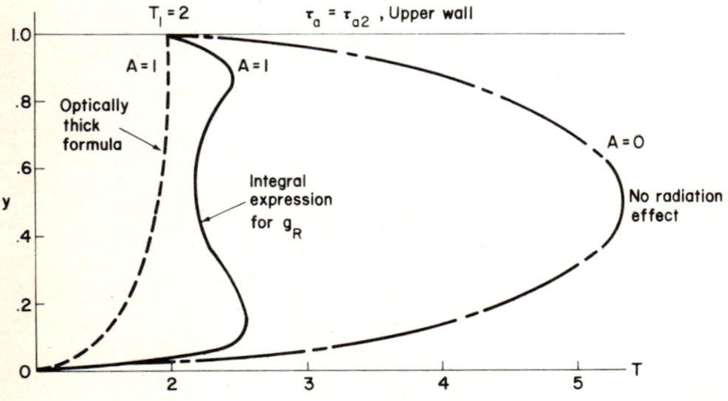

Fig. 7.9 Temperature distributions for plane Couette flow of MGD including thermal radiation effects.

thermal radiation, there is a tendency that the temperature in the central portion of the flow field becomes more or less uniform.

REFERENCES

1. S. I. Pai, "Magnetogasdynamics and Plasma Dynamics," Springer-Verlag, New York, 1962.
2. S. I. Pai, A Critical Survey of Magnetofluid Dynamics, parts I–IV, *Aerospace Res. Lab. Repts.* ARL 63-175 and ARL 66-0060, Wright-Patterson Air Force Base, Ohio, 1963 and 1966.
3. R. J. Rosa, Physical Principles of MHD Power Generation, *AVCO-Everett Res. Lab. Rept.* 69, 1960.
4. S. I. Pai, Wave Motion of Small Amplitude in a Fully Ionized Plasma Under Applied Magnetic Field, *Phys. Fluids*, vol. 5, no. 2, pp. 234–240, 1962.
5. B. S. Tanenbaum and D. Mintzer, Wave Propagation in a Partly Ionized Gas, *Phys. Fluids*, vol. 5, no. 10, pp. 1226–1237, 1962.
6. S. I. Pai and J. O. Powers, Nonequilibrium Effects on Energy Transfer in an Ionized Fluid Flow, S. I. Pai et al. (eds.), *Symposium on the Dynamics of Fluids and Plasmas* in honor of Professor J. M. Burgers, Academic Press, Inc., New York, 1967.
7. J. M. Burgers, The Application of Transfer Equations to the Calculation of Diffusion, Heat Conduction, Viscosity and Electric Conductivity, *Inst. Fluid Dynamics Appl. Math. Tech. Notes* BN124a and 124b, University of Maryland, 1958.
8. J. M. Burgers, The Bridge between Particle Mechanics and Continuum Mechanics, *Proc. Intern. Symposium on Plasma Dynamics, Woods Hole, Mass., 1958*, pp. 119–186, Addison-Wesley Publishing Company, Inc., Reading, Mass. 1960.
9. S. Chandrasekhar, "Plasma Physics," University of Chicago Press, Chicago, Ill., 1960.
10. S. I. Pai, "Radiation Gasdynamics," Springer-Verlag, New York, 1966.

APPENDICES

Appendix A

SYMBOLS, UNITS, AND DIMENSIONS

Quantity*	Symbol	Dimensions	Meter-kilogram-second (mks) system
Capacitance	**C**	$M^{-1}L^{-2}t^2q^2$	Farad
Charge	q	q	Coulomb
Charge density	ρ_e	$L^{-3}q$	Coulomb/cubic meter
Conductivity	σ	$M^{-1}L^{-3}tq^2$	Mho/meter
Current	I	$t^{-1}q$	Ampere
Current density	**J**	$L^{-2}t^{-1}q$	Ampere/square meter = coulomb/second-square meter
Density	ρ	ML^{-3}	Kilogram/cubic meter
Dielectric displacement	**D**	$L^{-2}q$	Coulomb/square meter
Electric field intensity	**E**	$MLt^{-2}q^{-1}$	Volt/meter = joules/coulomb-meter = newton-meter/coulomb-meter
Electric potential	ϕ, V	$ML^2t^{-2}q^{-1}$	Volt
Energy	u, W	ML^2t^{-2}	Joule
Enthalpy	h, I	L^2t^{-2}	Joule
Entropy	S	$L^2t^{-2}T^{-1}$	Joule/°K
Flux density	**B**	$Mt^{-1}q^{-1}$	Weber/square meter or newton-meter-second/square meter-coulomb
Force/unit mass	**F**	Lt^{-2}	Newton/kilogram
Gas constant	R	$L^2t^{-2}T^{-1}$	Newton meter/kilogram °K
Heat	Q	ML^2t^{-2}	Joule
Inductance	L	ML^2q^{-2}	Henry
Inductive capacity	ϵ	$M^{-1}L^{-3}t^2q^2$	Farad/meter
Magnetic dipole moment	m	$L^2t^{-1}q$	Ampere-square meter
Magnetic field intensity	**H**	$L^{-1}t^{-1}q$	Ampere-turn/meter = coulomb turns/second-meter
Magnetic flux	Φ	$ML^2t^{-1}q^{-1}$	Weber
Mass velocity	$\dot{m}$	Mt^{-1}	Kilogram/second
Permeability	μ_e	MLq^{-2}	Henry/meter
Power	P	ML^2t^{-3}	Watt
Pressure	p	$ML^{-1}t^{-2}$	Newton/square meter
Resistance	R	$ML^2t^{-1}q^{-2}$	Ohm
Specific heat	c	$L^2t^{-2}T^{-1}$	Joules/kilogram °K
Specific volume	v	L^3M^{-1}	Cubic meter/kilogram
Temperature	T	T	Degrees Kelvin
Thermal conductivity	k	$MLT^{-1}t^{-3}$	Joules/second meter °K
Velocity	V,q,w	Lt^{-1}	Meters/second
Viscosity	μ, β	$ML^{-1}t^{-1}$	Kilogram/meter second
Work	W	ML^2t^{-2}	Joules

*Unlisted quantities are either defined in the text or are combinations of those listed.

Appendix B

MAGNETOFLUID-DYNAMIC PARAMETERS

Quantity	*Comparison*	*Units*	*Symbol*	*Name*
Pressure	$\dfrac{\text{Magnetic pressure}}{\text{Dynamic pressure}}$	$\dfrac{\mu_e H_0^2/2}{\rho V^2/2}$	$= R_H$	Magnetic Pressure no.
Force	$\dfrac{\text{Magnetic force}}{\text{Viscous force}}$	$\dfrac{\sigma V \mu_e^2 H_0^2}{\mu V/L^2}$	$= H_a^2$	(Hartmann no.)^2
	$\dfrac{\text{Magnetic force}}{\text{Inertial force}}$	$\dfrac{\sigma V \mu_e^2 H_0^2}{\rho V^2/L}$	$= R_m^2$	(Magnetic no.)^2
Velocity	$\dfrac{\text{Flow velocity}}{\text{Alfvén's velocity}}$	$\dfrac{V^2}{V_H^2} = \dfrac{V^2}{\mu_e H^2/\rho}$	$= M_m^2 = \dfrac{1}{R_H}$	(Magnetic Mach no.)^2
Diffusivity	$\dfrac{\text{Viscous diffusivity}}{\text{Magnetic diffusivity}}$	$\dfrac{\nu}{\nu_H}$	$= P_m = \dfrac{R_\sigma}{R_e}$	(Magnetic Prandtl no.)
$\dfrac{\text{Dynamic}}{\text{Magnetic}}$	$\dfrac{\text{Dynamic}}{\text{Magnetic diffusivity}}$	$\dfrac{VL}{1/\sigma\mu_e}$	$= R_\sigma$	(Magnetic Reynolds no.)

Appendix C

PROPERTY DATA

Fluid	Temp. (deg.)	σ (mho/m)	μ (kg/m-sec)	ρ (kg/m^3)
Hg	20 C	1.05×10^6	0.00155	13,550
Na	100 C	1.03×10^7	0.000686	928
NaK	100 C	2.43×10^6	0.00047	847
Plasma	10,000 K/ 1 atm	3.73	0.000288	4.8×10^{-5}
Sea water	20 C	4.75	0.00102	1,020
Tap water	20 C	0.01	0.001	1,000
Pure water	20 C	4.0×10^{-6}	0.001	1,000
NaCl (25%)	20 C	21.6	0.00158	1,160
H_2SO_4 (con.)	20 C	73.5	0.00197	1,230
CuCl (20%)	20 C	10.0	0.0017	2,440

Appendix D

DIFFERENTIAL OPERATORS IN ORTHOGONAL COORDINATES

ψ = Scalar quantity; $\mathbf{V}$ = vector quantity

Rectangular coordinates (x, y, z)

$$\text{grad}_x\, \psi = \frac{\partial \psi}{\partial x}, \quad \text{grad}_y\, \psi = \frac{\partial \psi}{\partial y}, \quad \text{grad}_z\, \psi = \frac{\partial \psi}{\partial z}$$

$$\text{Div}\ \mathbf{V} = \frac{\partial V_x}{\partial x} + \frac{\partial V_y}{\partial y} + \frac{\partial V_z}{\partial z}$$

$$\nabla^2 \psi = \text{div grad}\ \psi = \frac{\partial^2 \psi}{\partial x^2} + \frac{\partial^2 \psi}{\partial y^2} + \frac{\partial^2 \psi}{\partial z^2}$$

$$\text{curl}_x\ \mathbf{V} = \frac{\partial V_z}{\partial y} - \frac{\partial V_y}{\partial z}, \quad \text{curl}_y\ \mathbf{V} = \frac{\partial V_x}{\partial z} - \frac{\partial V_z}{\partial x}, \quad \text{curl}_z\ \mathbf{V} = \frac{\partial V_y}{\partial x} - \frac{\partial V_x}{\partial y}$$

$$\frac{D}{Dt} = \frac{\partial}{\partial t} + V_x \frac{\partial}{\partial x} + V_y \frac{\partial}{\partial y} + V_z \frac{\partial}{\partial z}$$

Cylindrical coordinates (r, θ, z)

$$\text{grad}_r\, \psi = \frac{\partial \psi}{\partial r}, \quad \text{grad}_\theta\, \psi = \frac{1}{r}\frac{\partial \psi}{\partial \theta}, \quad \text{grad}_z\, \psi = \frac{\partial \psi}{\partial z}$$

$$\text{Div}\ \mathbf{V} = \frac{1}{r}\frac{\partial (rV_r)}{\partial r} + \frac{1}{r}\frac{\partial V_\theta}{\partial \theta} + \frac{\partial V_z}{\partial z}, \quad \nabla^2 \psi = \frac{\partial^2 \psi}{\partial r^2} + \frac{1}{r}\frac{\partial \psi}{\partial r} + \frac{1}{r^2}\frac{\partial^2 \psi}{\partial \theta^2} + \frac{\partial^2 \psi}{\partial z^2}$$

$$\text{curl}_r\, \mathbf{V} = \frac{1}{r}\frac{\partial V_z}{\partial \theta} - \frac{\partial V_\theta}{\partial z}, \quad \text{curl}_\theta\, \mathbf{V} = \frac{\partial V_r}{\partial z} - \frac{\partial V_z}{\partial r}, \quad \text{curl}_z\, \mathbf{V} = \frac{1}{r}\frac{\partial (rV_\theta)}{\partial r} - \frac{1}{r}\frac{\partial V_r}{\partial \theta}$$

$$\frac{D}{Dt} = \frac{\partial}{\partial t} + V_r \frac{\partial}{\partial r} + \frac{V_\theta}{r} \frac{\partial}{\partial \theta} + V_z \frac{\partial}{\partial z}$$

Spherical coordinates (r, θ, ϕ)

$$\text{grad}_r\, \psi = \frac{\partial \psi}{\partial r}, \quad \text{grad}_\theta\, \psi = \frac{1}{r}\frac{\partial \psi}{\partial \theta}, \quad \text{grad}_\phi\, \psi = \frac{1}{r \sin\theta}\frac{\partial \psi}{\partial \phi}$$

$$\text{Div}\ \mathbf{V} = \frac{1}{r^2}\frac{\partial (r^2 V_r)}{\partial r} + \frac{1}{r \sin\theta}\frac{\partial (\sin\theta V_\theta)}{\partial \theta} + \frac{1}{r \sin\theta}\frac{\partial V_\phi}{\partial \phi}$$

$$\nabla^2 \psi = \frac{1}{r^2}\frac{\partial [r^2(\partial\psi/\partial r)]}{\partial r} + \frac{1}{r^2 \sin\theta}\frac{\partial [\sin\theta(\partial\psi/\partial\theta)]}{\partial \theta} + \frac{1}{r^2 \sin^2\theta}\frac{\partial^2 \psi}{\partial \phi^2}$$

$$\text{curl}_r\, \mathbf{V} = \frac{1}{r \sin\theta}\left[\frac{\partial (\sin\theta V_\phi)}{\partial \theta} - \frac{\partial V_\theta}{\partial \phi}\right], \quad \text{curl}_\phi\, \mathbf{V} = \frac{1}{r}\frac{\partial (r V_\theta)}{\partial r} - \frac{1}{r}\frac{\partial V_r}{\partial \theta}$$

$$\text{curl}_\theta\, \mathbf{V} = \frac{1}{r \sin\theta}\left[\frac{\partial V_r}{\partial \phi} - \sin\theta \frac{\partial (r V_\phi)}{\partial r}\right]$$

$$\frac{D}{Dt} = \frac{\partial}{\partial t} + V_r \frac{\partial}{\partial r} + \frac{V_\theta}{r}\frac{\partial}{\partial \theta} + \frac{V_\phi}{r \sin\theta}\frac{\partial}{\partial \phi}$$

Appendix E

ADDITIONAL REFERENCES

Books

Cowling, T. G., "Magnetohydrodynamics," Intersciences Publisher, Inc., New York, 1957.

Hughes, W. F., and Young, F. J., "The Electromagnetodynamics of Fluids," John Wiley & Sons, Inc., New York, 1966.

Pai, S. I., "Magnetogasdynamics and Plasma Dynamics," Springer-Verlag, Berlin, 1962.

Shercliff, J. A., "A Textbook of Magnetohydrodynamics," The Commonwealth & International Library of Science, Technology, Engineering, & Liberal Studies, London, England, 1965.

Sherman, A., and Sutton, G., "Engineering Magnetohydrodynamics," McGraw-Hill, New York, 1965.

Symposium Proceedings

Symposium of Engineering Aspects of MHD

2nd—University of Pennsylvania (March 1961)

3rd—University of Rochester (March 1962)

4th—University of California (April 1963)

5th—Massachusetts Institute of Technology (April 1964)

6th—University of Pittsburgh (April 1965)

7th—Princeton University (March 1966)

8th—Stanford University (March 1967)

9th—University of Tennessee (April 1968)

10th—Massachusetts Institute of Technology (March 1969)

11th—California Institute of Technology (March 1970)

12th—Argonne National Laboratory (March 1972)

13th—Stanford University (March 1973)

Symposium on MHD Electrical Power Generation Jointly Organized by the International Atomic Energy Agency and the European Nuclear Energy Agency of the Office for Economic Cooperation & Development.

Newcastle-on-Tyne, England (Sept. 1962)
Paris, France (July 1964)
Salzburg, Austria (July 1966)
Warsaw, Poland (July 1968)
Munich, Germany (April 1971)

PROBLEMS

PROBLEMS

CHAPTER 1

1.1. Select and describe several MFD problems for astrophysics, power generation, accelerators, etc.

1.2. Calculate the various parameters for the examples in Prob. 1.1 with the property data in Appendix C.

CHAPTER 2

2.1. Using the Biot-Savart law, determine a relation for the magnetic field of a current-carrying, long, straight wire, and calculate the magnetic field strength of a welding arc of 1,000 amperes.

2.2. Determine the relation for the force between two straight wires carrying 1,000 amperes each and calculate the force for several wire spacings.

2.3. Calculate the electric field intensity E that would be just sufficient to balance the gravitational force on an electron. If this electric field were produced by a second electron located below the first one, what would be the distance between the two electrons? (The charge on an electron is $e = -1.6 \times 10^{-19}$ coul and its mass is $m = 9.1 \times 10^{-31}$ kg.)

2.4. An electric field exists everywhere in the z direction. What can you conclude about the values of the partial derivatives of E with respect to x, to y, and to z

(a) if the charge density ρ_e is zero,

(b) if the charge density ρ_e is not zero?

2.5. The axial electric field intensity E on the axis of the accelerating tube in a particular type of ion accelerator is given approximately by

$$E_z = E_{z0} + kz^2$$

where z is measured from the center of the tube along its axis. The azimuthal component E_ϕ is equal to zero.

(a) Show that the radial electric field intensity in the neighborhood of the axis is $E_r = -kzr$, assuming that the charge density is zero.

(b) Draw a rough sketch of the lines of force in a radial plane.

(c) What is the maximum charge density in coulombs per meter that can be tolerated if the value calculated above for the radial field is to be accurate within 5 percent at the ends of the tube? The accelerating tube is 1.00 m long, E_{z0} is 7.5×10^5 volts/m and k is 1.00×10^6 volts/m^3.

2.6. Calculate the force due to the earth's magnetic field on a wire 100 ft long carrying a current of 50 amperes due north (magnetic). Take the earth's magnetic field of 0.5×10^{-4} weber/m^2 at an angle of 70° with the horizontal.

2.7. A thin, flat, conducting disk of thickness h, diameter D, and resistivity $\frac{1}{\sigma}$ is placed in a uniform alternating magnetic field $B = B_0 \sin \omega t$ parallel to the axis of the disk. Find the induced current density as a function of distance from the axis of the disk. What is the direction of this current?

2.8. Show that both the normal and tangential components of the vector potential $\mathbf{A}$ must be continuous across the interface between two media if the currents are constant.

2.9. Compare the energies per unit volume in

(a) a magnetic field of 1.0 weber/m^2, and

(b) an electrostatic field of 10^6 volts/m.

2.10. Using the vector MFD equations, write the component equations in the cylindrical and spherical coordinate systems.

2.11. Discuss the physical effects of a nonlinear stress-strain relation.

2.12. Assume representative values for the various parameters and then estimate the size of the terms dropped on the basis of the MFD assumptions.

CHAPTER 3

3.1. Using the theory of Sec. 3.11.2 calculate the area and Mach number distribution for a constant velocity channel for several magnetic field strengths.

3.2. Choose an area distribution for an MFD nozzle and calculate the velocity variation for a subsonic and a supersonic flow.

CHAPTER 4

4.1. Sketch the velocity and current fields for Hartmann and Couette flows for the short circuit, open circuit, and pumping cases with $H_x = 0$ at the lower insulated surface. Compare these sketches with Figs. 4.4 and 4.5.

4.2. Using Eqs. (4.157)-(4.159), derive Eq. (4.160).

CHAPTER 5

5.1. From the fundamental equations of magnetofluid dynamics, derive the system of linearized equations under the conditions of Eq. (5.1) and show that the resultant equations may be divided into three independent groups.

5.2. Discuss the longitudinal waves given by Eq. (5.15). Find the speeds of propagation of the heat wave and the sound wave in a viscous and heat-conducting medium and their damping factors.

5.3. Calculate the total energy in the two transverse magnetogas-dynamic waves given by Eq. (5.17) and show how the energy is transferred by means of the magnetic field.

5.4. Draw in scale the Friedrichs diagram of Fig. 5.2 for the case (a) $V_H = \frac{1}{2} a$ and (b) $a = \frac{1}{2} V_H$. Draw also the wave patterns in these Friedrichs diagrams for the four cases shown in Fig. 5.3.

5.5. From Eqs. (5.27), show that the entropy along any flow line is a constant.

5.6. Discuss the system of characteristics of magnetogas dynamics of an ideal plasma for three-dimensional unsteady flow.

5.7. Calculate some typical cases of oblique shocks in magnetogas dynamics of an ideal plasma and show that the shock angle may be greater than 90°.

5.8. Find the Rankine-Hugoniot relations across a normal shock in radiation magnetogas dynamics of an ideal plasma including the effects of radiation pressure and radiation energy density.

5.9. Derive the expression for the critical magnetic field h_b discussed in Sec. 5.8 and show that when $h_1 = h_b$, a smooth solution may be obtained for the transitional region even when viscosity and heat conductivity are neglected.

CHAPTER 6

6.1. Derive a relation for the voltage as a function of the average velocity through a two-dimensional MFD channel and design the electrical circuit for a MFD flow meter.

6.2. Repeat problem 1 for a two-dimensional MFD pump.

6.3. Using the theory in Sec. 6.4, determine the magnetic field strength and power requirements for a 25-knot version of the model MFD submarine of Way and Devlin.

CHAPTER 7

7.1. Calculate the velocity and magnetic field distribution of MHD Rayleigh problem by including the Hall current in which an infinite flat insulated plate moving impulsively into a uniform velocity U in its own plane x direction in the presence of a transverse uniform magnetic field (in the y direction) of strength $H_y = H_0$ = constant.

7.2. Calculate the pressure distribution in the y direction and the magnetic field distribution for the MHD channel flow with Hall current discussed in Sec. 7.3.

7.3. By means of two-fluid theory, calculate the nozzle flow based on strictly one-dimensional-flow theory.

7.4. Assuming a uniform flow with velocity U passing an infinite sinusoidal surface under an applied magnetic field H_0, calculate the velocity and magnetic field distributions in the flow field if we assume that the linearized theory is applicable for the two special cases: (a) U and H_0 are parallel and (b) U and H_0 are perpendicular. In both cases, the uniform velocity U is parallel to the surface.

7.5. Derive the generalized Ohm's law if the pressure gradient of the electron is not negligible.

7.6. Discuss the wave motion of infinitesimal amplitude in an ideal plasma which consists of three species: singly charged ions, electrons, and atoms, assuming that there is no externally applied magnetic field.

7.7. What is the Saha relation for the degree of ionization of a plasma? Discuss its relation with the multifluid theory. (See Ref. 1.)

7.8. Discuss the Rankine-Hugoniot relations of a normal shock for a fully ionized plasma by two-fluid theory, assuming that there is no externally applied magnetic field.

7.9. Find the expression of the collision terms of Eq. (7.85) if (a) the main collision is due to binary collisions between the molecules only, and (b) the collision is mainly due to distant encounters by involving a great number of acts of very small momentum transfer, the so-called Fokker-Planck mechanism.

7.10. Derive the maxwellian molecular distribution function F_0 by assuming that the gas consists of one kind of molecules, that there is no body force on the molecules, that only binary collision is important, and that the state of the gas is uniform.

7.11. Study the plasma oscillation by means of the Boltzmann equation for a one-dimensional motion of electrons. (See Chap. XIV, Sec. 8 of Ref. 1.)

7.12. What are some subregions of the flow field of a rarefied plasma when the electric and magnetic Knudsen numbers are important.

7.13. Calculate the first-order motion of a charged particle if the applied electric and magnetic fields are slightly different from constant values, and assume that the basic electric and magnetic field are perpendicular.

7.14. Consider the case of thermodynamic equilibrium between matter and radiation in an enclosure. Calculate the thermodynamic relations between the interaction of matter and radiation in terms of the radiation pressure number R_p.

7.15. Derive the integral expression for the radiative flux between two parallel plates by assuming that the gas is gray and the flow is steady and one dimensional.

7.16. From the expression of radiative flux of Prob. 7.15, derive the corresponding expressions for (a) optically thick medium, and (b) optically very thin medium.

7.17. Discuss the flow field in front of a blunt body in a hypersonic flow by including the radiative flux.

SUBJECT AND AUTHOR INDEXES

AUTHOR INDEX

Numbers in italics refer to the pages on which the complete references are listed.

T

W

Y

SUBJECT INDEX

N

O

P

Q

R

S

T

U

V

W